Springer Series in Optical Sciences

Volume 215

Springer Series in Optical Sciences

The Springer Series in Optical Sciences, under the leadership of Editor-in-Chief William T. Rhodes, Georgia Institute of Technology, USA, provides an expanding selection of research monographs in all major areas of optics: lasers and quantum optics, ultrafast phenomena, optical spectroscopy techniques, optoelectronics, quantum information, information optics, applied laser technology, industrial applications, and other topics of contemporary interest.
With this broad coverage of topics, the series is of use to all research scientists and engineers who need up-to-date reference books.
The editors encourage prospective authors to correspond with them in advance of submitting a manuscript. Submission of manuscripts should be made to the Editor-in-Chief or one of the Editors. See also www.springer.com/series/624

More information about this series at http://www.springer.com/series/624

Wei Lu · Ying Fu

Spectroscopy of Semiconductors

Numerical Analysis Bridging Quantum Mechanics and Experiments

Springer

Wei Lu
Shanghai Institute of Technical Physics
Shanghai, China

Ying Fu
Department of Applied Physics
Royal Institute of Technology
Solna, Sweden

ISSN 0342-4111 ISSN 1556-1534 (electronic)
Springer Series in Optical Sciences
ISBN 978-3-030-06943-8 ISBN 978-3-319-94953-6 (eBook)
https://doi.org/10.1007/978-3-319-94953-6

Printed on acid-free paper

This Springer imprint is published by the registered company Springer Nature Switzerland AG
The registered company address is: Gewerbestrasse 11, 6330 Cham, Switzerland

Preface

The basic function of the eye is to distinguish light from dark. Human eye has been greatly evolved so that we can sense lights and discriminate colors in a broad wavelength range (known as the visible range) with a great sensitivity in low light levels, and an ability to detect motion and resolve objects for locomotion and navigation. The most fundamental of developing and evolving light sensing (also known as sight sensing) is the information richness of the optical properties of an object we observe when our eyes look at it. The other fundamental of light (sight) sensing is that we can look at the object without physically touching it. (The latter is also true concerning hearing and smell.)

The success of the human eye evolution is self-evident,[1] and the persistent development to assist, enable, and extend our light sensing is breathtaking. 2017 Nobel Prize in Physics was awarded to Rainer Weiss, Barry C. Barish, and Kip S. Thorne "for decisive contributions to the LIGO detector and the observation of gravitational waves." The LIGO detector is a laser-based interferometer consisting of two arms that form an L. A single laser beam is split at the corner of the L into two identical beams that travel along the arms and reflect back from the mirrors mounted at the ends of the arms and then meet again at the corner. When the lengths of the two arms are equal, the two beams meet at the corner synchronously. A passing gravitational wave will compress one arm while stretch the other, resulting in the loss of the synchronization where the beams meet, thereafter becoming captured by the LIGO detector. "The world's first captured gravitational waves were created in a violent collision between two black holes, 1.3 billion light years away."[2]

This identifying a material by measuring (synonyms: quantifying) optical properties of the material is the essence of the optical spectroscopy. Light looks so obvious but its quantitative description can be very mysterious. Though the

[1] **On the Origin of Species by Means of Natural Selection, or the Preservation of Favoured Races in the Struggle for Life**, Charles Darwin, published on 24 November 1859.

[2] http://www.nobelprize.org/nobel_prizes/physics/laureates/2017/.

mechanism of the LIGO detector can be fully described by the Maxwell's equations (Sect. 1.3) that have been thoroughly studied and validated for a very broad spectrum of physical phenomena, it is well known that the Maxwell's equations are limited. Quantum electrodynamics describes fully how light and matter interact to an unprecedented accuracy (1965 Nobel Prize in Physics was awarded to Sin-Itero Tomonaga, Julian Schwinger, and Richard P. Feynman "for their fundamental work in quantum electrodynamics, with deep-ploughing consequences for the physics of elementary particles"[3]). However, the microscopic equations of the quantum electrodynamics are difficult to solve as well as to understand in many cases. We take a look at the systems to be studied in this book and observe that we study in the realm of non-relativity. Since the Maxwell's equations can be derived from non-relativistic quantum electrodynamics, while the dynamics of the nonrelativistic light–matter interaction has been well-explained quantum mechanically, we describe the light in the following way in this book. Light in transport is an electromagnetic wave described by the Maxwell's equations in terms of the amplitude and the wavelength of the electromagnetic wave, while it is a photon when interacting with matter quantified by the number of photons and the energy carried per photon.

Optical spectroscopy of matter is the study of interaction between light and matter as a function of the wavelength of the light, which is commonly represented by the wavelength-dependent optical spectrum. Spectroscopy, i.e., recording and analyzing wavelength-dependent optical spectrum of an object (sample) under a probing light (in advanced setups, the object can be subjected simultaneously to various modulations such as an electric field or heat or mechanical force), is probably the best and most important tool for studying the world where we live. Our eyes sense the light around us, such as solar light transmitted through atmosphere, reflected and diffracted from various objects, in the visible range of ca 390–700 nm, which coincides of course with the wavelength range of the solar light most abundant on the earth surface. We have never stopped extending our light sensing, from the passive sensing, i.e., seeing the light available, to the active sensing such as night vision by lighting a fire. We have greatly extended the optical wavelength range of our light sensing down below 10^{-5} nm (γ rays) as well as up above 10^{12} nm (radio waves).

Spectroscopy is widely used in basic research in laboratory, in industrial sectors, e.g., for routine quality assessment, as well as in our daily life (e.g., radio-frequency identification and smartphone-based glucose monitor), least to mention to understand our universe. There has been an ever-growing demand to estimate optical spectra of various bulk and composite materials more accurately, which is made technically possible by the continuous development of novel optoelectronic devices as light sources and detectors. For instance, the scientific as well as industrial interest in the optical properties of novel nanomaterials comes from their widespread presence and relevant importance both in strategic engineering and life

[3]http://www.nobelprize.org/nobel_prizes/physics/laureates/1965/.

sciences sectors. Another nonconventional and very recent area of interest for the methodologies of optical investigation is in the field of bio-optical medicine, where the interaction of light with different biomaterials can be exploited in novel drug development.

The book focuses on the spectroscopy of semiconductors, i.e., unraveling the microscopic light–matter interactions in semiconductors, both bulk materials (including thin films) and nanostructures (quantum well, quantum wire, and quantum dot), by measuring and analyzing various optical spectra of the semiconductors. We first introduce physical parameters to define the spectroscopy in Chap. 1, discuss the microscopic light–matter interactions in semiconductors in Chap. 2, and then the fundamental principles of reflection, transmission, photoluminescence, modulation, and photocurrent spectroscopies in terms of the microscopic light–matter interactions in Chaps. 3–6. The reason of the variety of the spectroscopies is just that one spectral analysis cannot usually provide every aspect of the microscopic light–matter interaction of the sample under investigation. For example, the transmission spectroscopy based on the Beer–Lambert law is not proper to measure the refractive index of a sample. The combination of the various spectroscopies in studying one sample is highlighted in Chap. 6 where we learn the functioning of quantum-well-based infrared photodetector. We close the book by introducing the latest applications and developments of semiconductor spectroscopy in the field of bio-nanophotonics in Chap. 7 where semiconductor colloidal quantum dots are described for research and development of novel biomarkers using absorbance, fluorescence, time-resolved fluorescence (ns scale), fluorescence blinking (ms), and bioimaging.

The unique feature of the book is the two-way highway of numerical analysis bridging microscopic quantum physics and macroscopic spectroscopic setups and measuremental results, which distinguishes this book from many books on this topic. Built mostly on the knowledge of university physics (e.g., Young and Freedman, University Physics with Modern Physics, 13th edition. Addison-Wesley Boston 2012), the book is largely self-contained. Details of theoretical analysis of spectroscopy (physics analysis, mathematical operations, and physics-based numerical approximations) are clearly presented stepwise in light of experimental details and results.

The book is an introduction for studying the physics of spectroscopy and spectroscopic analysis of materials. All studied cases presented in the book are about real materials and devices with valid physical parameters so that the book can be further used as a toolbox for researching and developing nanomaterials based on spectroscopy.

Despite our best efforts, it is inevitable that there will be errors in the book. We encourage readers who find errors to let us know so that we can correct them in the future printing.

Shanghai, China

Wei Lu
Ying Fu

Contents

Chapter 1
Optical Spectral Measurement

Abstract A monochromatic light is quantified by its brightness and color, i.e., its optical power and wavelength. And a light beam is normally composed of many monochromatic lights of different powers and wavelengths. In this chapter we discuss three optical dispersive devices, namely, prism, diffraction grating, and Fourier transform spectroscopy, to resolve the compositing monochromatic lights of a light beam. Prism and diffraction grating disperse the compositing monochromatic lights in the light beam into different spatial regions, and are therefore denoted as spatial dispersion, while Fourier transform spectroscopy disperses the light beam in the time domain.

Just like human eye senses light by its color and brightness, we describe quantitatively a light beam as an electromagnetic field (or electromagnetic wave) using two physical parameters, one is the wavelength λ (color), measured in meters and the other the optical power of the electromagnetic field denoted as S (brightness), measured in watts per meter squared (W/m^2). Theoretically we can assign one wavelength to a light beam, resulting in the so-called monochromatic light denoted as $S(\lambda)$. A typical green light-emitting diode (LED) can be described as its emission wavelength $\lambda = 550$ nm and its light output power 20 mW. However no real light beam is purely monochromatic, therefore scientifically speaking, monochromatic light only refers to a light beam of a narrow band of wavelengths. A more precise description is that the green LED emits a light beam with a peak wavelength 550 nm, a wavelength bandwidth 20 nm and light output power 20 mW, or something similar. Mathematically, $S(\lambda)$ of the green LED is a function of λ which is maximal at $\lambda = 550$ nm. $S(\lambda = 540$ nm) is nonzero because of the finite bandwidth. And

$$\int \left[\int S(\lambda)\, \mathrm{d}\lambda \right] \mathrm{d}\boldsymbol{a} = 20\,\mathrm{mW} \tag{1.1}$$

where $\boldsymbol{a}$ is a geometric surface enclosing the green LED. Discussions about the spatial dependence of $S(\lambda)$ in the above equation may lead us to imaging, which will be looked at in Chaps. 4 and 7, while the dependence of $S(\lambda)$ on λ, known as the optical spectrum of the light beam, is the central theme of this book.

W. Lu and Y. Fu, *Spectroscopy of Semiconductors*, Springer Series in Optical Sciences 215, https://doi.org/10.1007/978-3-319-94953-6_1

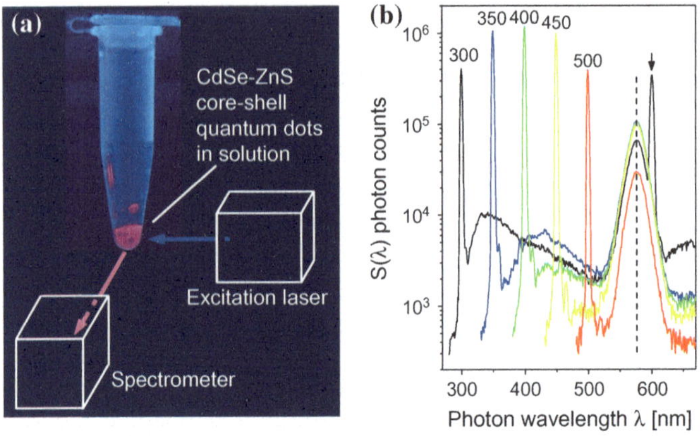

Fig. 1.1 **a** Schematics of quantum dot spectral characterization. **b** Optical spectra of the quantum dots under different excitations of different wavelengths. Note that the color in the figures does not represent the true color associated with the light in reality, it is only used for clear presentation

Figure 1.1 shows an optical spectrum obtained from the following experimental setup: There is a drop of water-soluble CdSe-ZnS core-shell quantum dots [1] in an Eppendorf tube, an excitation laser beam, represented by the blue arrow in Fig. 1.1a, incidents to the quantum dot solution in the Eppendorf tube and the quantum dot photoluminescence, the violet arrow, is collected and recorded by a spectrometer (we will extensively discuss these quantum dots in Chap. 7). The spectrometer is normally positioned at a 90° angle relative to the excitation laser to avoid directly reading the excitation laser which in many cases may be too strong for the spectrometer. The excitation laser light can reach the spectrometer via reflection, diffraction etc., which may be blocked by putting a proper optical filter in front of the spectrometer. Figure 1.1b is the readout from the spectrometer, i.e., the optical spectrum $S(\lambda)$ of the CdSe-ZnS core-shell quantum dots. The horizontal axis is the wavelength λ in unit of nm (10^{-9} m), and the vertical axis is $S(\lambda)$ in units of photon counts, i.e., the number of photons that are recorded per a certain wavelength interval. There is no optical filter in Fig. 1.1a so that the excitation laser light, mostly scattered by the quantum dots in the Eppendorf tube, is also detected and appeared in the optical spectrum of Fig. 1.1b. We use different excitations of different wavelengths, namely, 300, 350, 400, 450 and 500 nm, respectively, to excite the quantum dots. The resulting quantum dot photoluminescence peak is centered at 580 nm (vertical dash line) with a bandwidth of about 20 nm (which is known to be the full width at half maximum, FWHM), independent of the wavelength of the excitation laser. Quite interestingly we also observe a double-wavelength excitation light at 600 nm marked by a vertical arrow in Fig. 1.1b accompanying the 300 nm excitation laser light. We will discuss this accompanying light shortly in Sect. 1.2.

We can see directly from Fig. 1.1b that the quantum dot's photoluminescence spectrum characterized by the photoluminescence peak wavelength and FWHM of the photoluminescence peak does not depend on the wavelength of the excitation laser, at least within the excitation strength and wavelength range in Fig. 1.1b. Moreover, bandwidths (measured by FWHM) of the excitation laser lights are much narrower than that of the quantum dot's photoluminescence peak. We may ask why and how these optical properties are correlated with the electronic properties of quantum dots and later on how can we utilize these properties. Optical spectrum is the result of the interaction between material, i.e., quantum dots in Fig. 1.1, and light, normally referred to as the light-matter interaction, so that the precise and correct spectral measurement and thereafter the spectral analysis are two key factors to understand the light-matter interaction. The principal object of the spectral measurement and analysis is to extract the relationships between optical power S and wavelength λ before and after the light-matter interaction, and the experimental setup of the spectral measurement consists of a light source, its incidence to the sample (the material under investigation), and the recording of light before and after the light-matter interaction, schematically shown in Fig. 1.1.

The basic quantification of our visual perception is the brightness versus darkness, i.e., the optical power. Light is further categorized by its wavelength such as radio wave, microwave, infrared light, visible light (red, yellow, green, blue etc.), ultraviolet light, X ray, and Gamma ray. By these we define the optical spectrum $S(\lambda)$ of a light beam, e.g., Fig. 1.1b, showing quantitatively how the light beam is composed of monochromatic lights of different optical powers and wavelengths.

We normally need a device, generally denoted as an optical spectrometer, to decompose the light beam into monochromatic lights. Human eye, naturally one of the optical spectrometers, has three types of cones that response red (wavelength 500–700 nm), green (450–630 nm), and blue (400–500 nm) lights, the whole response wavelength range is 400–700 nm, which is referred to as the visible spectral range. Figure 1.1b display a spectrum in the wavelength range of 280–670 nm with a spectral resolution of 1 nm, i.e., $d\lambda$ in (1.1), which is defined as the smallest difference in wavelength that can be distinguished by the spectrometer. There are different spectrometers with different response wavelength ranges and spectral resolutions depending on the eventual applications of the spectrometers.

Now we know that the measurement of an optical spectrum is first to decompose the light beam into its constituent monochromatic lights then measure the optical powers of the monochromatic lights. Light decomposition is normally realized by an optical dispersive device. Optical filters selectively transmit monochromatic lights in particular ranges of wavelengths while blocking the remainder, they are generally referred to as direct optical dispersions.

Below we introduce three optical dispersive devices which can provide us with optical spectra in broad wavelength ranges, namely, prism dispersion, diffraction grating, and Fourier transform spectroscopy. Through the following sections we will see that the dispersing strategies of these dispersive devices are different, prism dispersion and diffraction grating disperse monochromatic lights in the light beam into different spatial regions, thus can be denoted as spatial dispersion, while Fourier transform spectroscopy disperses the light beams in the time domain.

1.1 Prism to Disperse Light

The most well-known optical dispersive device is the triangular prism with an isosceles triangular base and rectangular sides. It is also known as the Cornu prism, as shown schematically in Fig. 1.2a. Another commonly-used triangular prism is a half of the Cornu prism with a reflective mirror, known as the Littrow prism, see Fig. 1.2b.

Let us take a close look at the Cornu prism to understand its dispersing mechanism. Refer to Fig. 1.2a, the Cornu prism is described geometrically by a vertex angle α and a base length AC $= \ell$, and physically by the refractive index of the prism material as a function of the wavelength denoted as $n(\lambda)$. Let a monochromatic light beam of wavelength λ incident to facet ABB'A' of the prism of Fig. 1.2a from vacuum or air at an incident angle θ_i. The light will exit the prism from facet CBB'C' after two refractions. Here we implicitly assume that the prism has perfect geometric surfaces so that we apply Snell's law (see more in Sect. 3.1) to describe the optical path of the light beam, which is shown in Fig. 1.3.

Since the refractive index of air is 1 (very close to 1.0 in the optical range of our interest), the angles of the optical path in Fig. 1.3 are correlated by Snell's law of refraction

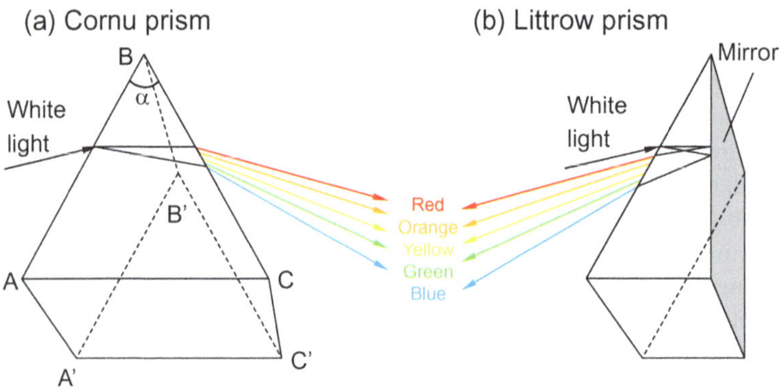

Fig. 1.2 Schematics of **a** Cornu prism and **b** Littrow prism that disperse light

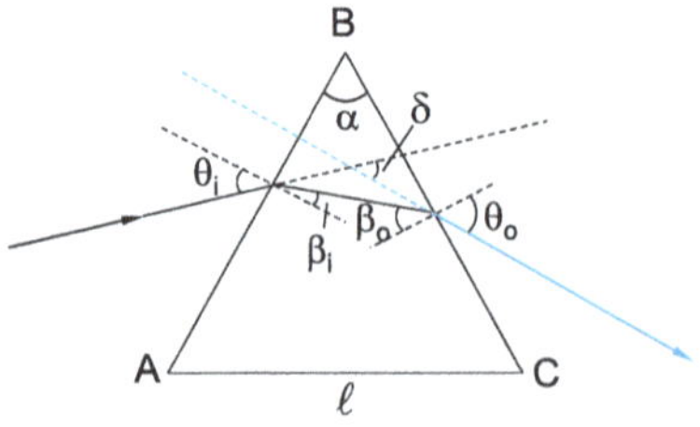

Fig. 1.3 Schematics of the light refraction in a Cornu prism described geometrically by its vertex angle α and base length AC $= \ell$. δ is the deflection angle between the directions of the incident light and the deflected light

$$\sin\theta_\mathrm{i} = n(\lambda)\sin\beta_\mathrm{i}, \quad \sin\theta_\mathrm{o} = n(\lambda)\sin\beta_\mathrm{o} \tag{1.2}$$

where θ_i is the incident angle and θ_o the output angle.

A deflection angle δ, also known as the deviation angle, is normally introduced here to describe how much the light is deflected by the prism, see Fig. 1.3. Since the refractive index $n(\lambda)$ of the prism material is a function of wavelength λ, deflection angle δ therefore also depends on λ, explicitly expressed as $\delta(\lambda)$, which represents the ability of the prism to break the light down into its constituent monochromatic lights, i.e., the light dispersion. In common applications, the geometric setup of the prism and the light beam is designed in such a way that $\theta_\mathrm{i} = \theta_\mathrm{o}$, which means $\beta_\mathrm{i} = \beta_\mathrm{o}$ as well. Under such a condition,

$$\theta_\mathrm{i} = \frac{\alpha + \delta(\lambda)}{2}, \quad \beta_\mathrm{i} = \frac{\alpha}{2} \tag{1.3}$$

Differentiate the first equation of (1.2) with respect to wavelength λ results in

$$\cos\left[\frac{\alpha + \delta(\lambda)}{2}\right]\frac{\mathrm{d}\delta(\lambda)}{\mathrm{d}\lambda} = 2\sin\left(\frac{\alpha}{2}\right)\frac{\mathrm{d}n(\lambda)}{\mathrm{d}\lambda} \tag{1.4}$$

from which we obtain the angular dispersion of the prism

$$\frac{\mathrm{d}\delta(\lambda)}{\mathrm{d}\lambda} = \frac{2\sin\left(\frac{\alpha}{2}\right)}{\sqrt{1 - n^2(\lambda)\sin^2\left(\frac{\alpha}{2}\right)}}\frac{\mathrm{d}n(\lambda)}{\mathrm{d}\lambda} \tag{1.5}$$

which shows that the angular dispersion of the prism is determined by vertex angle α and the dependence of the refractive index on wavelength λ, i.e., $\mathrm{d}n(\lambda)/\mathrm{d}\lambda$.

Thus far, we have studied the prism from the geometric-optics point of view, meaning that the optical path of a light beam in a uniform space is a straight line. Let us take a look at the implication of light's wave nature. When a light beam passes through a single slit of width d, the spatial distribution of the output light at an observation plane, which is also known as the image of the slit, forms the so-called diffraction pattern, as shown in Fig. 1.4a, which can be easily obtained by the following considerations. Refer to Fig. 1.4b, we divide the light (with a total electric field amplitude E_0) from the slit into sub-beams (amplitude $E_0\mathrm{d}y/d$) such as beam 1 with a width $\mathrm{d}y$ originating at y. The length of its optical path to the point y' at the observation plan is $\sqrt{D^2 + (y' - y)^2}$. In practice, however, the distance D between the slit and the observation position is much larger than d ($y' \gg d$), so that all beams between $y \in (-d/2, d/2)$ reach point y' with the same angle θ as beam 2. The electric field of the light reaching y' is then

$$E(\theta, t) = \frac{E_0}{d}\int_{-d/2}^{d/2}\sin\left(\omega t + \frac{2\pi y\sin\theta}{\lambda}\right)\mathrm{d}y \tag{1.6}$$

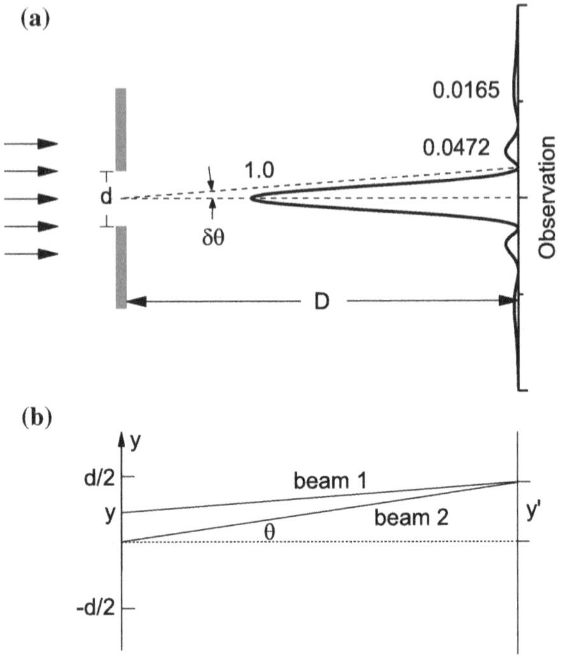

Fig. 1.4 **a** Schematics of the light diffraction through a single slit. Numerical values near peaks denote peak optical powers. **b** Optical paths. See text for details

where ω is the frequency of the light. Simple mathematical manipulations result in

$$E(\theta, t) = E_0 \sin\Big(\omega t + \beta\Big)\frac{\sin\beta}{\beta} \tag{1.7}$$

where $\beta = \pi d \sin\theta/\lambda$. Since the optical power of the light $S \propto \langle |E(t)|^2\rangle_t$, where $\langle \ldots \rangle$ denotes time average, we obtain

$$S(\theta) = S_0 \left[\frac{\sin\left(\frac{\pi d \sin\theta}{\lambda}\right)}{\frac{\pi d \sin\theta}{\lambda}}\right]^2 \tag{1.8}$$

In the observation plane the Fraunhofer diffraction pattern of the slit, i.e., the numerical profile of the above equation, appears as shown in Fig. 1.4a.

It is then easy to identify the first power minimum closest to the principal peak at an angle $\delta\theta$, see Fig. 1.4a,

$$\frac{\pi d \sin(\delta\theta)}{\lambda} = \pi \quad \rightsquigarrow \quad \sin(\delta\theta) = \frac{\lambda}{d}$$

which is well approximated as

$$\delta\theta = \frac{\lambda}{d} \tag{1.9}$$

since $d \gg \lambda$ in practice.

For the prism of Fig. 1.3, the incident light strikes side $\overline{\mathrm{AB}}$ of the prism with an equivalent slit width

$$d = \overline{\mathrm{AB}} \cos \theta_{\mathrm{i}}$$

Since the width of base $\overline{\mathrm{AC}}$ is

$$\ell = 2\overline{\mathrm{AB}} \sin\left(\frac{\alpha}{2}\right)$$

we re-formulate (1.5) to be

$$\frac{\mathrm{d}\delta(\lambda)}{\mathrm{d}\lambda} = \frac{\ell}{d} \frac{\mathrm{d}n(\lambda)}{\mathrm{d}\lambda} \tag{1.10}$$

By the well-established Rayleigh criterion, the minimum resolvable detail occurs when the first diffraction minimum of the output of one light beam coincides with the maximum of another. Thus, we need to have

$$\mathrm{d}\delta(\lambda) \geq \delta\theta \tag{1.11}$$

in order to distinguish two monochromatic lights whose wavelengths are separated by $\mathrm{d}\delta(\lambda)$. Insert (1.5) and (1.10) to (1.11),

$$\frac{\lambda}{\mathrm{d}\lambda} \leq \ell \frac{\mathrm{d}n(\lambda)}{\mathrm{d}\lambda} \tag{1.12}$$

The above equation shows that the maximal spectral resolution

$$R \equiv \frac{\lambda}{\mathrm{d}\lambda}$$

of a dispersing prism, known as the resolving power, is then given by

$$R = \ell \frac{\mathrm{d}n(\lambda)}{\mathrm{d}\lambda} \tag{1.13}$$

Thus, a broad base $\overline{\mathrm{AC}}$, i.e., a large ℓ, results in a large resolving power. However, (1.13) shows that the resolving power is also proportional to the dependence of the refractive index $n(\lambda)$ on wavelength λ, which unfortunately is quite small for most commonly used prism materials, which therefore is the critical limiting factor for the light-dispersing prism. Assume that we make a prism with $\ell = 15$ cm using glass N-BK7HT from SCHOTT [2]. Figure 1.5 show the refractive index $n(\lambda)$, resolving power $R(\lambda)$, and spectral resolution $\mathrm{d}\lambda$ as functions of λ. Here we see that depending on $\mathrm{d}n/\mathrm{d}\lambda$, we can reach a spectral resolution of ca 0.5 nm when $\lambda = 300$ nm, 5 nm

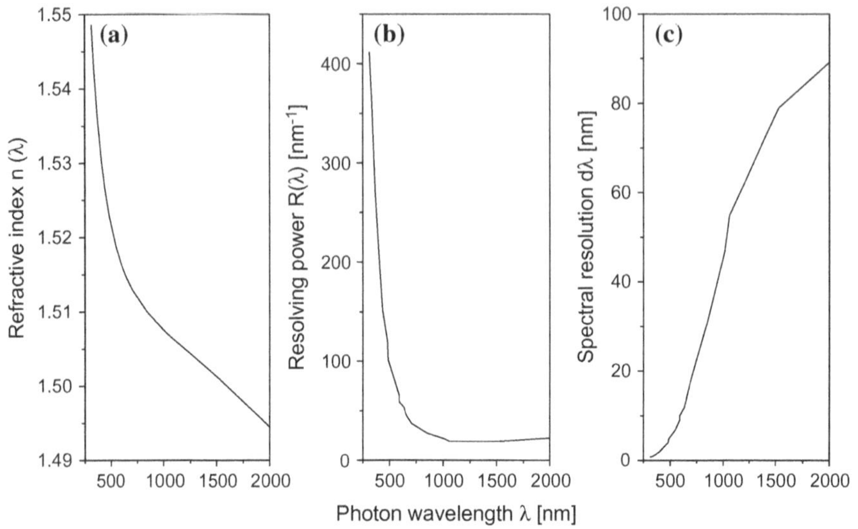

Fig. 1.5 **a** Refractive index $n(\lambda)$ of glass N-BK7HT from SCHOTT [2]. **b** Resolving power $R(\lambda)$ and **c** spectral resolution $d\lambda$ of a prism with $\ell = 15$ cm using glass N-BK7HT

at 500 nm, 10 nm at 600 nm within the visible range, but 100 nm at 2 μm in the short-wavelength infrared range.

The above analysis about the resolving power of the prism is based on the fundamental principles of physics, which sets the upper limit for the resolving power that can be reached by the prism. There are many other limiting factors in reality that can significantly deteriorate the performance of a dispersing prism. Two most important technical issues are the spatial uniformity of the prism material (which will cause the non-uniformity of the refractive index) and the surface roughness in the facets, especially when we want to increase the geometric size of the prism, i.e., ℓ, to compensate the weak dependence of the refractive index $n(\lambda)$ on wavelength λ of many materials that make the prisms.

The advantage of the dispersing prism is its capability of light dispersion over a broad wavelength range. The refractive indices of commonly used prisms are mostly monotonic functions of wavelength over a broad wavelength range, so, as long as the incident light is composed of monochromatic lights of different wavelengths, the monochromatic lights will, in principle, always exit at different deviation angles and thus are resolved in space (thus is referred to as spatial dispersion, as mentioned before).

1.2 Diffraction Grating

The principle of the dispersing prism is largely based on geometric optics that because the refractive index depends on the wavelength, the propagations of monochromatic lights of different wavelengths are dispersed into different spatial directions (the resolving power was however discussed in terms of the wave nature of light). Grating, also known as diffraction grating, disperses monochromatic lights to different exit directions by utilizing the interference effects of the diffracted lights. In other words, the principle of the diffraction grating is based totally on the wave optics of light.

There are two diffraction grating configurations, one is the transmission and the other the reflection. In the transmission grating configuration, the optical energy is largely concentrated in the zeroth order of the diffraction which is not dispersive, so that the reflection grating is commonly used to disperse lights. In the following we focus on the light dispersion of the reflective diffraction grating.

Figure 1.6 shows schematically a blazed grating which is one special type of the reflective diffraction grating. The grating has a constant line spacing d. The grating is alternatively defined by $N = 1/d$, called the grating constant, which is the number of lines per unit length. The grating lines normally possess a triangular, sawtooth-shaped cross-section, forming a step structure. The steps are tilted at the so-called blaze angle α with respect to the grating surface.

Let a light incident to the grating surface at an incident angel θ_i with respect to the normal direction $\boldsymbol{n}$ of the grating surface. The light will be diffracted by various steps in the step structure on the grating surface. Consider two diffracted beams, beam 1 and 2 in Fig. 1.6, outgoing at a diffraction angle θ_o. The two beams have a phase difference $d\left(\sin\theta_\mathrm{i} - \sin\theta_\mathrm{o}\right)$ after diffraction as a function of the diffraction angle θ_o. When the phase difference satisfies the following equation

$$d\left[\sin\theta_\mathrm{i} - \sin\theta_\mathrm{o}^{(m)}\right] = m\lambda \tag{1.14}$$

the two beams will interfere constructively at the observation plane perpendicular to the outgoing direction. Equation (1.14) is normally referred to as the grating equation,

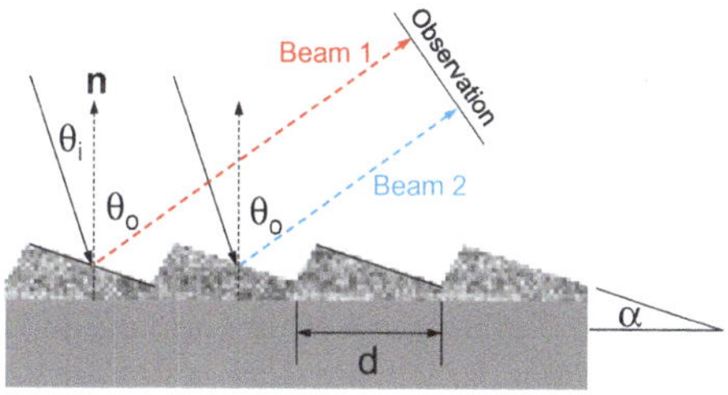

Fig. 1.6 Schematics of a reflective diffraction grating specified by its grating constant $N = 1/d$ and blaze angle α. A light incidents to the grating at an incident angle θ_i and becomes diffracted. The two diffracted beams, beam 1 and 2 interfere constructively at a certain outgoing angle θ_o

where m, a positive integer, is the diffraction order, and $\theta^{(m)}$ is the diffraction angle of the constructive interference of order m.

Equation (1.14) shows that for a fixed θ_i and m, $\theta_o^{(m)}$ is a function of λ. Now, when a light beam, a mixture of two monochromatic lights of two different wavelengths λ_1 and λ_2, strikes to the grating surface with the same incident angle $\theta_i = \theta_{i1} = \theta_{i2}$, the two monochromatic lights are spatially separated via the two different diffraction angles of the constructive interference

$$\theta_{o1}^{(m)} \neq \theta_{o2}^{(m)} \tag{1.15}$$

This is the dispersing mechanism of the diffraction grating to disperse lights of different wavelengths.

Note that by (1.14), when a light of λ interferes constructively at θ_o, lights of $\lambda_\ell = \lambda/\ell$, where $\ell = 1, 2, \ldots$, also interfere constructively at the same diffraction angle (of different diffraction orders which are $m\ell$, of course). Let us denote the optical power $S^{(m)}(\lambda)$ which is constructive at θ_o of a diffraction order m. What we observe at θ_o is the sum of lights

$$\sum_{\ell=1}^{\infty} S^{(m\ell)}\left(\frac{\lambda}{\ell}\right) \tag{1.16}$$

A more thorough analysis shows that the above equation is not complete. If the diffraction order m is larger than 1, say $m = 2$, we expect constructive interferences of lights of $\lambda_\ell = \lambda/\ell$ with $\ell = 1/2, 1, 2, \ldots$ by (1.14). Thus, the total sum of lights observed at θ_o is

$$S(\theta_o) = \sum_{m\ell=1}^{\infty} S^{(m\ell)}\left(\frac{\lambda}{\ell}\right) \tag{1.17}$$

where m and $m\ell$ are integers.

This is commonly referred to as the overlapping effects of diffraction. Recall Fig. 1.1 that we use excitation lights of different wavelengths to excite quantum dots. When an excitation light of 300 nm is selected, a light of 600 nm is also recorded by the spectrometer, see the vertical arrow in Fig. 1.1b. In the optical spectrometer (FluoroMax-3, HORIBA JOBIN YVON in the case of Fig. 1.1b) that excites the quantum dots then records the quantum dot photoluminescence, the selection of the excitation light is made by a diffraction grating from a broad light source of a xenon arc lamp. A monochromatic light of 300 nm is then expected to be accompanied by a monochromatic light of 600 nm when the diffraction order m is larger than 1, and the 600 nm is exactly recorded by the spectrometer in the wavelength range of Fig. 1.1b.

By the way, in practice, the grating operates normally at the so-called blazing order such that

$$\theta_i + \theta_o = 2\alpha \tag{1.18}$$

For a certain diffraction order of m, (1.14) shows that the diffraction angle $\theta_o^{(m)}$ is a function of wavelength λ. For a fixed incident angle θ_i, differentiating (1.14) with respect to wavelength λ results in

$$\frac{d\theta_o^{(m)}}{d\lambda} = -\frac{m}{d\cos\theta_o^{(m)}} = -\frac{\sin\theta_i - \sin\theta_o^{(m)}}{\lambda\cos\theta_o^{(m)}} \tag{1.19}$$

which is known as the angular dispersion of the grating, representing the ability of the grating to disperse lights of different wavelengths. A larger angular dispersion means a better ability to separate lights of different wavelengths.

A good angular dispersion can be achieved by optimizing the geometric structure of the grating. Take a close look at (1.19), a large angular dispersion can be obtained by a large diffraction order m and a small d (i.e., a large grating constant $N = 1/d$). At the blazing order of (1.18) and when the light incidents normal to the grating ($\theta_i = 0$), $\theta_o = 2\alpha$ so that

$$\frac{d\theta_o^{(m)}}{d\lambda} = -\frac{m}{d\cos(2\alpha)} \tag{1.20}$$

which shows that a larger blazing angle α means a better angular dispersion.

However, the dispersion capability of the diffraction grating is limited by not only its angular dispersion but also the diffraction limit because of the wave nature of light, very much alike the prism discussed in the previous section. When a light of wavelength λ strikes the grating of a total width W at an incident angle θ_i, the grating is equivalent to a single slit of width $W\cos\theta_i$. By (1.11), the angle between the two power minima close to the principal power peak is

$$\delta\theta = \frac{\lambda}{W\cos\theta_i} \tag{1.21}$$

Applying the Rayleigh criterion, we need to have

$$d\theta_o^{(m)} \geq \delta\theta$$

in order to be able to distinguish the two diffracted monochromatic lights. By (1.19) and (1.21),

$$\frac{\lambda}{d\lambda} \leq \frac{W\cos\theta_i\left[\sin\theta_o^{(m)} - \sin\theta_i\right]}{\lambda\cos\theta_o^{(m)}} \tag{1.22}$$

Similar to the previous section, we define the resolving power R

$$R = \frac{W\cos\theta_i\left[\sin\theta_o^{(m)} - \sin\theta_i\right]}{\lambda\cos\theta_o^{(m)}} \tag{1.23}$$

showing that a broad grating (i.e., a large grating width W) will result in a good resolving power R. In other words, we need a broader grating to achieve a better dispersion capability. In reality, the fabrication of a broad grating with uniform and periodic lines is not trivial.

In comparison with the prism, the diffraction grating can reach a much higher resolving power. However, it has the overlapping effect of (1.17) so that supplemental optical filters are commonly used in the diffraction grating applications (such as to filter away the 600 nm light in Fig. 1.1b) to realize light dispersion in a restricted optical wavelength range. Another different aspect about the prism dispersion and the diffraction grating is that the diffraction grating is normally made of metallic materials with high refractive indices, while the materials of prism are normally highly transparent.

To obtain a basic picture about a realistic diffraction grating, we go to thorlabs.com and look at the typical 500 nm blaze wavelength reflective diffraction grating, say Item nr GR25-0305: width × height × depth $= 25 \times 25 \times 6\,\text{mm}^3$, blaze wavelength $= 500\,\text{nm}$ (the wavelength of light that yields the greatest absolute efficiency of the diffraction grating), $N = 300$ per mm, blaze angle $\alpha = 4°18'$, and dispersion $= 3.32$ nm/mrad (which is the inverse of the angular dispersion defined by (1.19)).

1.3 Fourier Transform Spectroscopy

The principle of the Fourier transform spectroscopy is totally different from the ones of the dispersing prism and the diffraction grating that decompose a multi-colored light into its compositing monochromatic lights along different spatial output directions. Fourier transform spectroscopy is realized through the Michelson interferometer shown schematically in Fig. 1.7.

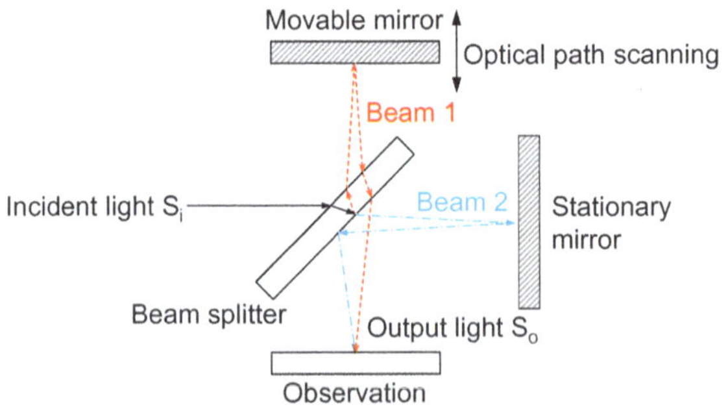

Fig. 1.7 Schematics of the Michelson interferometer. An incident light is split into two beams by the beam splitter then merged after the optical path of beam 1 is modified by a movable mirror. Mathematical analysis, i.e., Fourier transform, of the output light $S_o(\Delta)$ in terms of the optical path difference Δ between beam 1 and 2 gives us information of $S_i(\lambda)$

When an incident monochromatic light $S_i(\lambda)$ of wavelength λ is introduced into the Michelson interferometer, it is split into two beams by the beam splitter, one of them, beam 1, strikes to the reflecting movable mirror, and the other, beam 2 goes to the stationary mirror. When the movable mirror is controlled in such a way to introduce an optical path difference Δ to the two beams, the output light $S_o(\Delta)$ at the observation plane is a function of Δ

$$S_o(\Delta) = S_i(\lambda)\left[1 + \cos\left(\frac{2\pi\Delta}{\lambda}\right)\right] \tag{1.24}$$

Since λ in the above equation is a denominator, we commonly introduce $1/\lambda = \nu$ as the wavenumber of the monochromatic light in order to simplify mathematical expressions. When the incident light is composed of many monochromatic lights, i.e., a polychromatic light, the output light is

$$S_o(\Delta) = \sum_{\nu} S_i(\nu)\left[1 + \cos\left(2\pi\nu\Delta\right)\right] \tag{1.25}$$

For a distributed polychromatic light such that $S_i(\nu)$ is a continuous function of ν, the summation in the above equation becomes integration

$$S_o(\Delta) = \int_0^{\infty} S_i(\nu)\left[1 + \cos\left(2\pi\nu\Delta\right)\right] d\nu \tag{1.26}$$

Figure 1.8a shows a typical relationship between the output light and the optical path difference Δ, normally known as the interferogram, while (b) shows the part of the interferogram in the vicinity of the interference maximum, and (c) displays the situation when Δ is very large. Zero optical path difference means that the optical paths of the two beams are equal after passing through the Michelson interferometer, and the output light is the most strongest, since all monochromatic lights are at the constructive interference state. The output signal oscillates as a function of Δ, its optical power also decreases gradually until the signal is eventually drowned by the noise. We observe strong oscillations of the output light in the vicinity of the interference maximum. The interference effect persists even the optical power oscillation is weak when the optical path difference is long. Figure 1.8a is obtained by measuring $S_o(\Delta)$ as a function of Δ, which is deduced from the speed of the movable mirror and the measurement time.

The optical spectrum of the original incident light $S_i(\lambda)$ is finally extracted from $S_o(\Delta)$ by the mathematic Fourier transform of either (1.25) when the spectrum is discrete or (1.26) for the continuous spectrum. Consider only the modulated part of the continuous spectrum, (1.26) becomes

$$S_o(\Delta) = \int_0^{\infty} S_i(\nu)\cos\left(2\pi\nu\Delta\right) d\nu \tag{1.27}$$

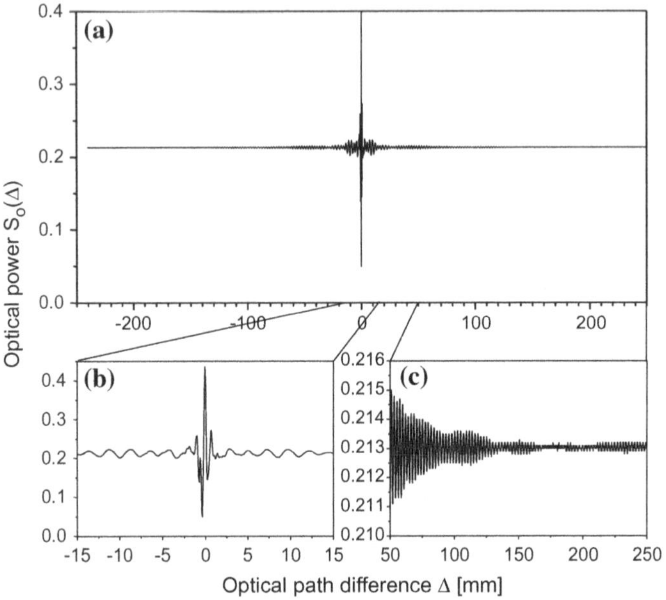

Fig. 1.8 **a** A typical interferogram $S_o(\Delta)$; **b** Interferogram around interference maximum; **c** Interferogram when the optical path difference is large

by which the spectrum of the incident light beam is obtained by the inverse Fourier transform

$$S_i(\nu) = 4\int_{-\infty}^{\infty} S_o(\Delta)\,\cos\left(2\pi\nu\Delta\right)\mathrm{d}\Delta \tag{1.28}$$

therefore the name of Fourier transform spectroscopy. The above mathematic operation can be easily performed via either software or hardware.

It is easily realized that $S_i(\lambda)$ depends on the whole data set of $S_o(\Delta)$ as a function of Δ, which is the speed of the movable mirror multiplied by the measurement time. In other words, during the Fourier transform spectroscopy measurement, any error or mistake at any time will affect the determination of the whole spectrum $S_i(\lambda)$. This is rather different from the dispersing prism and diffraction grating measurements where we measure directly as well as independently $S_i(\lambda)$ as a function of λ. One error or mistake at one wavelength can be corrected by a new measurement at that specific wavelength, while the correction or revision of a Fourier transform spectroscopic spectrum is much more involved. Therefore, the measurement of the Fourier transform spectroscopy is in general much more complicated and demanding as compared with the prism dispersion and diffraction grating measurements.

The optical spectrum $S_i(\lambda)$ after Fourier transforming the interferogram $S_o(\Delta)$ of Fig. 1.8 is presented in Fig. 1.9. This is an optical spectrum in the infrared range, consisting of discrete monochromatic lights below 1000 cm^{-1} as well as a broad

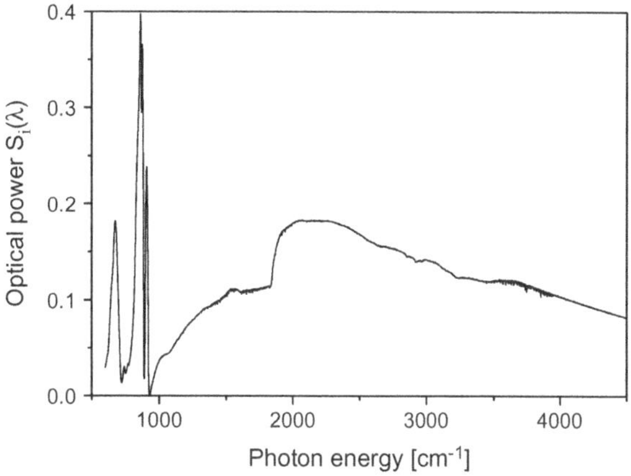

Fig. 1.9 Optical spectrum $S_i(\lambda)$ obtained by Fourier transforming the interferogram $S_o(\Delta)$ of Fig. 1.8

continuous band above 1000 cm^{-1}. As implicated by the Fourier transform, a broad spectrum will be reflected in the interferogram as a fast decay of the interference oscillation as a function of the optical path difference. Conversely, a narrow peak in the optical spectrum will be reflected as a slow decay in the interferogram. At the extreme situation, a monochromatic light, i.e., a single δ function of wavelength corresponds to a monotonic cosine function in the interferogram. All these are clearly reflected in Figs. 1.8 and 1.9. In Fig. 1.9, the broad and continuous spectrum above 1000 cm^{-1} is related to the fast oscillating and quick decaying feature in Fig. 1.8b in the vicinity of the zero optical path difference. The periodic and slow decaying feature in Fig. 1.8c results in the discrete and sharp peaks below 1000 cm^{-1} in Fig. 1.9.

The resolving power of the Fourier transform spectroscopy is determined by the maximal optical path difference Δ_{max} (unit = cm) in the Michelson interferometer that its inversion $1/\Delta_{max}$ (unit = cm^{-1}) is denoted as the spectral resolution $\delta\nu$ of the Fourier transform spectroscopy. Physically, the spectral resolution $\delta\nu$ means to introduce a phase difference of 2π between two monochromatic lights of ν_1 and ν_2, $\nu_1 - \nu_2 = \delta\nu$, i.e.,

$$2\pi\left(\nu_2 - \nu_1\right)\Delta_{max} = 2\pi\delta\nu\Delta_{max} = 2\pi \tag{1.29}$$

so that

$$\delta\nu = \frac{1}{\Delta_{max}} \tag{1.30}$$

Thus, for a Fourier transform spectroscopy with a spectral resolution of 2 cm^{-1}, the maximal optical path difference between the two split beams is 0.5 cm; For a high resolution of 0.01 cm^{-1}, Δ_{max} needs to be 100 cm, i.e., $\Delta_{max} = 1$ m! For whatever the optical path difference, (1.27) and (1.28) requires an absolute error less than half the wavelength since

$$\cos\left(2\pi\nu\Delta\right) = \cos\left[2\pi\nu(\Delta + \lambda)\right]$$

For a monochromatic light of 2 μm wavelength (short-wavelength infrared), the mechanical motion and the position of the movable mirror is to be controlled within an error range of less than 1 μm. We understand immediately the meaning of this 1 μm in terms of 1 m for the Fourier transform spectroscopy, both the mechanical precision control as well as the resistance to the disturbance from environment during the measurement.

Note that in a common interferogram such as Fig. 1.8, the amplitude of the optical power oscillation decreases following the increase of the optical path difference. Let us denote the value of the optical path difference Δ_{noise} when the oscillation amplitude equals the noise level of the spectroscopy. Further increasing Δ will not give us new information. Thus, the noise level of the spectroscopy also sets an upper limit for the spectral resolution. When $\Delta_{\text{noise}} < \Delta_{\max}$, the final spectral resolution of the spectroscopy is determined by Δ_{noise}. In other words, the spectral resolution will be reduced when the measurement sensitivity becomes deteriorated because of the decrease of the sensitivity of the photodetector, the efficiency of the light source, the reflectivity of the mirrors along the optical path, the efficiency of the beam splitting, etc.

2017 Nobel Prize in Physics was awarded to Rainer Weiss, Barry C. Barish, and Kip S. Thorne "for decisive contributions to the LIGO detector and the observation of gravitational waves." The LIGO detector is a laser-based interferometer consisting of two arms that form an L. A single laser beam is split at the corner of the L into two identical beams that travel along the arms and reflect back from the mirrors mounted at the ends of the arms then meet again at the corner. When the lengths of the two arms are equal, the two beams meet at the corner synchronously. A passing gravitational wave will compress one arm while stretch the other, resulting in the loss of the synchronization where the beams meet, thereafter becoming captured by the LIGO detector. With various careful noise considerations and because the changes of the arm length caused by the gravitational wave was estimated to be very small and proportional to the arm length, the arm length in the LIGO detector was 4 km long, and the light beam originated from a pre-stabilized Nd:YAG laser with a wavelength of 1064 nm [3].

1.4 Modulation Spectroscopy Based on Fourier Transform

When measuring an optical spectrum in the infrared range, especially in the long-wavelength (8–15 μm) and far infrared range (15–1000 μm), many factors start to affect the measurement capability as well as the sensitivity. The factors in the top list include light sources, photodetectors and various optical components. Prism dispersion and diffraction grating are to be replaced by the Fourier transform spectroscopy in order to reach the necessary spectral resolution in the infrared range since we simply cannot use a dispersing prism or a diffraction grating with a geometric size of 1 m in order to reach a spectral resolution of 0.01 cm^{-1}, see (1.13) and (1.23). Fourier transform spectroscopy is thus developed to circumvent this issue in space

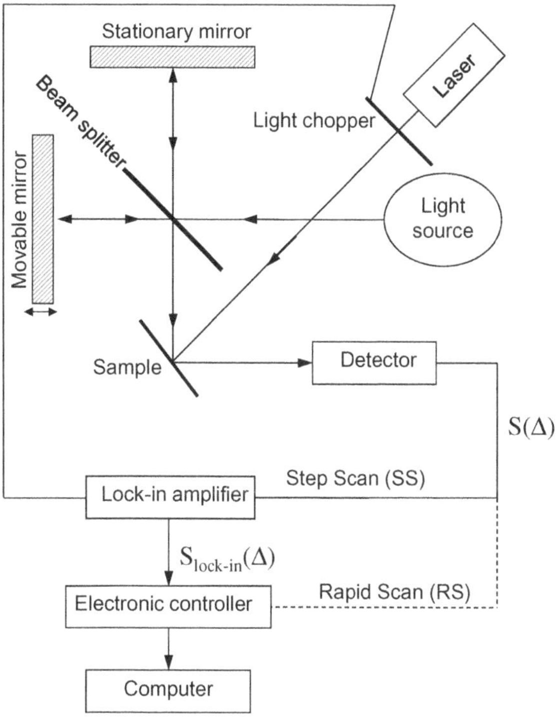

Fig. 1.10 Schematic optical setup of the modulation spectroscopy based on the Fourier transform spectroscopy

(large device size) by obtaining the optical spectrum in the time domain (optical path modulation by the movable mirror in the Michelson interferometer).

We briefly discussed the noise issue in the previous section which can be a dominant limiting factor in determining the spectral resolution of the Fourier transform spectroscopy. In order to detect weak signals, modulation spectroscopy is developed that a periodically modulated signal is added to the sample, which will induce a periodic response(s) from the sample. The periodic response(s), which may be rather weak, can be detected via frequency and/or phase lock-in techniques that suppress noises which are in principle random in both the space and time domains. We shall learn the modulation spectroscopy in details in Chap. 5.

Figure 1.10 shows schematically the optical setup of a modulation spectroscopy based on the Fourier transform spectroscopy. A probing continuous light beam from the light source, normally polychromatic, is split into two beams that strike to the stationary and movable mirrors, respectively. The reflected beams are then introduced to the sample. At the same time, a monochromatic light from a laser is introduced to the sample through a light chopper (thus periodical modulation). Light from the sample is collected by the detector, forming $S(\Delta)$. Both $S(\Delta)$ and the modulating light are fed into the lock-in amplifier which extracts the signal $S_{\mathrm{lock-in}}(\Delta)$ from $S(\Delta)$ that has the same frequency and same phase as the modulating light.

Let us be a bit more specific in order to understand the functioning mechanism of the modulation spectroscopy. We say that we want to measure the reflectivity of the material, which is the response of electrons inside the material to the external light, i.e., the so-called light-matter interaction. $S_{\text{lock-in}}(\Delta)$ from the modulation spectroscopy is then the change of the reflectivity of the material caused by the modulating light, which can be further traced back to the most fundamental light-matter interaction inside the material.

Fourier transform modulation spectroscopy can work in two modes. One is the so-called rapid scan (RS) mode. In this RS mode, the movable mirror moves at a constant velocity v, and the optical path difference at time t is given by $\Delta = 2vt$. The interferogram, always an explicit function of time t, is now determined by both the speed of the movable mirror and the frequency of the modulating light, for which, the RS mode is also known as the double-modulation mode. The RS mode works well when the optical signal does not change or changes very slowly relative to the time needed to complete the measurement. In most practical infrared applications, the optical response frequency is not high, normally only tens kHz. Therefore, a proper choice of scanning speed is very critical for the RS operation [4, 5].

The other operation mode is the step scan (SS) mode. Here the movable mirror is moved incrementally stepwise. When the sample is modulated, the temporal dependence of the response signal, i.e., the interferogram, is totally determined by the modulating light. Fourier transform spectroscopy commonly uses highly-stable He-Ne laser to calibrate the optical path difference Δ. And during the operation, the same laser light may be scattered to the sample and subsequently acquired by the photodetector. Moreover, background radiation from ambient environment may also be acquired, contributing to the output signal $S(\Delta)$. In the SS mode, real signal from the sample is expected to have the same temporal dependence as the modulating light, while all other signals do not. A lock-in amplifier can therefore easily extract the true sample signal from others such as due to background radiation or the calibration He-Ne laser [6].

1.5 A Few Key Notes in Spectral Measurement

In the previous two sections we have listed a few critical factors in spectral measurements. As we all know, there are millions of details in any experiment that need to be totally under control. For spectral measurements, the photodetector acquires any light that reaches the photodetector without any discrimination, including the light from the sample, as well as lights from the light source via reflection and diffraction etc. Moreover, the geometric size of the photodetector is normally small so that light collection/focus is always a key issue. To stably keep the whole measurement setup at its optimal operation status, especially for field operations in various and varying environments, is a daunting task. It is generally known that the absolute optical power is the most difficult parameter to be properly assessed, while the wavelength is almost the easiest to be absolutely measured, and the next easiest is the geometric shape of an optical spectrum such as the width of the quantum dot photoluminescence peak in Fig. 1.1b. This is normally why most spectral analysis focuses on the peak wavelength and the spectral peak width, while the analysis of the optical power is secondary.

The measurement of a reflection spectrum is much vulnerable than the transmission spectrum since it depends much on the surface roughness of the sample as well as the geometric control of the incident and reflected lights with respect to the sample surface. The sample positioning in the reflection spectrum needs further care such that its reflection surface should be aligned with the calibration mirror (to acquire the background signal of the system). The calibration mirror is highly reflective with a metal coating for a reflectivity of as close to 100% as possible, free of dust, and most importantly, free of diffraction.

In infrared spectral measurement, the whole measurement setup is commonly put into a vacuum chamber to avoid absorption by CO_2 and H_2O in the atmosphere, which can significantly modify the optical path of the measurement setup if the setup was previously optimized in another ambient environment. Therefore, the calibration of the reference spectrum is to be performed in the same environment as acquiring the spectrum of the sample. This can be illustrated by the following spectral analysis procedures.

By impinging a probing light $S_\mathrm{i}(\lambda)$ on a sample we measure the optical response such as reflection spectrum $S_\mathrm{sample}(\lambda)$ of the sample. Ideally, it is only $S_\mathrm{i}(\lambda)$ that strikes the sample to induce $S_\mathrm{sample}(\lambda)$ and the optical signal we measure is

$$D_\mathrm{sample}^\mathrm{ideal}(\lambda) = S_\mathrm{sample}(\lambda)\ S_\mathrm{i}(\lambda) \tag{1.31}$$

Here we take the linear optics as an example where the output signal is proportional to the probing light and the optical response of the sample. Happily we deduce $S_\mathrm{sample}(\lambda)$ as

$$S_\mathrm{sample}(\lambda) = \frac{D_\mathrm{sample}^\mathrm{ideal}(\lambda)}{S_\mathrm{i}(\lambda)} \tag{1.32}$$

However, $S_\mathrm{i}(\lambda)$ can induce many secondary lights before striking the sample which will also contribute to the output signal. In real measurement therefore,

$$D_\mathrm{sample}^\mathrm{real}(\lambda) = S_\mathrm{sample}(\lambda)\ S(\lambda) \tag{1.33}$$

where $S(\lambda)$ includes all the lights that strike the sample. The determination of $S(\lambda)$ is not trivial and experimentally the problem is circumvented by calibrating the measurement setup through measuring the signal from a reference sample such that $D_\mathrm{reference}(\lambda) \approx S(\lambda)$ and

$$S_\mathrm{sample}(\lambda) \approx \frac{D_\mathrm{sample}^\mathrm{real}(\lambda)}{D_\mathrm{reference}(\lambda)} \tag{1.34}$$

Many spectrometers output $S_{sample}(\lambda)$ directly from their softwares. However we notice by (1.34) that $D_{reference}(\lambda)$ is the denominator so that even a very small error or fluctuation in $D_{reference}(\lambda)$ in some frequencies or frequency ranges where $D_{reference}(\lambda)$ is close to zero, such as absorption lines of CO_2 and H_2O in the infrared range, will result in a huge peak in $S_{sample}(\lambda)$ which is not necessarily originated from the sample. Thus, a proper reference about $D_{reference}(\lambda)$ when analyzing $S_{sample}(\lambda)$ is of critical importance. We will present an example of the reference measurement in Sect. 6.3.

As we have learned before, Fourier transform spectroscopy measures indirectly the optical spectrum. It first measures the interferogram then converts the interferogram into the optical spectrum through mathematical Fourier transform. When environmental perturbations, such as electromagnetic radiations from a nearby AC power source, are accidentally acquired, they will appear in the output spectrum from the Fourier transform. The simplest way to check and double check the final result is to run repeatedly the measurement by varying the scanning speed. Normally, true spectral features do not depend on the scanning speed, while any features that depend on the scanning speed should be deleted.

The vertical axis of an optical spectrum is normally the optical power (equivalently, photon count), while the horizontal axis differs in different measurements, which can be either the photon energy or the wavenumber (such as Fig. 1.9 from Fourier transform spectroscopy), or the wavelength (such as Fig. 1.1 from a grating). When analyzing optical spectra from semiconductors we normally prefer to use the photon energy as the unit for the horizontal axis, since physical models are commonly based on the energy band structures of the semiconductors. For example, absorption peaks are normally correlated to the electron transition from one energy level to another so that the photon energy of an absorption peak corresponds directly to the energy difference between the two energy levels. Another example for using the photon energy as the unit of the horizontal axis is that optical interference induces a series of peaks equally distanced in energy. This is very useful when analyzing optical spectrum of a thin film since the film's geometrical structure can easily induce light interference effects. A further example about the unit is that the absorption and emission peaks are commonly described by either Lorentz or Gaussian peaks, depending on the dominant energy-relaxation processes, when the photon energy is used as the unit of the horizontal axis. On the contrary, the energy feature of the electron transition becomes vague when wavelength is used as the unit; or we cannot assess directly the interference feature; or the absorption and/or emission peak will not be symmetric in a horizontal axis using the unit of wavelength.

The photon energy is normally in the unit of electron volt (eV) or wavenumber (cm^{-1}), while the wavelength can be nm or μm. The conversion between photon energy E and wavelength λ is straightforward:

$$E\,[\mathrm{eV}] = \frac{1.2398}{\lambda\,[\mu\mathrm{m}]} \tag{1.35}$$

and $1\,[\mathrm{eV}] = 8.066 \times 10^3$ wavenumber $[\mathrm{cm}^{-1}]$.

References

1. H. Xu, H. Brismar, Y. Fu, Influence of surface states on blinking characteristics of single colloidal CdSe-CdS/ZnS core-multishell quantum dot. J. Colloid Interface Sci. **505**, 528–536 (2017)
2. www.schott.com. Data sheet: schott_-2017-01-20.pdf
3. www.nobelprize.org/nobel_-prizes/physics/laureates/2017/
4. A.R. Reisinger, R.N. Roberts, S.R. Chinn, T.H. Myers II, Photoluminescence of infrared-sensing materials using an FTIR spectrometer. Rev. Sci. Instrum. **60**, 82–86 (1989)
5. F. Fuchs, A. Lusson, J. Wagner, P. Koidl, Double modulation techniques in Fourier transform infrared photoluminescence. Proc. SPIE **1145**, 323–326 (1989)
6. J. Shao, W. Lu, F. Yue, X. Lü, W. Huang, Z. Li, S. Guo, J. Chu, Photoreflectance spectroscopy with a step-scan Fourier-transform infrared spectrometer: technique and applications. Rev. Sci. Instrum. **78**, 013111(8pp) (2007)

Chapter 2
Introduction to Physics and Optical Properties of Semiconductors

Abstract As the theoretical basis for the whole book, we introduce fundamental physics and optical properties of semiconductors in this chapter. We start the chapter by describing the electrons and their energy band structures in semiconductors based on quantum mechanics, then the electromagnetic field by the Maxwell's equations. The focus of the chapter, and actually the central theme of the book, namely, the light-matter interaction, is studied in terms of first quantizing the electromagnetic field in terms of the number of photons and the energy carried per photon then calculating the optical transition of the electron between energy bands in semiconductors after either absorbing or emitting one photon. We also describe the phonon spectrum of the semiconductor which is responsible for many key features in the infrared spectra of the semiconductors.

Semiconductors are materials that have moderately good conductivities, which are higher than those of insulators and lower than those of metals, resulting in the name of semi-conductor. Various semiconductors exhibit different optical properties. Here are just a few of these optical properties: Irradiation with light can transform the semiconductor from insulator-like behavior to metal-like behavior. The optical absorption spectra of semiconductors normally exhibit a threshold. Below the threshold frequency, light can pass through with practically no losses, whereas above it the light is strongly absorbed. Silicon is the main-stream material in semiconductor electronics industry. Its optical applications are however limited since it is an indirect energy bandgap material; While III–V semiconductors are main actors in photonics, crystalline silicon solar panels dominate worldwide markets.

All these macroscopic properties of a semiconductor can be traced back to a common microscopic origin: its energy band structure of electrons and electron-photon interaction. In this chapter we present a brief introduction and quantum mechanical description of electrons and photons in semiconductor materials. The chapter starts with the basic electronic energy band structure of a single atom, the modifications of the energy band structure when more atoms are brought together to form bulk materials, i.e., solid states. We study the light-matter interaction in the following way. Quantum electrodynamics describes fully the light-matter interact to an unprecedented accuracy. It is a relativistic theory. Since our study is non-relativistic and the

W. Lu and Y. Fu, *Spectroscopy of Semiconductors*, Springer Series in Optical Sciences 215, https://doi.org/10.1007/978-3-319-94953-6_2

classical Maxwell's equations can be derived from the quantum electrodynamics in the realm of non-relativity, light in transport is described as an electromagnetic field by the Maxwell's equations in terms of the amplitude and the wavelength of the electromagnetic field, while it is a photon when interacting with matter quantified by the number of photons and the energy carried per photon. The focus of the chapter, and actually the central theme of the book, namely, the interaction between the electron and the photon, the so-called light-matter interaction, is studied in terms of the optical transition of the electron between energy bands in semiconductors after either absorbing or emitting one photon, at the end of the chapter. In the following chapters, we will introduce other physical processes related to various optical spectra such as plasmon oscillation of free electrons in Sect. 3.5, multiphoton excitation in Sect. 4.6 and ultra-fast (femtosecond) optical transitions in Sect. 6.4.

The mathematical equations presented in the chapter look tough. They are however straightforward results of careful (and many times, tedious) systematic analysis, mathematical operations and numerical considerations. For example, we present the Bloch theorem about the energy band structures of electrons in a bulk semiconductor by discussing the periodic lattice structure. The Maxwell's equations are studied in very detail in order to introduce a complete picture, as fundamental as possible, about the light-matter interaction in semiconductors of our interest. Many realistic physical parameters are also considered and discussed so that the pre-conditions of our studies can be checked and consistent with the final conclusions.

2.1 Energy Band Structure of Electron State

A pure semiconductor, also referred to as an intrinsic semiconductor, has an electrical conductivity whose value is normally between the ones of a metal (conductor) and an insulator. The wide applications of the semiconductor are mainly due to the tunability of its electrical conductivity by many means including modulating temperature, illumination and doping. The electrical conductivity of an extremely pure semiconductor decreases following the decrease of its temperature, it may reach zero at zero Kelvin, while a highly-doped semiconductor can have an electrical conductivity as high as that of a metal. Furthermore, a semiconductor will absorb light that impinges on it when the photon energy is higher than a certain threshold value. Photons whose energies are smaller than the threshold value will pass through the semiconductor without significant loss.

Ge and Si, i.e., group IV materials, are dominating semiconductor materials mostly for electronics ever since the beginning of the semiconductor industry marked by the invention of PNP point-contact Ge transistor in 1947 by John Bardeen, William Shockley, and Walter Brattain in Bell Labs. GaAs and other III–V materials have now been widely used because of their direct energy bandgaps and high electron mobilities. At room temperature, the direct bandgap of GaAs is 1.42 eV and its electron mobility is 8500 $cm^2/V{\cdot}s$, while Si is an indirect-bandgap material with a typical value of the electron mobility about 1400 $cm^2/V{\cdot}s$.

All these properties and applications of semiconductors are solely determined by a common microscopic origin: the energy band structures of electrons in the periodic lattice structures of atoms in semiconductors. Today, we study and develop semiconductor nanotechnology within which we position atoms of semiconductor materials in constrained three-dimensional spaces to modify/tailor the energy band structures as well as construct totally novel energy band structures for various applications. For example, [1] reports in 2016 nanostructured amorphous In-Ga-Zn-O field-effect network to sense glucose, while colloidal semiconductor quantum dots, to be closely studied in Chap. 7, have been extensively studied and developed (many of these quantum dots have already been commercialized) in the past 20 years.

Let us focus on the energy band structure of electrons in a semiconductor. In a semiconductor, atoms, either single atoms or groups of different atoms, are positioned in a periodic lattice in space. Energy levels of electrons in a single atom in vacuum are discrete and very narrow (infinitely narrow in theory). Moving two single atoms close to each other will make the wave functions of energy levels of the electrons in the two individual atoms overlap, resulting in interactions between electrons originally confined to their parental atoms. Quantum mechanically, one energy level of the electron in a single atom and the corresponding level in another identical atom (the two energy levels are degenerate when the two identical atoms are well separated in space) will split into two energy levels separated in the energy domain when the two atoms are brought together in the space domain. And the energy difference between the two split energy levels represents the interaction strength between the two energy levels of the individual atoms, it increases following the decrease of the spatial distance between the two atoms. Pure theoretically, degenerate energy levels of individual atoms will split into a number of energy levels when atoms are placed closely with each other. And the number of split energy levels equals the number of atoms, and the energy separations between the split energy levels are inversely proportional to the space distances between atoms. At room temperature, the lattice constant (see below about its precise definition) of GaAs is $a = 5.65\,\text{Å}$, it is 5.43 Å in Si. The density of atoms in these materials is therefore as high as $1/a^3 \sim 10^{28}$ atoms/m^3. At such a high density of atoms, the split energy levels are so densely populated in the energy domain that they form continuous energy bands.

Table 2.1 lists the atomic structures of elements that compose common semiconductors. Electrons in one individual atom are normally categorized into core electrons and valence electrons. Core electrons are tightly bound to the nucleus so that they normally do not interact significantly with valence electrons and electrons of other atoms, while valence electrons interact with each other in the same atom, they also interact with valence electrons of other atoms such as forming chemical bonds between atoms. Core electrons are normally not responsible for the unique properties of semiconductors so are not directly relevant for semiconductor applications. Lattice structures, electronic and optical properties of semiconductors and devices are mostly determined by the valence electrons. Table 2.1 shows that the valence electrons of commonly used semiconductor elements are all either s or p type, which is the major reason that electron states close to the valence bandedges and conduction bandedges in semiconductors are similarly s or p type.

Table 2.1 Atomic structures of elements that compose common semiconductors

Element	Core electrons	Valence electrons
IV-Si	$1s^2 2s^2 2p^6$	$3s^2 3p^2$
IV-Ge	$1s^2 2s^2 2p^6 3s^2 3p^6 3d^{10}$	$4s^2 4p^2$
III-Ga	$1s^2 2s^2 2p^6 3s^2 sp^6 3d^{10}$	$4s^2 4p^1$
V-As	$1s^2 2s^2 2p^6 3s^2 sp^6 3d^{10}$	$4s^2 4p^3$
II-Cd	$1s^2 2s^2 2p^6 3s^2 3p^6 3d^{10} 4s^2 4p^6 4d^{10}$	$5s^2$
VI-Se	$1s^2 2s^2 2p^6 3s^2 3p^6 3d^{10}$	$4s^2 4p^4$

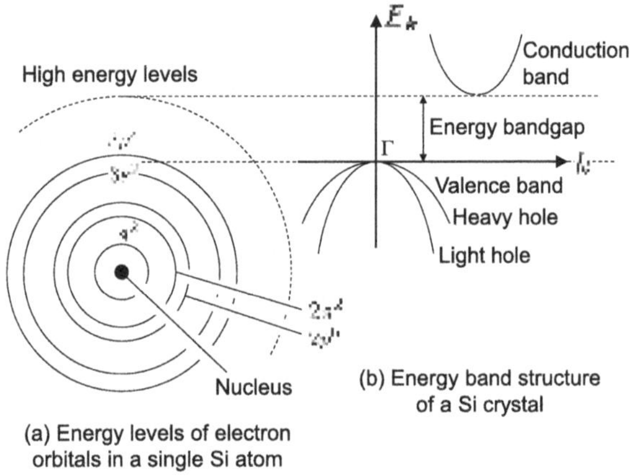

Fig. 2.1 **a** Schematics of energy levels of electron orbitals in an individual Si atom. Electrons in $1s$, $2s$ and $2p$ levels are categorized as core electrons and electrons in $3s$ and $3p$ are valence electrons. **b** Schematic energy band structure of a Si crystal in the $\boldsymbol{k}$ space

In general, when atoms exchange or share valence electrons so that the outmost shells are completely filled, they have a lower electrostatic energy than when they are spatially separated. For example, an individual Si atom has four valence electrons, 2 electrons occupying the single $3s$ energy level and 2 electrons occupying the three degenerate $3p$ energy levels, constituting the outmost shell, see Table 2.1 and Fig. 2.1.

In an infinitely large Si bulk material, the four valence electrons in the s and p levels of one Si atom first hybrid to form four degenerate sp-hybrid electrons then share with four neighboring Si atoms, resulting in a three-dimensional cubic atomic lattice structure in which the outmost sp hybrid states of every Si atom are completely occupied, see Fig. 2.2a. Most semiconductors used in electronics and photonics have this cubic crystal structure. This crystal structure can be decomposed into two interlaced face-centered cubic (FCC) lattices, see Fig. 2.2b, with a spatial displacement of one quarter of the cubic diagonal. The structure can also be viewed as one FCC lattice with a basis composed of two atomic sites at

$$d_1 = (000), \quad d_2 = \frac{a}{4}(111)$$

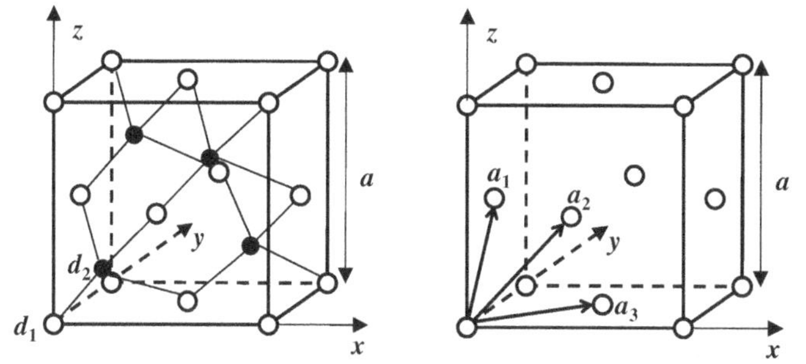

Fig. 2.2 **a** Zincblende crystal structure. d_1 and d_2 may be occupied by two different types of atoms, such as Ga and As, respectively, in GaAs; When they are of same type of atoms, e.g., two Si atoms in a Si crystal, the crystal structure is called the diamond structure. **b** Face-centered cubic (FCC) crystal. $\boldsymbol{a}_1$, $\boldsymbol{a}_2$, and $\boldsymbol{a}_3$ are basis vectors of the crystal. a is the so-called lattice constant

respectively, see Fig. 2.2a, where a is the so-called lattice constant. When two atoms sitting at d_1 and d_2 are identical, such as Si atoms in a pure Si crystal, also other group IV materials, the structure is called the diamond structure. If the two atoms are different, such as GaAs, AlAs and CdS, the structure is referred to as the zincblende structure. Note that the crystal structures of group II–VI materials such as CdSe are normally more complicated than group IV and III–V materials, e.g., β-CdSe is zincblende shown in Fig. 2.2a, while α-CdSe is wurtzite.

In short, atoms in semiconductors are periodically positioned at lattice points in space, which can be mathematically expressed

$$\boldsymbol{R}_m = m_1\boldsymbol{a}_1 + m_2\boldsymbol{a}_2 + m_3\boldsymbol{a}_3 \tag{2.1}$$

where $\boldsymbol{R}_m$ is the space position of lattice point m, $m = (m_1, m_2, m_3)$ are three integers (may be negative), and $(\boldsymbol{a}_1, \boldsymbol{a}_2, \boldsymbol{a}_3)$ are basis vectors of the semiconductor crystal structure. The three basis vectors of the face-centered cubic lattice is shown in Fig. 2.2b.

We now turn to quantum mechanics to study the valence electrons in semiconductor crystals. The Hamiltonian of one electron (we focus on the valence electrons which are simply referred to as electrons from now on), say electron i denoted by its spatial position $\boldsymbol{r}_i$, in a semiconductor whose crystal lattice structure is described by (2.1), is

$$H(\boldsymbol{r}_i) = \frac{-\hbar^2\nabla_i^2}{2m_0} - \sum_m \frac{Z_m^* e^2}{4\pi\epsilon_0|\boldsymbol{r}_i - \boldsymbol{R}_m|} + \sum_{j\neq i} \frac{e^2}{4\pi\epsilon_0|\boldsymbol{r}_i - \boldsymbol{r}_j|} \tag{2.2}$$

where Z_m^* is the effective atomic number of atom m at lattice position R_m defined as the atomic number minus the number of core electrons. $Z^* = 4$ for Si atom, $Z^* = 3$ for Ga atom, $Z^* = 5$ for As atom, and so on. m_0 is the free electron mass. The first term on the right side of the above equation represents the kinetic energy of electron i, the second term is the Coulombic potential energy due to its interactions with nuclei plus core electrons, i.e., Z^*e, and the last term is the Coulombic potential energy due to its interactions with other valence electrons. The next step is to solve the following Schrödinger equation

$$\left[\sum_i H(\boldsymbol{r}_i)\right]\Psi(\boldsymbol{r}_1,\boldsymbol{r}_2,\ldots\boldsymbol{r}_i\ldots) = E\,\Psi(\boldsymbol{r}_1,\boldsymbol{r}_2,\ldots\boldsymbol{r}_i\ldots) \tag{2.3}$$

which in practice is very difficult. What we actually do now is to carefully analyze the situation: Atoms are positioned periodically in space, see (2.1), which implicates that the spatial distributions of electrons, both the core electrons and valence electrons, are also periodically distributed in space! Electron i therefore will "feel" the Coulombic fields from Z^*e which are periodic in space as well as the Coulombic fields of other valence electrons which are also periodic in space. We therefore can formulate the following single-electron Schrödinger equation

$$\left[\frac{-\hbar^2\nabla^2}{2m_0} + V_0(\boldsymbol{r})\right]\Psi(\boldsymbol{r}) = E\,\Psi(\boldsymbol{r}) \tag{2.4}$$

where the second term $V_0(\boldsymbol{r})$ on the left side of the above equation is the effective potential of the crystal which is periodic in space

$$V_0(\boldsymbol{r}) = V_0(\boldsymbol{r} + \boldsymbol{R}_m) \tag{2.5}$$

Note that now we work on the single electron, we drop off subscript i and j that distinguish electrons in the crystal in (2.2) and (2.3).

Consider a one-dimensional model crystal in Fig. 2.3. Atoms are periodically positioned along the z axis at a spatial interval a, which is the lattice constant of this one-dimensional crystal. There are totally N atoms so that the size of the crystal is Na. The whole crystal can be viewed as a composition of N unit cells centered at atoms with a size of a. Naturally there can be more than one atom in each cell, such as the zincblende lattice in Fig. 2.2. Back to Fig. 2.3, the crystal is defined within $z \in (0, Na)$, and the effective periodic potential $V_0(z + R_m) = V_0(z)$, $m = 1, 2, \ldots N$, $R_m = ma$ is the spatial position of lattice site m (it becomes vector $\boldsymbol{R}_m$ of (2.1) in a three-dimensional crystal). In practice the crystal is very large so that we can extend Na along the z axis such that $V_0(z + ma) = V_0(z)$ with any integer m.

A real crystal extends in the three dimensional space. Moreover, the three basis vectors $(\boldsymbol{a}_1, \boldsymbol{a}_2, \boldsymbol{a}_3)$ are not necessarily orthogonal to each other, similar to the case of Fig. 2.2b. Let us simplify the situation by assuming a basis set of orthogonal vectors $(a_x, 0, 0)$, $(0, a_y, 0)$, $(0, 0, a_z)$, along the three axes of a Cartesian coordinate, and the numbers of unit cells are N_x, N_y and N_z. We then extend the three dimensional crystal into the whole mathematical space so that the crystal is now translational symmetric,

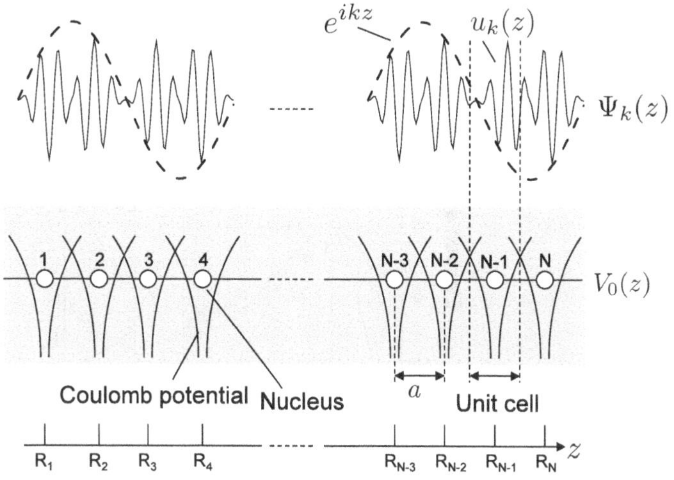

Fig. 2.3 A one-dimensional model crystal

rigorously, $V_0(\boldsymbol{r} + \boldsymbol{R}) = V_0(\boldsymbol{r})$, where $\boldsymbol{R}$ is the spatial position of any lattice site, i.e., (2.5). Here we drop off the subscript "m" in $\boldsymbol{R}$ to simplify mathematical expressions.

Refer to Fig. 2.3. Since $V_0(\boldsymbol{r})$ is periodic in space, the distribution of an electron in this periodic potential must be periodic too. Without losing generality, the solution of (2.4) can be expressed as

$$\Psi_k(\boldsymbol{r}) = u_k(\boldsymbol{r})\, e^{i\boldsymbol{k}\cdot\boldsymbol{r}}, \quad u_k(\boldsymbol{r}) = u_k(\boldsymbol{r} + \boldsymbol{R}) \tag{2.6}$$

so that the electron distribution, which is $|\Psi_k(\boldsymbol{r})|^2$ quantum mechanically, satisfies the requirement to be periodic. This is the well-known Bloch theorem by which we describe the wave function of the electron in a crystal by a plan wave $e^{i\boldsymbol{k}\cdot\boldsymbol{r}}$ modulated by a periodic function $u_k(\boldsymbol{r})$, see Fig. 2.3. Here $\boldsymbol{k}$ is normally known as the wave vector of the electron in a crystal.

Insert (2.6) into (2.4) to obtain the following equation for $u_k(\boldsymbol{r})$

$$\left[\frac{\hbar^2\boldsymbol{k}^2}{2m_0} + \frac{-i\hbar^2\boldsymbol{k}\cdot\nabla}{m_0} + \frac{-\hbar^2\nabla^2}{2m_0} + V_0(\boldsymbol{r})\right] u_k(\boldsymbol{r}) = E_k\, u_k(\boldsymbol{r}) \tag{2.7}$$

which shows that $u_k(\boldsymbol{r})$ depends on $\boldsymbol{k}$. This is very important, see more below when we study the light-matter interaction. This is why we denote $u_k(\boldsymbol{r})$ by subscript "$\boldsymbol{k}$", so is the case for the eigen energy $E = E_k$ that is $\boldsymbol{k}$-dependent. $u_k(\boldsymbol{r})$ is commonly called the periodic function, or Bloch function. Possible numerical values of $\boldsymbol{k}$ depend on the lattice structure which form the so-called $\boldsymbol{k}$ space, which was already introduced in Fig. 2.1 where a schematic energy band structure E_k is shown in the $\boldsymbol{k}$ space.

Because of the large number of unit cells existed in a crystal, there are a lot of eigen solutions from the above equation (recall the discussion of energy level split

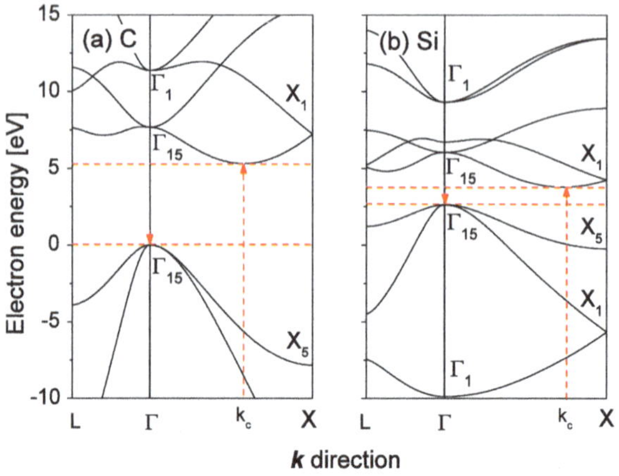

Fig. 2.4 Energy band structures of crystal C (**a**) and Si (**b**) calculated from the sp^3s^* tight-binding theory. $\boldsymbol{k}_c$ denotes the wave vector of the conduction bandedge

when atoms are brought together). It is found from numerical calculations that the eigen solutions form energy bands, which can be denoted by band index ℓ such that

$$\Psi_{\ell \boldsymbol{k}}(\boldsymbol{r}) = u_{\ell \boldsymbol{k}}(\boldsymbol{r})\, e^{i\boldsymbol{k}\cdot\boldsymbol{r}}, \quad u_{\ell \boldsymbol{k}}(\boldsymbol{r}) = u_{\ell \boldsymbol{k}}(\boldsymbol{r}+\boldsymbol{R}) \tag{2.8}$$

with eigen energy $E_{\ell \boldsymbol{k}}$. The relationship between $E_{\ell \boldsymbol{k}}$ and $\boldsymbol{k}$ is known to be the energy dispersion relationship, or energy band structure. Moreover,

$$E_{\ell,\boldsymbol{k}} = E_{\ell,-\boldsymbol{k}} \tag{2.9}$$

because of the periodicity of the crystal lattice in space. Figure 2.4 shows the energy band structures of crystal C and Si calculated by using the so-called sp^3s^* tight-binding theory [2].

A little explanation is needed to understand Fig. 2.4. The vertical axis is the electron energy in unit of eV. The horizontal axis is "$\boldsymbol{k}$ direction". For the FCC lattice of Fig. 2.2b, the $\boldsymbol{k}$ space, i.e., the numerical values of $\boldsymbol{k}$ in (2.6), are denoted by points of symmetry such as

$$\Gamma = (0,0,0), \quad \mathrm{X} = (1,0,0), \quad \mathrm{L} = \left(\frac{1}{2}, \frac{1}{2}, \frac{1}{2}\right) \tag{2.10}$$

in terms of unit vectors along k_x, k_y and k_z directions (whose lengths are $2\pi/a$, where a is the lattice constant of the FCC lattice). Thus in Fig. 2.4, the $\boldsymbol{k}$ value from Γ to X symmetry point changes linearly

$$(0,0,0) \rightarrow \left(\frac{2\pi}{a}, 0, 0\right)$$

From Γ to L,

$$(0, 0, 0) \rightarrow \left(\frac{\pi}{a}, \frac{\pi}{a}, \frac{\pi}{a}\right)$$

Now we know the energy band structure, the next step is to fill electrons of the crystal to the energy levels in the band structure, starting from the lowest energy level. According to the Pauli exclusion principle, each energy level can be maximally occupied by two electrons, one spin up and one spin down. It just happens that for intrinsic semiconductors and insulators at a not-too-high temperature, valence electrons fill completely energy bands up to the so-called the valence band (VB). In other words, the valence band is the highest occupied band. All those energy bands which are higher than the valence band are totally empty. The lowest unoccupied band is called the conduction band (CB). In a metal, however, the conduction band is partially filled. Note that the temperature of the semiconductor affects the electron occupation of the energy level. We put a condition of a not-too-high temperature here temporarily.

By these we are able to understand, at least qualitatively, the property differences among an insulator, a semiconductor and a metal that we read at the beginning of this chapter: First of all, the electron transport is described quantum mechanically by its electric current density vector

$$\boldsymbol{j}_k(\boldsymbol{r}) = \frac{-e\hbar}{2im_0}\left[\Psi_k^*(\boldsymbol{r})\nabla\Psi_k(\boldsymbol{r}) - \Psi_k(\boldsymbol{r})\nabla\Psi_k^*(\boldsymbol{r})\right] \tag{2.11}$$

which for electron state $\Psi_k(r)$ of (2.6) becomes

$$\boldsymbol{j}_k = \int \boldsymbol{j}_k(\boldsymbol{r})\,\mathrm{d}\boldsymbol{r} = \frac{-e\hbar\boldsymbol{k}}{m_0} \tag{2.12}$$

after a few simple but necessary mathematical operations. The above equation indicates that the electrical conductivities of insulators and intrinsic semiconductors are in theory zero since their energy bands up to the valence bands are completely filled. $\boldsymbol{k}$ and $-\boldsymbol{k}$ states, whose energies are equal, see (2.9), are equally occupied but their contributions to the electrical conduction cancel with each other, i.e.,

$$\boldsymbol{j}_k + \boldsymbol{j}_{-k} = 0 \tag{2.13}$$

Metal, on the other hand, has electrons that partially fill the conduction band, thus is highly conducting.

The valence band and the conduction band is separated in the energy domain by the so-called energy bandgap, see the energy ranges between horizontal dashed lines in Fig. 2.4. The energy bandgaps of common semiconductors are about 1.0 eV, while they are normally higher than 3.0 eV for insulators. Electrons that originally occupy the valence band can be thermally excited to the conduction band if the energy bandgap is not too wide, which is the case for a semiconductor (think about the thermal excitation energy of k_BT which is 25 meV at room temperature of

$T = 300$ K), therefore, an intrinsic semiconductor can conduct at a temperature with a corresponding thermal excitation energy comparable to the energy bandgap, while an insulator needs to be heated up to an extremely high temperature (may already exceeds its melting temperature) in order to conduct. Doping can induce impurity levels inside the energy gap that narrows the energy bandgap that contributes to the electrical conductivity. Furthermore doping a certain amount Si atoms to replace Ga atoms in GaAs will introduce extra valence electrons which occupy partially the conduction band, making the Si-doped GaAs conducting.

One property about the semiconductor we introduced at the beginning of this chapter is that impingement of light with an energy larger than a certain value on a semiconductor makes the semiconductor conducting and light below this certain photon energy will pass through the semiconductor without significant loss. We now understand this that this certain photon energy value corresponds to the energy bandgap. When the photon energy is larger than the energy bandgap, the photon will be absorbed by the electron initially occupying the valence band, exciting the electron to jump to the conduction band. The process is called an optical transition, which makes the conduction band partially filled and leaves some valence-band states unoccupied, resulting in the absorption of the photon and an increased electrical conductivity. Electron cannot absorb photon whose energy is smaller than the energy bandgap because there is no available state in the energy bandgap for the electron to stay (this is the so-called one-photon excitation and we shall discuss multiphoton excitation later).

We need to dig a bit deeper about the energy band structure in order to understand the optical spectrum in the coming chapters. Quantum mechanically, the amplitude of an electron wave function represents the distribution of the electron in space. Thus, the wave function should be normalized in space. Consider a three-dimensional crystal defined by lattice vectors $(a_x, 0, 0)$, $(0, a_y, 0)$, $(0, 0, a_z)$, and the numbers of unit cells are N_x, N_y and N_z along the x, y, and z directions, respectively. The volume of the unit cell is $\Omega_{\text{cell}} = a_x a_y a_z$, and $N = N_x N_y N_z$ is the total number of unit cells in the crystal with a total crystal volume $\Omega = N\Omega_{\text{cell}}$. The spatial integration of the electron distribution in space is

$$\begin{aligned}\int_{\Omega} |\Psi_{\ell\boldsymbol{k}}(\boldsymbol{r})|^2 \mathrm{d}\boldsymbol{r} &= \int_{\Omega} |u_{\ell\boldsymbol{k}}(\boldsymbol{r})|^2 \mathrm{d}\boldsymbol{r} = \sum_{\boldsymbol{R}} \int_{\text{cell}} |u_{\ell\boldsymbol{k}}(\boldsymbol{r}+\boldsymbol{R})|^2 \mathrm{d}\boldsymbol{r} \\ &= \sum_{\boldsymbol{R}} \int_{\text{cell}} |u_{\ell\boldsymbol{k}}(\boldsymbol{r})|^2 \mathrm{d}\boldsymbol{r} = N \int_{\text{cell}} |u_{\ell\boldsymbol{k}}(\boldsymbol{r})|^2 \mathrm{d}\boldsymbol{r} \end{aligned} \tag{2.14}$$

where we first insert (2.6), convert the integration over the crystal space $\int_{\Omega}$ to a summation over unit cells and an integration over one unit-cell space $\int_{\text{cell}}$, then utilize the translational symmetry of the crystal lattice, i.e., (2.8), and the final equality is due to the fact that there are N unit cells (we neglect difference between surface unit cells and unit cells deep in the crystal since the number of surface cells is relatively very small as compared with the number of cells deep in the crystal).

Normally we set the Bloch function as normalized within the unit cell

$$\int_{\text{cell}} |u_{\ell k}(r)|^2 \mathrm{d}r = 1 \tag{2.15}$$

so that the result of (2.14) equals N, resulting in the textbook expression of the Bloch theorem

$$\Psi_{\ell k}(r) = \frac{1}{\sqrt{N}} u_{\ell k}(r)\, e^{ik\cdot r}, \quad u_{\ell k}(r) = u_{\ell k}(r + R) \tag{2.16}$$

As we now understand, an intrinsic semiconductor has a completely filled valence band and a totally empty conduction band at low temperature. It can be thermally or optical excited, or doped for various electronics and photonics operations so that one or both of the valence and conduction bands become partially occupied (for n-type doping, only the conduction band becomes partially occupied). The populations of electrons in the conduction band and missing electrons (called holes) in the valence band are normally low compared with a metal (the electron density in a highly doped GaAs for Ohmic contact with a metal electrode is about 10^{18} cm^{-3}, while it is about 10^{22} cm^{-3} in gold). When we either take away these electrons in a filled valence band starting from the valence bandedge (marked by vertical down arrows in Fig. 2.4), or fill in these electrons to an empty conduction band from the conduction bandedge, we will find that the $\boldsymbol{k}$ values of these electrons are either very small, such as the ones near the valence bandedge, or close to $\boldsymbol{k}_c$ of the conduction bandedge (see vertical up arrows in Fig. 2.4). Note here that in Fig. 2.4, $\boldsymbol{k}_c \neq 0$ of the conduction bandedge and $\boldsymbol{k} = 0$ of the valence bandedge are not equal, the corresponding material is denoted as an indirect energy bandgap material. Figure 2.5 shows the energy band structures of GaAs and InAs. Quite differently from C and Si in Fig. 2.4, both the conduction band and valence band of GaAs and InAs have their optimal points at the Γ symmetry points in the $\boldsymbol{k}$ space, so that GaAs and InAs are called direct energy bandgap materials. We will discuss shortly the impact of this energy band character on the optical properties. For the moment let us discuss the consequences of small $\boldsymbol{k}$ around either Γ or $\boldsymbol{k}_c$.

Equation (2.16) shows that the wave function consists of two parts. The first part describes the periodicity of the electron wave function in the crystal in the form of

$$e^{ik\cdot r}$$

around Γ or

$$e^{i(k+k_c)\cdot r}$$

around $\boldsymbol{k}_c$. Both $\boldsymbol{k}$ in the above two expressions are very small so that

$$e^{ik\cdot r}$$

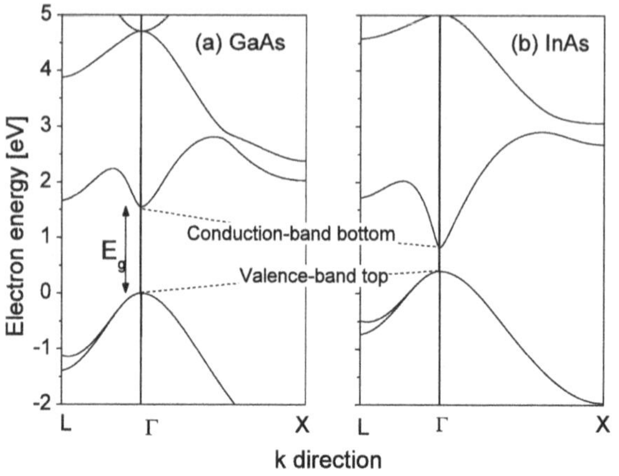

Fig. 2.5 Energy band structures of GaAs (**a**) and InAs (**b**) calculated from the sp^3s^* tight-binding theory

is a slow-varying function in space. The second part in the electron wave function, $u_{\ell k}(\boldsymbol{r})$ represents the orbital nature of atoms in the crystal, see its schematic representation in Fig. 2.3. Its spatial variations are limited within one unit cell. It is normally difficult to describe numerically the two parts with the same numerical precisions (the lattice constant of Si is 5.4 Å and 5.6 Å for GaAs while the geometric sizes of Si and GaAs semiconductors in applications are surely much large). One solution is to separate the two parts. Consider the wave function of (2.16) and its corresponding Schrödinger equation (2.4)

$$\left[\frac{\boldsymbol{p}^2}{2m_0} + V_0(\boldsymbol{r})\right] \Psi_{\ell k}(\boldsymbol{r}) = E_{\ell k} \Psi_{\ell k}(\boldsymbol{r}) \tag{2.17}$$

Let us assume $\boldsymbol{k}_c = 0$ while the reader is encouraged to follow the coming discussion for the case of a nonzero $\boldsymbol{k}_c$ to see its effects. Since quantum mechanically $\boldsymbol{p} = -i\hbar\nabla$ and mathematically $\nabla e^{i\boldsymbol{k}\cdot\boldsymbol{r}} = i\boldsymbol{k}\, e^{i\boldsymbol{k}\cdot\boldsymbol{r}}$,

$$-i\hbar\nabla\left[e^{i\boldsymbol{k}\cdot\boldsymbol{r}} u_{\ell k}(\boldsymbol{r})\right] = -i\hbar\, e^{i\boldsymbol{k}\cdot\boldsymbol{r}}\left(\nabla + i\boldsymbol{k}\right) u_{\ell k}(\boldsymbol{r}) = e^{i\boldsymbol{k}\cdot\boldsymbol{r}}\left(\boldsymbol{p} + \hbar\boldsymbol{k}\right) u_{\ell k}(\boldsymbol{r}) \tag{2.18}$$

The result of the second operation of the momentum operator is $e^{i\boldsymbol{k}\cdot\boldsymbol{r}}\left(\boldsymbol{p} + \hbar\boldsymbol{k}\right)^2 u_{\ell k}(\boldsymbol{r})$ and the Schrödinger equation becomes

$$e^{i\boldsymbol{k}\cdot\boldsymbol{r}}\left[\frac{\left(\boldsymbol{p} + \hbar\boldsymbol{k}\right)^2}{2m_0} + V_0(\boldsymbol{r})\right] u_{\ell k}(\boldsymbol{r}) = e^{i\boldsymbol{k}\cdot\boldsymbol{r}} E_{\ell k} u_{\ell k}(\boldsymbol{r}) \tag{2.19}$$

By removing $e^{i\boldsymbol{k}\cdot\boldsymbol{r}}$ from the two sides of the above equation and expanding $\left(\boldsymbol{p} + \hbar\boldsymbol{k}\right)^2$ we obtain

$$\left[\frac{\boldsymbol{p}^2}{2m_0} + V_0(\boldsymbol{r}) + \frac{\hbar \boldsymbol{k} \cdot \boldsymbol{p}}{m_0} + \frac{\hbar^2 \boldsymbol{k}^2}{2m_0}\right] u_{\ell \boldsymbol{k}}(\boldsymbol{r}) = E_{\ell \boldsymbol{k}} u_{\ell \boldsymbol{k}}(\boldsymbol{r}) \tag{2.20}$$

By comparing the above equation with (2.17), we see that the first two terms remain the same, which is denoted as

$$H_0 = \frac{\boldsymbol{p}^2}{2m_0} + V_0(\boldsymbol{r})$$

Since $\boldsymbol{k}$ is small (this is our current interest that we focus on the energy states close to the band edges), the last two terms can be treated as perturbations so that (2.20) becomes

$$\Big(H_0 + H_1 + H_2\Big) u_{\ell \boldsymbol{k}}(\boldsymbol{r}) = E_{\ell \boldsymbol{k}} u_{\ell \boldsymbol{k}}(\boldsymbol{r}) \tag{2.21}$$

where

$$H_1 = \frac{\hbar \boldsymbol{k} \cdot \boldsymbol{p}}{m_0}, \quad H_2 = \frac{\hbar^2 \boldsymbol{k}^2}{2m_0}$$

are the first-order and the second-order perturbation, respectively.

At the symmetry point Γ where $\boldsymbol{k} = 0$, only H_0 is nonzero. We denote its solutions as $u_\ell(\boldsymbol{r})$, where ℓ is the energy band index of H_0, i.e.,

$$H_0 u_\ell(\boldsymbol{r}) = E_\ell u_\ell(\boldsymbol{r}) \tag{2.22}$$

Surprisingly (or maybe not at all!), it is found that for most commonly used semiconductors, there are three $u_\ell(\boldsymbol{r})$ at Γ in the valence band which are largely p-type, and one $u_\ell(\boldsymbol{r})$ at the conduction bandedge (remember that the conduction bandedge does not always sit at Γ) is s-type. Recall Table 2.1 when we are bringing together individual atoms to form a crystal, eigen energy levels of the valence electrons of an individual atom are no longer eigen because of the interactions among valence electrons, but the s and p characters of the valence electrons of the semiconductor atoms are well preserved in the Bloch functions in crystal.

Back to (2.21). The solutions $u_{\ell \boldsymbol{k}}(\boldsymbol{r})$ for nonzero $\boldsymbol{k}$ are in principle linear combinations of $u_m(\boldsymbol{r})$

$$u_{\ell \boldsymbol{k}}(\boldsymbol{r}) = \sum_m c_{\ell m}(\boldsymbol{k}) u_m(\boldsymbol{r}) \tag{2.23}$$

In the above equation we used m as the band index (the same ℓ in (2.22)) in order to distinguish the band index of finite $\boldsymbol{k}$ from the band index at $\boldsymbol{k} = 0$. Insert the above equation in (2.20),

$$H \sum_m c_{\ell m}(\boldsymbol{k}) u_m(\boldsymbol{r}) = \sum_m c_{\ell m}(\boldsymbol{k}) H \; u_m(\boldsymbol{r}) = E_{\ell \boldsymbol{k}} \sum_m c_{\ell m}(\boldsymbol{k}) u_m(\boldsymbol{r}) \tag{2.24}$$

Multiply $u_\ell^*(\boldsymbol{r})$ then integrate over space

$$\sum_m c_{\ell m}(\boldsymbol{k})\langle u_\ell|H|u_m\rangle = E_{\ell\boldsymbol{k}} \sum_m c_{\ell m}(\boldsymbol{k})\langle u_\ell|u_m\rangle = E_{\ell\boldsymbol{k}} c_{\ell\ell}(\boldsymbol{k}) \tag{2.25}$$

Here we utilize the relation $\langle u_\ell|u_m\rangle = \delta_{\ell m}$, i.e., the Bloch functions are orthogonal and normalized. Note that $\delta_{\ell m} = 1$ when $\ell = m$, it is zero otherwise.

Insert the Hamiltonian of (2.20),

$$\left(E_\ell + \frac{\hbar^2\boldsymbol{k}^2}{2m_0}\right) c_{\ell\ell}(\boldsymbol{k}) + \sum_m \frac{\hbar}{m_0}\langle u_\ell|\boldsymbol{k}\cdot\boldsymbol{p}|u_m\rangle c_{\ell m}(\boldsymbol{k}) = E_{\ell\boldsymbol{k}} c_{\ell\ell}(\boldsymbol{k}) \tag{2.26}$$

Since u_ℓ and ∇u_ℓ have opposite parities, the spatial integration of their product is zero, i.e., $\langle u_\ell|\boldsymbol{p}|u_\ell\rangle = -i\hbar\langle u_\ell|\nabla|u_\ell\rangle = 0$. Thus,

$$\left(E_\ell + \frac{\hbar^2 k^2}{2m_0}\right) c_{\ell\ell}(\boldsymbol{k}) + \sum_{m\neq\ell} \frac{\hbar}{m_0}\langle u_\ell|\boldsymbol{k}\cdot\boldsymbol{p}|u_m\rangle c_{\ell m}(\boldsymbol{k}) = E_{\ell\boldsymbol{k}} c_{\ell\ell}(\boldsymbol{k}) \tag{2.27}$$

Neglect the $\boldsymbol{k}\cdot\boldsymbol{p}$ terms, we have

$$\left(E_\ell + \frac{\hbar^2\boldsymbol{k}^2}{2m_0}\right) c_{\ell\ell}(\boldsymbol{k}) = E_{\ell\boldsymbol{k}} c_{\ell\ell}(\boldsymbol{k}) \tag{2.28}$$

which implies

$$E_{\ell\boldsymbol{k}} = E_\ell + \frac{\hbar^2\boldsymbol{k}^2}{2m_0}, \quad c_{\ell\ell}(\boldsymbol{k}) = 1 \tag{2.29}$$

Now include the $\boldsymbol{k}\cdot\boldsymbol{p}$ terms, it is easy to obtain by the first-order perturbation theory of quantum mechanics that the eigen energy is

$$E_{\ell\boldsymbol{k}} = E_\ell + \frac{\hbar^2\boldsymbol{k}^2}{2m_0} + \frac{\hbar^2}{m_0^2}\sum_{m\neq\ell}\frac{|\langle u_m|\boldsymbol{k}\cdot\boldsymbol{p}|u_\ell\rangle|^2}{E_\ell - E_m} \tag{2.30}$$

and the corresponding wave function is

$$u_{\ell\boldsymbol{k}}(\boldsymbol{r}) = u_\ell(\boldsymbol{r}) + \frac{\hbar}{m_0}\sum_{m\neq\ell}\frac{\langle u_m|\boldsymbol{k}\cdot\boldsymbol{p}|u_\ell\rangle}{E_\ell - E_m} u_m(\boldsymbol{r}) \tag{2.31}$$

Here we observe the relationship between $u_{\ell\boldsymbol{k}}(\boldsymbol{r})$ and $\boldsymbol{k}$, which has a deep impact on the optical transition (see below).

We study more carefully the eigen energy $E_{\ell\boldsymbol{k}}$ of (2.30). To mimic (2.29), we introduce the so-called effective mass m_ℓ^* so that

$$E_{\ell k} = E_\ell + \sum_{ij} \frac{\hbar^2}{2m^*_{\ell,ij}} k_i k_j \tag{2.32}$$

where $i, j = x, y, z$, and

$$\frac{m_0}{m^*_{\ell,ij}} = \delta_{ij} + \frac{2}{m_0} \sum_{m \neq \ell} \frac{\langle m|p_i|\ell\rangle\langle\ell|p_j|m\rangle}{E_\ell - E_m} \tag{2.33}$$

It is very important to observe that the effective mass can be anisotropic, which is closely related to the operation of photodetection using quantum well structures [3].

Normally the interaction between two energy levels is small when the energy separation between the two levels is large. Thus, focus only on the conduction-band state $u_c(\boldsymbol{r})$ and the valence-band state $u_v(\boldsymbol{r})$,

$$\frac{m_0}{m^*_{c,ij}} \approx \delta_{ij} + \frac{2}{m_0} \frac{\langle u_v|p_i|u_c\rangle\langle u_c|p_j|u_v\rangle}{E_c - E_v} \tag{2.34}$$

Moreover, since $E_c - E_v = E_g$ is the energy bandgap,

$$\begin{aligned} E_{ck} &= E_c + \sum_{ij} \frac{\hbar^2}{2m^*_{c,ij}} k_i k_j \\ \frac{m_0}{m^*_{c,ij}} &\approx \delta_{ij} + \frac{2}{m_0} \frac{\langle u_v|p_i|u_c\rangle\langle u_c|p_j|u_v\rangle}{E_g} \end{aligned} \tag{2.35}$$

The above equations show that the effective mass of the conduction-band electron is small when the energy bandgap is narrow. This is exactly so in reality. For commonly used semiconductors, the effective mass of the conduction-band electron and the energy bandgap of InSb are all small. Note that the valence band is more complicated due to interactions among the valence-band levels at the Γ point so that there is not a clear dependence of the effective masses of the valence band levels on the energy bandgap.

It is of interest to take an extra look at the wave function of (2.31) which, after careful study, is actually

$$u_{\ell k}(\boldsymbol{r}) \approx u_\ell(\boldsymbol{r}) + i\boldsymbol{k} \cdot \boldsymbol{r} u_\ell(\boldsymbol{r}) \tag{2.36}$$

Recall that we have so far restrained us to small $\boldsymbol{k}$ so that one more step further from the above equation is

$$u_{\ell k}(\boldsymbol{r}) \approx \left(1 + i\boldsymbol{k} \cdot \boldsymbol{r}\right) u_\ell(\boldsymbol{r}) \approx e^{i\boldsymbol{k}\cdot\boldsymbol{r}} u_\ell(\boldsymbol{r}) \tag{2.37}$$

which is exactly the Bloch theorem when $\boldsymbol{k}$ is very small.

We have studied thus far the energy states of electrons close to the conduction bandedge and the valence bandedge, which are basically all we need to know about semiconductors for utilizing semiconductors. More specifically, the relevant physical parameters that we need to know are: lattice constant a, energy bandgap E_g, effective masses of conduction-band electrons and valence-band holes.

A few further mathematical derivations and physical discussions lead us to the well-established effective mass approximation which is important for studying semiconductor nanostructures. Though details are beyond the scope of this book, the reader can refer to literatures including many monographs in the field, e.g., [4]. One simple way of reasoning is that we replace $\boldsymbol{k}$ in (2.35) by $-i\nabla$ such that

$$E_c + \sum_{ij} \frac{\hbar^2}{2m^*_{c,ij}} k_i k_j \rightarrow E_c + \sum_{ij} \frac{-\hbar^2}{2m^*_{c,ij}} \frac{\partial^2}{\partial x_i \partial x_j} \tag{2.38}$$

When we adopt it as a Hamiltonian and try to find its eigen solutions

$$\left(E_c + \sum_{ij} \frac{-\hbar^2}{2m^*_{c,ij}} \frac{\partial^2}{\partial x_i \partial x_j} \right) \psi_c(\boldsymbol{r}) = E\, \psi_c(\boldsymbol{r}) \tag{2.39}$$

for a bulk material, i.e., a single crystal that extends over the whole space, the solution of $\psi_c(\boldsymbol{r})$ is in the form of $e^{i\boldsymbol{k}\cdot\boldsymbol{r}}$ since E_c is constant in space. $\boldsymbol{k}$ is the wave vector which may depend on E in a complicated manner because of the anisotropic effective mass.

By referring to (2.39) we construct the well-known effective mass approximation for nanostructures: We call $\psi_c(\boldsymbol{r})$ from (2.39) the envelope wave function of the conduction-band electron, or simply the envelope function. The total wave function of E is $\Psi_c(\boldsymbol{r}) = \psi_c(\boldsymbol{r})u_c(\boldsymbol{r})$. For a nanostructure composed of one semiconductor in one space region and another semiconductor in another space region, E_c varies from one space region to another. Thus, in addition to the lattice constant a, energy bandgap E_g and effective masses that we have already specified, one extra parameter is needed that defines the relative energy positions of either the conduction bandedge or the valence bandedge. References [5–7] list the energies of valence bandedges of commonly used semiconductors, known as the valence-band offsets (VBOs). The conduction bandedge is simply VBO + E_g.

The Schrödinger equations of the effective mass approximation for various materials including nanostructures are therefore

$$\begin{aligned}
&\left[\frac{-\hbar^2\nabla}{2m^*_c} + E_g(\boldsymbol{r}) + \mathrm{VBO}(\boldsymbol{r})\right] \psi_c(\boldsymbol{r}) = E_c \psi_c(\boldsymbol{r}) \\
&\left[\frac{\hbar^2\nabla}{2m^*_{hh}} + \mathrm{VBO}(\boldsymbol{r})\right] \psi_{hh}(\boldsymbol{r}) = E_{hh} \psi_{hh}(\boldsymbol{r}) \\
&\left[\frac{\hbar^2\nabla}{2m^*_{\ell h}} + \mathrm{VBO}(\boldsymbol{r})\right] \psi_{\ell h}(\boldsymbol{r}) = E_{\ell h} \psi_{\ell h}(\boldsymbol{r})
\end{aligned} \tag{2.40}$$

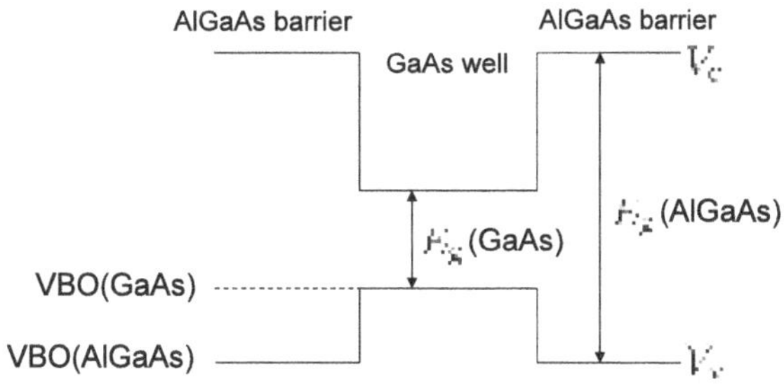

Fig. 2.6 Spatial variations of VBO, energy bandgap E_g, conduction bandedge $V_c = \text{VBO} + E_g$ and valence bandedge $V_v = \text{VBO}$ in an AlGaAs/GaAs quantum barrier/well nanostructure

where $E_g(\boldsymbol{r})$ and VBO($\boldsymbol{r}$) can be functions of space $\boldsymbol{r}$ such as in an AlGaAs/GaAs quantum well of Fig. 2.6, E_g and VBO assume the values of bulk AlGaAs in AlGaAs barriers, they assume the values of bulk GaAs in the GaAs quantum well. The above expressions show that in a crystal that extends over the whole three-dimensional space, the potential energy is constant so that the carrier, either an electron in the conduction band or a hole (either a heavy hole denoted as "hh" or a light hole denoted as "ℓh") in the valence band, moves as a free carrier with a mass m_c^* or m_{hh}^* or $m_{\ell h}^*$ in the form of $e^{i\boldsymbol{k}\cdot\boldsymbol{r}}$, its kinetic momentum is $\hbar\boldsymbol{k}$. Thus, $\hbar\boldsymbol{k}$ is further known as the quasi momentum of the electron motion in crystal ("quasi" since the real wave function of the carrier also contains the Bloch function $u_c(\boldsymbol{r})$ or $u_v(\boldsymbol{r})$).

In (2.40), m_c^*, m_{hh}^* and $m_{\ell h}^*$ are effective masses of electrons, heavy holes and light holes, respectively. In nanostructures, the values of these effective masses are determined in the similar manner of $E_g(\boldsymbol{r})$ and VBO($\boldsymbol{r}$) but not exactly the same. When the effective mass becomes spatial dependent, (2.40) is no longer Hermitian [8, 9]. In practice, the effective mass can be evaluated by the energy level and the wave function. For example, in an AlGaAs/GaAs quantum well, the effective mass of an electron in a state confined in the GaAs region adopts the one of bulk GaAs.

The conduction bandedges of Si do not locate at the Γ point in the $\boldsymbol{k}$ space, see Fig. 2.4. Electrons in the conduction bandedges in Si are commonly described by 6 ellipsoids, with a longitudinal effective mass $m_\ell^* = 0.9163$ and a transverse effective mass $m_t^* = 0.1905$ in terms of free electron mass m_0 [10]. The 6 ellipsoids are grouped into 3 groups, each 2-fold degenerate:

1. $(m_x^*, m_y^*, m_z^*) = (m_t^*, m_t^*, m_\ell^*)$
2. $(m_x^*, m_y^*, m_z^*) = (m_\ell^*, m_t^*, m_t^*)$
3. $(m_x^*, m_y^*, m_z^*) = (m_t^*, m_\ell^*, m_t^*)$

where m_x^*, m_y^* and m_z^* are effective masses along x, y and z direction, respectively. By (2.38) we then obtain the following effective-mass Hamiltonian for the Si conduction-band electron

$$H = -\frac{\hbar^2}{2m_x^*}\frac{\partial^2}{\partial x^2} - \frac{\hbar^2}{2m_y^*}\frac{\partial^2}{\partial y^2} - \frac{\hbar^2}{2m_z^*}\frac{\partial^2}{\partial z^2} + V_c(\boldsymbol{r}) \tag{2.41}$$

Back to (2.27): This is actually the well-established $\boldsymbol{k} \cdot \boldsymbol{p}$ theory. Since we care mostly electrons and holes in the close vicinities of the conduction- and valence bandedges, and the interaction between two bandedge states that are well separated in energy is small, we follow Kane [11] to choose only eight $u_\ell(\boldsymbol{r})$ to formulate the eight-band $\boldsymbol{k} \cdot \boldsymbol{p}$ theory, i.e., the eigen functions of (2.22), as the basis functions to describe conduction- and valence-bandedge electrons and holes, which are normally denoted as

$$|S \uparrow\rangle, |X \uparrow\rangle, |Y \uparrow\rangle, |Z \uparrow\rangle, |S \downarrow\rangle, |X \downarrow\rangle, |Y \downarrow\rangle, |Z \downarrow\rangle$$

where the up and down arrows denote spin up and down, respectively. $|X, Y, Z\rangle$ are the p-type states and $|S\rangle$ is s-type. Interactions between these basis functions, i.e., $\langle u_\ell|\boldsymbol{p}|u_m\rangle$, are obtained by fitting various experimentally data such as energy bandgap and effective mass. Energy band structure parameters of various group IV, III–V and II–VI semiconductors can be found in review books and articles, e.g., [5–7]. The characteristic s- and p-type features of these basis functions are also the foundations of the sp^3s^* tight-binding theory [2] used to calculate Figs. 2.4 and 2.5.

2.2 Lattice Vibration and Phonon Spectrum

In Sect. 3.3 we will try to extract information about the interaction between an electromagnetic field and a lattice vibration (phonons, their frequencies, strengths and damping rates of interactions with the electromagnetic field, etc.) from an optical spectrum in the infrared range. Let us get know the lattice vibration and phonon spectrum in this section.

In the study of energy levels of electrons in individual atoms and the energy band structure of valence electrons in a semiconductor crystal we implicitly assume that the nuclei and core electrons sit motionlessly at the crystal lattice sites, which is valid as long as the motions of the nuclei and core electrons are much smaller than the valence electrons.

In reality, the lattice constant of a semiconductor material varies as a function of external environmental condition, most noticeable, the temperature due to the thermal motion of the atoms that the lattice constant becomes large when the material temperature is high. At normal device operation state, the lattice positions defined by the lattice constant, i.e., (2.1), are statistic average positions of atoms in the time domain, commonly referred to as equilibrium positions. In other words, atoms move, or oscillate, around these equilibrium positions. Let the instantaneous spatial position of an atom, denoted by its equilibrium position $\boldsymbol{R}$, at time t be $\boldsymbol{r}(\boldsymbol{R}, t)$ so that

$$\langle \boldsymbol{r}(\boldsymbol{R}, t)\rangle_t = \boldsymbol{R} \tag{2.42}$$

where $\langle \ldots \rangle_t$ represents the time average operation. We can also use the displacement $\boldsymbol{u}(\boldsymbol{R}, t)$ to represent the motion of the atom

$$\boldsymbol{r}(\boldsymbol{R},t)=\boldsymbol{R}+\boldsymbol{u}(\boldsymbol{R},t),\quad \langle\boldsymbol{u}(\boldsymbol{R},t)\rangle_t=0 \tag{2.43}$$

Note that we use the conventional notation $\boldsymbol{u}$ to represent the displacement, which should not be too difficult to get distinguished from the notation of the Bloch function by context.

We now study $\boldsymbol{u}(\boldsymbol{R},t)$ which is known to be the lattice vibration. In common semiconductors, the nucleus-nucleus interaction $\phi[\boldsymbol{r}(\boldsymbol{R},t)-\boldsymbol{r}(\boldsymbol{R}',t)]$ between atoms located at $\boldsymbol{R}$ and $\boldsymbol{R}'$ dominantly determines the lattice vibration (note that the energy band structure of valence electrons is determined jointly by the electron-nucleus and electron-electron interactions). The total potential energy of the lattice atoms is

$$U=\frac{1}{2}\sum_{\boldsymbol{R}\boldsymbol{R}'}\phi[\boldsymbol{r}(\boldsymbol{R},t)-\boldsymbol{r}(\boldsymbol{R}',t)] \tag{2.44}$$

By including the kinetic energies of atoms we obtain the Hamiltonian of the lattice vibration

$$H_{\text{lattice}}=\sum_{\boldsymbol{R}}\frac{\boldsymbol{P}_{\boldsymbol{R}}^2}{2M_{\boldsymbol{R}}}+U \tag{2.45}$$

where $M_{\boldsymbol{R}}$ is the mass of the atom located at $\boldsymbol{R}$ and $\boldsymbol{P}_{\boldsymbol{R}}$ is the momentum.

Normally $\boldsymbol{u}(\boldsymbol{R},t)$ is small compared with the lattice constant so that we can do the following Taylor expansion of U with respect to the equilibrium position of the lattice site

$$\begin{aligned}U&=\frac{1}{2}\sum_{\boldsymbol{R}\boldsymbol{R}'}\phi[\boldsymbol{r}(\boldsymbol{R},t)-\boldsymbol{r}(\boldsymbol{R}',t)]=\frac{1}{2}\sum_{\boldsymbol{R}\boldsymbol{R}'}\phi[\boldsymbol{R}-\boldsymbol{R}'+\boldsymbol{u}(\boldsymbol{R},t)-\boldsymbol{u}(\boldsymbol{R}',t)]\\&=\frac{1}{2}\sum_{\boldsymbol{R}\boldsymbol{R}'}\phi(\boldsymbol{R}-\boldsymbol{R}')+\frac{1}{2}\sum_{\boldsymbol{R}\boldsymbol{R}'}\Big[\boldsymbol{u}(\boldsymbol{R},t)-\boldsymbol{u}(\boldsymbol{R}',t)\Big]\cdot\nabla\phi(\boldsymbol{R}-\boldsymbol{R}')\\&\quad+\frac{1}{4}\sum_{\boldsymbol{R}\boldsymbol{R}'}\Big\{\Big[\boldsymbol{u}(\boldsymbol{R},t)-\boldsymbol{u}(\boldsymbol{R}',t)\Big]\cdot\nabla\Big\}^2\phi(\boldsymbol{R}-\boldsymbol{R}')+\ldots\end{aligned} \tag{2.46}$$

The first term is a constant while the second term can be expressed as

$$\begin{aligned}&\frac{1}{2}\sum_{\boldsymbol{R}\boldsymbol{R}'}\Big[\boldsymbol{u}(\boldsymbol{R},t)\cdot\nabla\phi(\boldsymbol{R}-\boldsymbol{R}')-\boldsymbol{u}(\boldsymbol{R}',t)\cdot\nabla\phi(\boldsymbol{R}-\boldsymbol{R}')\Big]\\&=\frac{1}{2}\sum_{\boldsymbol{R}}\boldsymbol{u}(\boldsymbol{R},t)\left[\sum_{\boldsymbol{R}'}\nabla\phi(\boldsymbol{R}-\boldsymbol{R}')\right]-\frac{1}{2}\sum_{\boldsymbol{R}'}\boldsymbol{u}(\boldsymbol{R}',t)\left[\sum_{\boldsymbol{R}}\nabla\phi(\boldsymbol{R}-\boldsymbol{R}')\right]\end{aligned} \tag{2.47}$$

Since $\nabla\phi(\boldsymbol{R}-\boldsymbol{R}')$ is the force applied on the atom at $\boldsymbol{R}'$ by the atom at $\boldsymbol{R}$,

$$\sum_{\boldsymbol{R}'}\nabla\phi(\boldsymbol{R}-\boldsymbol{R}')$$

is the sum of forces on the atom at $\boldsymbol{R}$ from all other atoms, which should be zero at equilibrium. In other words, the second term in (2.46) is zero.

By neglecting the first constant potential term we obtain the harmonic oscillator model for the lattice vibration

$$U_{\text{harm}} = \frac{1}{4} \sum_{ij\boldsymbol{R}\boldsymbol{R}'} \Big[\boldsymbol{u}_i(\boldsymbol{R}, t) - \boldsymbol{u}_i(\boldsymbol{R}', t)\Big] \phi_{ij}(\boldsymbol{R} - \boldsymbol{R}') \Big[\boldsymbol{u}_j(\boldsymbol{R}, t) - \boldsymbol{u}_j(\boldsymbol{R}', t)\Big] \quad (2.48)$$

under the second-order approximation, i.e., including only the second-order expansion term in (2.46). In the above equation, $i, j = x, y, z$, and

$$\phi_{ij}(\boldsymbol{R} - \boldsymbol{R}') = \frac{\partial^2 \phi(\boldsymbol{R} - \boldsymbol{R}')}{\partial x_i \partial x_j} \quad (2.49)$$

Equation (2.48) shows that by the harmonic oscillator model, the force that induces the lattice vibration is proportional to the lattice displacement. We thus can apply Newtonian mechanics to study the lattice vibration. The acceleration of the atom equals to the sum of external forces

$$M_{\boldsymbol{R}} \frac{\partial^2 u_i(\boldsymbol{R}, t)}{\partial t^2} = \sum_{j\boldsymbol{R}'} \phi_{ij}(\boldsymbol{R} - \boldsymbol{R}') u_j(\boldsymbol{R}', t) \quad (2.50)$$

When we include only nearest neighbour interactions, the above equation will be significantly simplified. Consider a one-dimensional monatomic chain composed of N atoms ($n = 1, 2, \ldots N$) with a lattice constant a, see Fig. 2.7a, the potential energy of the lattice vibration is

$$U_{\text{harm}} = \frac{\alpha}{2} \sum_{n} \Big[u(na, t) - u(na + a, t)\Big]^2 \quad (2.51)$$

under the harmonic oscillator approximation, (2.50) becomes

$$M \frac{\partial^2 u(na, t)}{\partial t^2} = -\alpha \Big[2u(na, t) - u(na - a, t) - u(na + a, t)\Big] \quad (2.52)$$

Here we denote $\alpha = \phi_{ij}(\boldsymbol{R} - \boldsymbol{R}')$.

Very much alike the wave function of an electron in a translationally symmetric crystal, the lattice displacement is also spatially periodic

$$u(Na + a, t) = u(a, t), \quad u(0, t) = u(Na, t) \quad (2.53)$$

Note that there are only N atoms in the crystal, i.e., $n = 1, 2, \ldots N$, Atoms at 0 and $Na + a$ are hypothetical only for the periodic boundary condition. One critical point as compared with the Bloch theorem is that the lattice displacement here is a physical parameter, while the wave function in (2.6) is not, so there is an extra phase factor existed in the wave function which disappears when we try to obtain the electron distribution which is physically periodic.

The general form of the solution of (2.52) is

$$u(na, t) = s\, e^{i(kna-\omega t)} \tag{2.54}$$

where s is the amplitude of the lattice vibration. As expected, the displacement $\boldsymbol{u}$ of atom $\boldsymbol{R}$ is not only a function of space $\boldsymbol{R}$, it is also a function of time t. Insert the above expression into (2.52),

$$\begin{aligned} k &= \frac{2\pi}{a}\frac{m}{N}, \quad m = \text{integer} \\ \omega_k &= 2\sqrt{\frac{\alpha}{M}}\left|\sin\left(\frac{ka}{2}\right)\right| \\ u(na, t) &= s\cos\left(kna - \omega_k t\right) \end{aligned} \tag{2.55}$$

The second expression is normally known as the dispersion relationship of the lattice vibration in the one-dimensional atomic chain.

Consider a one-dimensional diatomic chain of Fig. 2.7b where the equilibrium positions of the two atoms are na and $na + d$, respectively, where $d \le a/2$. Let the atomic displacements of the two atoms be $u_1(na, t)$ and $u_2(na, t)$. Moreover, we assume that the interaction between atom na and its left-side atom $na - a + d$ is α and the interaction with its right-side atom $na + d$ is β. There are two lattice vibration equations

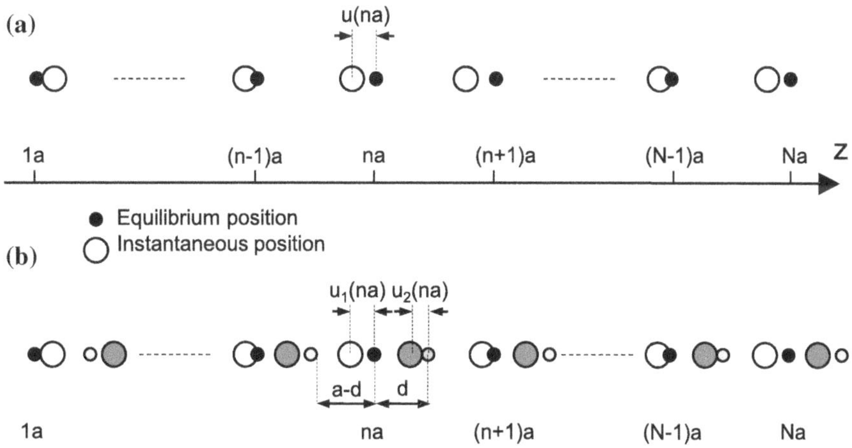

Fig. 2.7 **a** One-dimensional monatomic chain. **b** One-dimensional diatomic chain. a is the lattice constant, N is the number of unit cells contained in the crystal

$$M\frac{\partial^2 u_1(na,t)}{\partial t^2} = -\beta\left[u_1(na,t) - u_2(na,t)\right] - \alpha\left[u_1(na,t) - u_2(na-a,t)\right]$$
$$M\frac{\partial^2 u_2(na,t)}{\partial t^2} = -\beta\left[u_2(na,t) - u_1(na,t)\right] - \alpha\left[u_2(na,t) - u_1(na+a,t)\right] \quad (2.56)$$

Since we already assume that $d \le a/2$, i.e., the distance to the right-side atom is shorter than the left-side atom, $\alpha \le \beta$. Again we assume the general solution

$$u_1(na,t) = s_1\, e^{i(kna-\omega t)}, \quad u_2(na,t) = s_2\, e^{i(kna-\omega t)} \quad (2.57)$$

so that there are two solutions of (2.56), one is

$$\omega_k^2 = \frac{\alpha+\beta}{M} - \frac{1}{M}\sqrt{\alpha^2+\beta^2+2\alpha\beta\cos(ka)}, \quad \frac{s_2}{s_1} = \frac{\beta+\alpha e^{ika}}{|\beta+\alpha e^{ika}|} \quad (2.58)$$

and the other is

$$\omega_k^2 = \frac{\alpha+\beta}{M} + \frac{1}{M}\sqrt{\alpha^2+\beta^2+2\alpha\beta\cos(ka)}, \quad \frac{s_2}{s_1} = -\frac{\beta+\alpha e^{ika}}{|\beta+\alpha e^{ika}|} \quad (2.59)$$

Figure 2.8a shows the dispersion relationships (2.58) and (2.59) of a one-dimensional diatomic chain, and Fig. 2.8b is its corresponding density of vibrational states. Equations (2.55), (2.58) and (2.59) show that the eigen solutions of the lattice vibrations are wave-like characterized by wave vector $\boldsymbol{k}$. Analogous to the Bloch electron state, the eigen lattice vibrational state is known as a phonon, and the density of vibrational states is thereafter denoted as the density of phonon states.

We observe two solutions, also known as phonon bands, or phonon branches, in (2.58), (2.59), and in Fig. 2.8. The low-energy one, i.e., (2.58), is similar to the dispersion relationship (2.55) of the monatomic chain. Most importantly, when k is relatively small such that

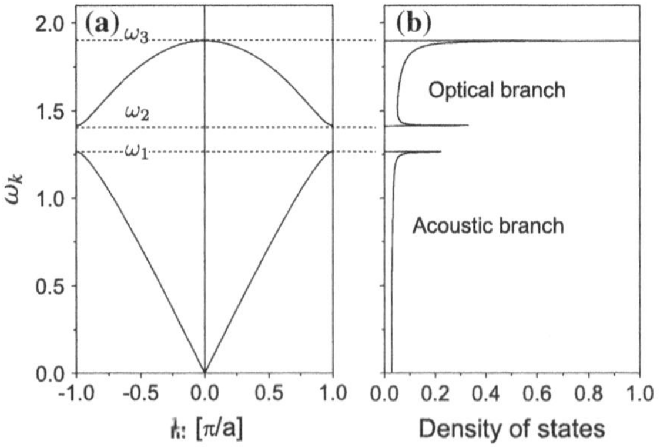

Fig. 2.8 Dispersion relationship ω_k (**a**) and density of phonon states (**b**) of a one-dimensional diatomic chain. $\omega_1 = \sqrt{2\alpha/M}$, $\omega_2 = \sqrt{2\beta/M}$, $\omega_3 = \sqrt{2(\alpha+\beta)/M}$

$$\cos(ka) \approx 1 - \frac{(ka)^2}{2} \tag{2.60}$$

Equation (2.58) become

$$\omega_k = ka\sqrt{\frac{\alpha\beta}{2(\alpha+\beta)M}}, \quad \frac{s_2}{s_1} = 1 \tag{2.61}$$

namely, ω_k is linearly proportional to k, and the amplitudes of displacements of adjacent atoms are the same. These are actually the acoustic waves. The above two relationships remain valid numerically in a quite wide range of k until k reaches π/a (and $\omega_k \to \omega_1$, see Fig. 2.8), so that this whole low-energy phonon branch is called acoustic.

As we shall see in the following section, the wave vector of a photon of our interest is quite small. If a semiconductor lattice is to absorb a photon thereafter emit a phonon (i.e., to start a lattice vibration), or convert a phonon into a photon, the momentum of the phonon must be very small in order to fulfill the momentum conservation requirement. The k value of the acoustic phonon in Fig. 2.8 spans quite widely between 0 and π/a, see Fig. 2.8. However, the corresponding density of the acoustic phonon states when $k \to 0$ is also small so that the interaction between acoustic phonons and photons is negligible.

On the other hand, we notice that the density of phonon states of the high-energy branch with small k, i.e., $\omega_k \to \omega_3$, is very high, see Fig. 2.8b. More critically, the sign of s_2/s_1 is negative in (2.59). In plain words, the motions of neighboring atoms either move towards each other or away from each other, forming electric dipole waves, which can be rather strong especially in ionic crystals such as III–V compound semiconductors, and therefore interact strongly with electromagnetic fields. Thus, the high-energy phonon branch is called optical due to its strong interaction with electromagnetic fields and thus is very important in many device applications. We come back to this in Sect. 3.3.

In the one-dimensional models of Fig. 2.8, displacements u of atoms are constrained along the z direction, parallel to wave vector k. Such a vibration mode is called longitudinal so that phonons are designated as either longitudinal acoustic (LA) phonons or longitudinal optical (LO) phonons. In three dimensional structures, displacements $\boldsymbol{u}$ can be perpendicular to the phonon propagation direction, i.e., $\boldsymbol{k}$. Such vibrations are called transverse and there are transverse acoustic (TA) and transverse optical (TO) phonons.

Expressions for interactions of (2.49) are complicated for three-dimensional lattice. In a zincblende lattice, the interaction between atom at $(0, 0, 0)$ and atom $(a/4, a/4, a/4)$, i.e., the two atoms at d_1 and d_2 in Fig. 2.2, is in the following matrix form

$$\begin{bmatrix} \phi_{xx} & \phi_{xy} & \phi_{xz} \\ \phi_{yx} & \phi_{yy} & \phi_{yz} \\ \phi_{zx} & \phi_{zy} & \phi_{zz} \end{bmatrix} = \begin{bmatrix} \alpha & \beta & \beta \\ \beta & \alpha & \beta \\ \beta & \beta & \alpha \end{bmatrix} \tag{2.62}$$

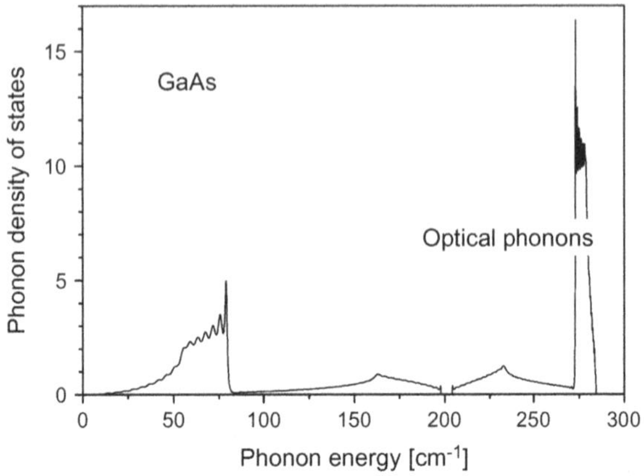

Fig. 2.9 The density of phonon states of bulk GaAs using the local Green's function theory

where α and β can be obtained by fitting the the theoretical model with experimentally determined energies of phonons at symmetry points, see [7, 12, 13]. To solve (2.50) for a three-dimensional lattice with an interaction matrix of (2.62), we can do the following mathematical transform. Let ω be the vibration frequency as before,

$$u_i(\boldsymbol{R}, t) = \frac{\psi_i(\boldsymbol{R})}{\sqrt{M_{\boldsymbol{R}}}} e^{-i\omega t}, \quad H_{ij}(\boldsymbol{R} - \boldsymbol{R}') = \frac{\phi_{ij}(\boldsymbol{R} - \boldsymbol{R}')}{\sqrt{M_{\boldsymbol{R}} M_{\boldsymbol{R}'}}} \tag{2.63}$$

Equation (2.50) then becomes an eigen-value problem

$$H\,|\psi\rangle = \omega^2 |\psi\rangle \tag{2.64}$$

There are many different numerical methods to solve the above eigen-value equation. One may try periodic trial functions for phonons in isotropic and translational symmetric bulk materials. Calculations of phonons in a nanostructure can be difficult. One numerical method is called the recursion method based on the local Green's function theory, see for example [14–17], which can solve the above equation indirectly but obtain directly the density of phonon states. Figure 2.9 shows the calculation result of the density of phonon states of bulk GaAs using the local Green's function theory, where we notice the massive density of phonon states of transverse optical (TO) phonons in the energy range of 270–290 cm^{-1}.

2.3 Light-Matter Interaction and Optical Spectrum

We first describe light as an electromagnetic field by Maxwell's equations. Table 2.2 lists principal wave bands, radio waves, microwaves, terahertz waves, infrared, visible, ultraviolet, X rays and γ rays, in the electromagnetic spectrum and typical physical processes of our interest. Wave bands can be categorized in more details

Table 2.2 Electromagnetic spectrum

Wave band	Wavelength λ (m)	Frequency $\omega/2\pi$ (Hz)	Typical physical processes
Radio waves	10^3–0.3	10^5–10^9	Electron spin, nuclear spin
Microwaves	0.3–10^{-3}	10^9–10^{11}	Molecular motion
Terahertz waves	10^{-3}–3.3×10^{-5}	3×10^{11}–10×10^{12}	Macromolecular motion, rotation and vibration
Infrared	10^{-3}–7.8×10^{-7}	3×10^{11}–3.8×10^{14}	Lattice vibration, free carrier transitions and electron transition in impurity levels
Visible	7.8–3.8×10^{-7}	3.8–7.8×10^{14}	Optical transitions of electron between valence- and conduction bands
Ultraviolet	10^{-7}–10^{-9}	3×10^{15}–10^{17}	Optical transitions between valence- and conduction bands in wide-band material
X rays	10^{-9}–10^{-12}	3×10^{17}–10^{20}	Optical transitions of core electrons
γ rays	10^{-11}–10^{-14}	3×10^{19}–10^{22}	Nuclei decays from high energy states

such as near infrared (750–1400 nm), short-wavelength infrared (1.4–3 μm), middle-wavelength infrared (3–8 μm), long-wavelength infrared (8–15 μm), and far infrared (15–1000 μm). Moreover, the divisions between wave bands are not sharp, there can be overlaps. For example, terahertz (1 THz = 10^{12} Hz) waves span from far infrared to millimeter waves which are normally categorized as microwaves. They become a distinguished hot spot in research and technical development since the terahertz band is a transition region from macroscopic classical theory to microscopic quantum theory as well as from electronics to photonics [18].

The wavelength of light spans from zero to infinity, and light generation, propagation and interactions with matter are vastly different. Divisions of wave bands in Table 2.2 are largely characterized by physical processes that are predominant in the corresponding wavelength ranges. In radio and microwave bands, light-matter interactions are mostly in the form of polarizations that light induces or modifies motions (vibrations and/or rotations) of electrons, and/or atoms, and/or ions in the matter that in their turns induce or modify electric dipoles and thereafter modify the incident light. Common semiconductors interact with light mostly via optical transitions of electrons between different energy bands (inter-band transition) and within the same energy band (intra-band transition) by either absorbing or emitting photons in infrared and visible ranges, which is the central theme of this section.

Light can be used to transfer information, the mostly well-known application is surely the radio, it is however the mobile communication that is dominating today. Light is also used to obtain information. Our eyes collect light around us that lets us know the world around us. We may see the world directly by naked eyes, we also invent tools such as telescopes to see things indirectly that are far away or microscopes to observe tiny things. We want to see not only the surface (color, geometric size, geometric structure etc.) of a matter, we also want to see microscopic structures (lattice structure of crystal, how atoms and electrons interact with each other to form the lattice etc.) inside the matter.

The electromagnetic field is described by the Maxwell's equations

$$\begin{aligned} \nabla \times \boldsymbol{H} &= \boldsymbol{J} + \frac{\partial \boldsymbol{D}}{\partial t} \\ \nabla \times \boldsymbol{E} &= -\frac{\partial \boldsymbol{B}}{\partial t} \\ \nabla \cdot \boldsymbol{D} &= \rho \\ \nabla \cdot \boldsymbol{B} &= 0 \end{aligned} \tag{2.65}$$

where $\boldsymbol{H}$ is the magnetic field, $\boldsymbol{J}$ is the electric current density, $\boldsymbol{D}$ electric displacement field, $\boldsymbol{E}$ electric field, $\boldsymbol{B}$ magnetic flux, and ρ charge density. Charge conservation law

$$\nabla \cdot \boldsymbol{J} = -\frac{\partial \rho}{\partial t} \tag{2.66}$$

is implicit in the Maxwell's equations. In a linear, isotropic and non-dispersive material,

$$\begin{aligned} \boldsymbol{D} &= \epsilon \boldsymbol{E} = \epsilon_0 \boldsymbol{E} + \boldsymbol{P} \\ \boldsymbol{B} &= \mu \boldsymbol{H} = \mu_0 \boldsymbol{H} + \boldsymbol{M} \\ \boldsymbol{J} &= \sigma \boldsymbol{E} \end{aligned} \tag{2.67}$$

where $\boldsymbol{P}$ is the polarization, i.e., the density of electric dipoles, $\boldsymbol{M}$ is the magnetization. ϵ_0 and μ_0 are free-space permittivity and permeability, respectively. In the International System of Units (SI), $\epsilon_0 = 10^7/4\pi c_0^2 = 8.854 \times 10^{-12}$ F/m, $\mu_0 = 4\pi \times 10^{-7} = 1.257 \times 10^{-6}$ H/m. $\epsilon_0\mu_0 = 1/c_0^2$, $c_0 = 2.997925 \times 10^8$ m/s is the speed of light in vacuum. ϵ, μ and σ are the dielectric coefficient, permeability and electrical conductivity of the material.

Three principal physical processes are explicit in the Maxwell's equations:

1. The electric charge density ρ is the source of the electric displacement field $\boldsymbol{D}$, while there is no magnetic monopole so that $\nabla \cdot \boldsymbol{B} = 0$;
2. The temporal variation of $\boldsymbol{D}$ ($\boldsymbol{B}$) accompanies the spatial variation in $\boldsymbol{H}$ ($\boldsymbol{E}$), forming the propagating electromagnetic field;
3. Light-matter interaction induces polarization $\boldsymbol{P}$ and/or magnetization $\boldsymbol{M}$ which modify electric displacement field $\boldsymbol{D}$ and magnetic flux $\boldsymbol{B}$ and thereafter the light.

This book focuses on "seeing", both directly and indirectly, the microscopic structures of semiconductors by comparing the optical spectra of the incident light and the light after light-matter interaction. For the majority of semiconductor photonic applications, the semiconductor materials are nonmagnetic so $\boldsymbol{M} = 0$ and thereafter $\mu = \mu_0$.

We introduce a vector field $\boldsymbol{A}$ and a scalar field ϕ such

$$\begin{aligned} \boldsymbol{E} &= -\nabla\phi - \frac{\partial \boldsymbol{A}}{\partial t} \\ \boldsymbol{B} &= \nabla \times \boldsymbol{A} \end{aligned} \tag{2.68}$$

and in the Coulomb gauge

$$\nabla \cdot \boldsymbol{A} = 0 \tag{2.69}$$

we obtain the following equations for the vector and scalar fields

$$\begin{aligned} \nabla^2 \boldsymbol{A} - \epsilon\mu_0 \frac{\partial^2 \boldsymbol{A}}{\partial t^2} - \epsilon\mu_0 \frac{\partial \nabla\phi}{\partial t} &= -\mu \boldsymbol{J} \\ \nabla^2\phi = -\frac{\rho}{\epsilon} \end{aligned} \tag{2.70}$$

The charge density of conducting (freely moving) electrons in the conduction band of a typical metal is about 10^{22} cm^{-3}, while even in heavily-doped semiconductors, the corresponding charge density is well below 10^{20} cm^{-3}. At 300 K, the resistivity of GaAs is about 10^{-4} Ω·m when GaAs is n-type doped to a doping concentration of 10^{20} cm^{-3}, it drastically reaches 0.15 Ω·m when the doping level is reduced to 10^{16} cm^{-3} [19]. The electrical conductivity of a semiconductor is therefore very low compared with those of metals. Moreover, ρ in (2.65) represents the net charge density, which is the sum of charges of free carriers and charges of ions that contribute the free carriers. Thus, ρ can be well approximated as 0 for most materials. With all these considerations we may simply set $\rho = 0$ and $\boldsymbol{J} = 0$ so that (2.70) become

$$\begin{aligned} \nabla^2 \boldsymbol{A} - \epsilon\mu_0 \frac{\partial^2 \boldsymbol{A}}{\partial t^2} - \epsilon\mu_0 \frac{\partial \nabla\phi}{\partial t} &= 0 \\ \nabla^2\phi &= 0 \end{aligned} \tag{2.71}$$

By a simple gauge transformation, we can eliminate the scalar potential then reduce the above first equation to the wave equation of $\boldsymbol{A}$

$$\nabla^2 \boldsymbol{A} - \epsilon\mu_0 \frac{\partial^2 \boldsymbol{A}}{\partial t^2} = 0 \tag{2.72}$$

with a general solution

$$\boldsymbol{A}(\boldsymbol{r}, t) = \boldsymbol{A}_0 e^{i(\boldsymbol{s}\cdot\boldsymbol{r} - \omega t)} + \text{c.c.} \tag{2.73}$$

where c.c. in the above expression represents the complex conjugate of the first term on the right side of the expression to make sure that vector field $\boldsymbol{A}$, which is a physical parameter, is real.

As we will see shortly, the amplitude of the electric field of the electromagnetic field of our interest is very small as compared with the electric field we apply externally to drive a semiconductor device in many applications. This leads to the negligence of the roles of the scalar field ϕ.

Insert (2.73) into (2.72) and we obtain

$$\boldsymbol{s} \cdot \boldsymbol{s} = \epsilon \mu_0 \omega^2 \tag{2.74}$$

which is the dispersion relationship of light, $\omega/2\pi$ is the frequency of light, ω is the angular frequency, $\boldsymbol{s}$ is the wave vector. We use $\boldsymbol{s}$ to denote the wave vector of an electromagnetic field to differentiate it from the wave vector $\boldsymbol{k}$ of an electron.

In general, the dielectric coefficient ϵ is a function of the angular frequency ω of the electromagnetic field. And most critically, it can be complex such that

$$\epsilon(\omega) = \epsilon'(\omega) + i\epsilon''(\omega) \tag{2.75}$$

where $\Re\epsilon = \epsilon'$, $\Im\epsilon = \epsilon''$. In the following we only write ϵ instead of its explicit ω-dependent expression $\epsilon(\omega)$ to simplify equations. In an anisotropic material, ϵ is a tensor, while in commonly used semiconductors, it is usually a scalar.

The wave vector $\boldsymbol{s}$ in (2.74) in general is complex because of the complex ϵ. By introducing a complex refractive index

$$\tilde{n} \equiv n + i\kappa = \sqrt{\frac{\epsilon}{\epsilon_0}} \tag{2.76}$$

where the real part of the complex refractive index n is called the refractive index (which was introduced in Chap. 1), while the imaginary part κ is known as the extinction coefficient.

A relative dielectric coefficient is commonly introduced

$$\epsilon_{\mathrm{r}} \equiv \frac{\epsilon}{\epsilon_0}$$

Many times, people do not express the distinction between ϵ_{r} and ϵ clearly, especially in equations. Reader can check their physical units or numerical values in order to differentiate the two. For example, ϵ_{r} is numerically around 12 (unit-less) for common semiconductors; while $\epsilon_0 = 8.85 \times 10^{-12}$ F/m so that $\epsilon = 12 \times 8.85 \times 10^{-12}$ F/m when $\epsilon_{\mathrm{r}} = 12$. It is easy to see the follow relations

$$\epsilon_{\mathrm{r}}' = n^2 - \kappa^2, \quad \epsilon_{\mathrm{r}}'' = 2n\kappa \tag{2.77}$$

After the introduction of the complex refractive index, the dispersion relationship of light becomes

$$s = \frac{\omega\tilde{n}}{c_0}s_0 = \left(\frac{\omega n}{c_0} + i\frac{\omega\kappa}{c_0}\right)s_0 \tag{2.78}$$

where s_0 is the unit vector of wave vector s. Insert the above dispersion relationship into (2.73),

$$\begin{aligned} A(r,t) &= A_0 \exp\left[i\left(\frac{\omega\tilde{n}}{c_0}s_0\cdot r - \omega t\right)\right] + \text{c.c.} \\ &= A_0 \exp\left[i\left(\frac{\omega n}{c_0}s_0\cdot r - \omega t\right)\right]\exp\left(-\frac{\omega\kappa}{c_0}s_0\cdot r\right) + \text{c.c.} \end{aligned} \tag{2.79}$$

The above expression shows that light propagates in an infinitely extended material at a speed of $c = c_0/n$. At the same time, the amplitude of the vector field gradually decays along its propagation due to the extinction coefficient κ.

Inserting (2.79) into the Coulomb gauge (2.69) results in

$$A_0 \cdot s_0 = 0 \tag{2.80}$$

since $\phi = 0$. This indicates that the wave vector of the photon, i.e., the propagation direction of the photon, is perpendicular to the vector field.

By (2.68),

$$\begin{aligned} E &= -\frac{\partial A}{\partial t} = \omega A_0 \exp\left[i\left(\frac{\omega\tilde{n}}{c_0}s_0\cdot r - \omega t\right)\right] + \text{c.c.} \\ B &= \nabla\times A = \frac{-i\omega\tilde{n}}{c_0}A_0\times s_0 \exp\left[i\left(\frac{\omega\tilde{n}}{c_0}s_0\cdot r - \omega t\right)\right] + \text{c.c.} \end{aligned} \tag{2.81}$$

so that

$$H = \frac{B}{\mu_0} = \frac{-i\omega\tilde{n}}{\mu_0 c_0}A_0\times s_0 \exp\left[i\left(\frac{\omega\tilde{n}}{c_0}s_0\cdot r - \omega t\right)\right] + \text{c.c.} \tag{2.82}$$

We may also write

$$\begin{aligned} E(r,t) &= E_0 \exp\left[i\left(\frac{\omega\tilde{n}}{c_0}s_0\cdot r - \omega t\right)\right] + \text{c.c.} \\ H(r,t) &= H_0 \exp\left[i\left(\frac{\omega\tilde{n}}{c_0}s_0\cdot r - \omega t\right)\right] + \text{c.c.} \\ E_0 &= \omega A_0, \quad H_0 = \frac{-i\tilde{n}}{\mu_0 c_0}E_0\times s_0 \end{aligned} \tag{2.83}$$

Note that H_0 in the above expression is complex while E_0 is real. One way of understanding the above equation is that E_0 and H_0 in a lossy medium (with a nonzero κ) are out of phase. It is anyway easy to see that $\boldsymbol{H}(\boldsymbol{r}, t)$ in the above equation is real.

By a straightforward but careful mathematical operations, we obtain the Poynting vector $\boldsymbol{S}$ that represents the directional energy flux density of light

$$\langle \boldsymbol{S} \rangle_t = \langle \boldsymbol{E} \times \boldsymbol{H} \rangle_t = 2n|\boldsymbol{A}_0|^2 \frac{\omega^2}{\mu_0 c_0} \exp\left(-\frac{2\omega\kappa}{c_0} \boldsymbol{s}_0 \cdot \boldsymbol{r}\right) \boldsymbol{s}_0 \tag{2.84}$$

In the above expression, $\langle \boldsymbol{S} \rangle_t$ denotes the time average of $\boldsymbol{S}$. Since light travels at the speed of $c = c_0/n$ in the material, the optical energy density is

$$\frac{\langle \boldsymbol{S} \rangle_t}{c} = 2\epsilon_0 n^2 \omega^2 |\boldsymbol{A}_0|^2 \exp\left(-\frac{2\omega\kappa}{c_0} \boldsymbol{s}_0 \cdot \boldsymbol{r}\right) \boldsymbol{s}_0 \tag{2.85}$$

which propagates along $\boldsymbol{s}_0$. Note that vector field $\boldsymbol{A}$, electric field $\boldsymbol{E}$, as well as magnetic field $\boldsymbol{H}$ all propagate along the direction of $\boldsymbol{s}_0$, which is the unit vector of wave vector $\boldsymbol{s}$. Moreover, $\boldsymbol{A}$ and $\boldsymbol{E}$ are perpendicular to $\boldsymbol{s}_0$, $\boldsymbol{H}$ is perpendicular to both $\boldsymbol{E}$ and $\boldsymbol{s}_0$, and the Poynting vector $\boldsymbol{S}$ and the optical energy density $\langle \boldsymbol{S} \rangle_t / c$ propagate along $\boldsymbol{s}_0$. The spatial orientations of these vectors are schematically presented in Fig. 2.10, which is elementary but critical for understanding the light-matter interaction.

Equation (2.84) is normally expressed as the Beer–Lambert law

$$S = S_0 \, e^{-\alpha(\omega) z} \tag{2.86}$$

where S is the amplitude of the Poynting vector, also known as the optical power after the light (with an initial power S_0) transmits through a layer of the material of thickness z. Here

$$\alpha = \frac{2\omega\kappa}{c_0}$$

is known to be the absorption coefficient.

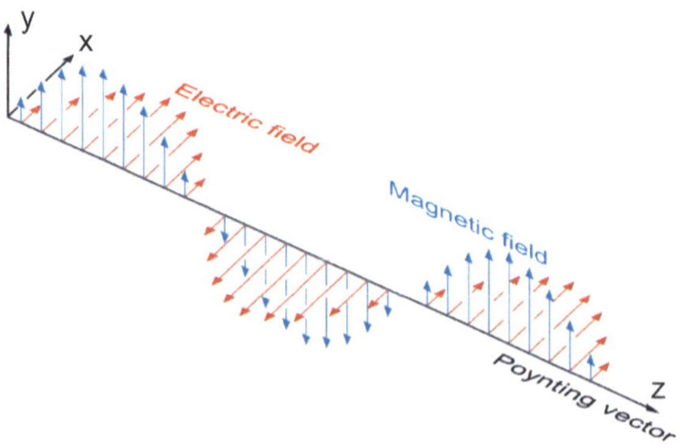

Fig. 2.10 Electromagnetic field in a homogeneous medium. The directions of the electric field (red arrows), magnetic field (blue arrows), and the Poynting vector (black arrow along the z axis) form a Cartesian coordinate

Next we quantize the electromagnetic field by the number of photons and the energy carried per photon to study the dynamics of the non-relativistic light-matter interaction. Let Ω be the space where the light-matter interaction occurs, N_ω is the number of the photons inside Ω and the photon energy is $\hbar\omega$, so that the optical energy density in Ω is

$$\frac{N_\omega \hbar\omega}{\Omega}$$

Because of the decay along $\boldsymbol{s}_0$ in (2.85), we expect that N_ω decreases in the same manner. Not getting into details right now (see more below) let us simply write the optical energy density as

$$\frac{N_\omega \hbar\omega}{\Omega} \exp\left(-\frac{2\omega\kappa}{c_0}\boldsymbol{s}_0 \cdot \boldsymbol{r}\right)\boldsymbol{s}_0 \tag{2.87}$$

By combining the above equation with (2.85) we obtain the amplitude of the vector field

$$|\boldsymbol{A}_0|^2 = \frac{N_\omega \hbar}{2\epsilon_0 n^2 \omega \Omega} \tag{2.88}$$

The density of photons, i.e., N_ω/Ω, is measurable from which we can determine the vector field. Insert the above expression into (2.73)

$$\boldsymbol{A}(\boldsymbol{r}, t) = \sqrt{\frac{N_\omega \hbar}{2\epsilon_0 n^2 \omega \Omega}} \left[e^{i(\boldsymbol{s}\cdot\boldsymbol{r}-\omega t)} + e^{-i(\boldsymbol{s}\cdot\boldsymbol{r}-\omega t)}\right] \boldsymbol{e}_\mathrm{s} \tag{2.89}$$

$\boldsymbol{e}_\mathrm{s}$ is the unit vector of $\boldsymbol{A}$. And $\boldsymbol{e}_\mathrm{s} \perp \boldsymbol{s}_0$ by (2.80).

In quantizing the electromagnetic field, the state of the photon, $|N_\omega\rangle$ is denoted by the angular frequency ω and the number of photons N_ω. Photon can be created or annihilated so we introduce the corresponding creation operator $\boldsymbol{b}^+$ and annihilation operator $\boldsymbol{b}$

$$\boldsymbol{b}^+|N_\omega\rangle = \sqrt{N_\omega + 1}|N_\omega + 1\rangle, \quad \boldsymbol{b}\,|N_\omega\rangle = \sqrt{N_\omega}|N_\omega - 1\rangle \tag{2.90}$$

The Hamiltonian of the photon state is

$$H_\omega = \hbar\omega\left(\boldsymbol{b}^+\boldsymbol{b} + \frac{1}{2}\right) \tag{2.91}$$

By (2.90), it is easy to obtain

$$H_\omega|N_\omega\rangle = \hbar\omega\left(N_\omega + \frac{1}{2}\right)|N_\omega\rangle \tag{2.92}$$

For polychromatic light composed of photons with different angular frequencies $\omega_1, \omega_2 \ldots$, the photon state is expressed as $|N_{\omega_1}, N_{\omega_2} \ldots\rangle$ and the corresponding Hamiltonian

$$\sum_i \hbar\omega_i \left(\boldsymbol{b}^+_{\omega_i} \boldsymbol{b}_{\omega_i} + \frac{1}{2} \right)$$

For simplicity, we focus on a monochromatic light, i.e., Hamiltonian of (2.91). The vector field of (2.89) now becomes a quantum mechanical operator

$$\boldsymbol{A} = \sqrt{\frac{\hbar}{2\epsilon_0 n^2 \omega \Omega}} \left[\boldsymbol{b}^+ e^{i(\boldsymbol{s}\cdot\boldsymbol{r}-\omega t)} + \boldsymbol{b}\, e^{-i(\boldsymbol{s}\cdot\boldsymbol{r}-\omega t)} \right] \boldsymbol{e}_s \tag{2.93}$$

The correspondence between (2.89) and (2.93) is found in the expectation value of quantum mechanical operator $\boldsymbol{A}$ in (2.93) to be compared with the physical observable $\boldsymbol{A}(\boldsymbol{r}, t)$ in (2.89). It is easy to show that in the eigen state of (2.92), $\langle \boldsymbol{A} \rangle = 0$ and $\langle \boldsymbol{A}^2 \rangle = N_\omega \hbar / 2\epsilon_0 n^2 \omega \Omega$. The same results are obtained for $\langle \boldsymbol{A}(\boldsymbol{r}, t) \rangle = 0$ and $\langle \boldsymbol{A}(\boldsymbol{r}, t)^2 \rangle$ by (2.89).

The electron is a fermion restrained by the Pauli exclusion principle. The average number of electrons at equilibrium occupying a single-particle state $|\boldsymbol{k}\rangle$ is given by the Fermi-Dirac distribution function

$$f_{\mathrm{FD}}(E_{\boldsymbol{k}}) = \frac{1}{e^{(E_{\boldsymbol{k}}-E_f)/k_B T} - 1} \tag{2.94}$$

where E_f is the Fermi level. The photon, on the other hand, is a boson obeying the Bose-Einstein distribution statistics

$$f_{\mathrm{BE}}(\hbar\omega) = \frac{1}{e^{\hbar\omega/k_B T} - 1} \tag{2.95}$$

such that there can more than one photon occupying the same photon state, i.e., N_ω can be any positive integer.

We are now ready to study the light-matter interaction in semiconductor materials and devices. Light can be transmitted, reflected, refracted, diffracted, and absorbed when impinging on a semiconductor. Moreover, light can be emitted from a semiconductor. All these are generally referred to optical properties, which can be quantified by reflectance, transmittance, absorption coefficient, etc. And their dependences on the wavelength of light are our optical spectra such as the reflection or transmission spectrum. Different optical spectrum reflects different aspect of the light-matter interaction thus constitutes optical fingerprint features of the material under investigation. Figure 2.11 shows schematically the three principal optical transitions of an electron from its initial electron state $\boldsymbol{k}$ to a final state $\boldsymbol{q}$, (a) absorption, (b) stimulated emission, and (c) spontaneous emission. At resonant steady-state processes, the photon energy equals the energy difference between the two electron states.

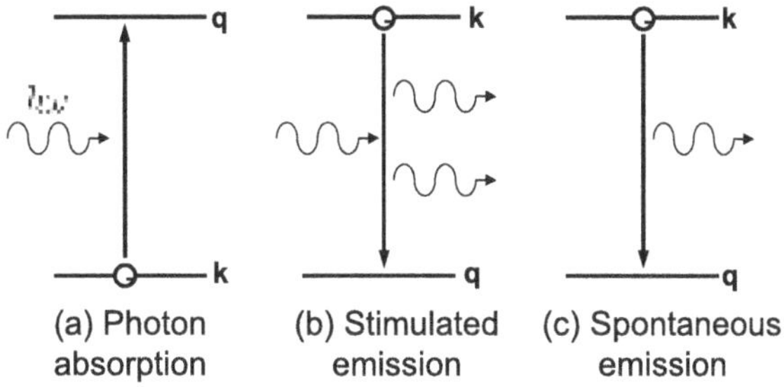

Fig. 2.11 Principal light-matter interactions due to the optical transition of an electron from its initial state $\boldsymbol{k}$ to the final state $\boldsymbol{q}$. **a** Photon absorption, **b** stimulated photon emission, and **c** spontaneous photon emission

In the presence of an electromagnetic field (interchangeably we may also say a photon field), the electron momentum $\boldsymbol{p}$ becomes $\left(\boldsymbol{p}+e\boldsymbol{A}\right)$, where $\boldsymbol{A}$ is the vector field of the electromagnetic field. Refer to (2.4), the Hamiltonian of the electron now becomes

$$\begin{aligned}&\frac{1}{2m_0}(\boldsymbol{p}+e\boldsymbol{A})^2+V_0(\boldsymbol{r})\\&=\frac{\boldsymbol{p}^2}{2m_0}+\frac{e}{m_0}\boldsymbol{A}\cdot\boldsymbol{p}-\frac{ie\hbar}{2m_0}\nabla\cdot\boldsymbol{A}+\frac{e^2}{2m_0}\boldsymbol{A}^2+V_0(\boldsymbol{r})\end{aligned}\tag{2.96}$$

where m_0 is the free electron mass. There should be a potential term $-e\phi$ on the right side of the equals sign in the above equation, where ϕ is the scalar potential of the electromagnetic field, but is neglected here based on the numerical estimation to be presented shortly. Note that the expansion of $\left(\boldsymbol{p}+e\boldsymbol{A}\right)$ in the above equation is not trivial. $\boldsymbol{p}=-i\hbar\nabla$ is an operator not only on a wave function ψ but also on $\boldsymbol{A}$ such that

$$\left(\boldsymbol{p}+e\boldsymbol{A}\right)\psi=\boldsymbol{p}\psi+e\boldsymbol{A}\psi$$

$$\left(\boldsymbol{p}+e\boldsymbol{A}\right)^2\psi=\left(\boldsymbol{p}+e\boldsymbol{A}\right)\cdot\left(\boldsymbol{p}\psi+e\boldsymbol{A}\psi\right)=\boldsymbol{p}^2\psi+e\boldsymbol{p}\cdot\left(\boldsymbol{A}\psi\right)+e\boldsymbol{A}\cdot\boldsymbol{p}\psi+e^2\boldsymbol{A}^2\psi$$

$$\boldsymbol{p}\cdot\left(\boldsymbol{A}\psi\right)=\psi\boldsymbol{p}\cdot\boldsymbol{A}+\boldsymbol{A}\cdot\boldsymbol{p}\psi$$

so that

$$\left(\boldsymbol{p}+e\boldsymbol{A}\right)^2\psi=\boldsymbol{p}^2\psi+e\psi\boldsymbol{p}\cdot\boldsymbol{A}+2e\boldsymbol{A}\cdot\boldsymbol{p}\psi+e^2\boldsymbol{A}^2\psi$$

The third term in (2.96) equation is zero under the Coulomb gauge of (2.69). Let us estimate the ratios between the second and first term and between the fourth and the second term:

$$\frac{\left|\frac{e\boldsymbol{A}\cdot\boldsymbol{p}}{m_0}\right|}{\left|\frac{\boldsymbol{p}^2}{2m_0}\right|} \approx \frac{2eA}{p}, \quad \frac{\left|\frac{e^2A^2}{2m_0}\right|}{\left|\frac{e\boldsymbol{A}\cdot\boldsymbol{p}}{m_0}\right|} \approx \frac{eA}{2p} \tag{2.97}$$

Consider a light beam with an optical power of 1.0 W/cm^2 (this is a very strong light source! Consider a 60-W light bulb in our daily life which illuminates at a distance $r = 1$ m away from you, the optical power that reaches you is only 60 W/4$\pi r^2 = 4.8 \times 10^{-4}$ W/cm^2) and a photon energy $\hbar\omega = 2.0$ eV (wavelength equals 620 nm, i.e., a red light), the photon density by (2.85) is

$$\frac{N_\omega}{\Omega} = \frac{|\langle \boldsymbol{S}\rangle_t|}{c_0\hbar\omega} = \frac{1.0\ \mathrm{W/cm^2}}{2.997925 \times 10^8\ \mathrm{m/s} \times 2 \times 1.60219 \times 10^{-19}\ \mathrm{J}} = 1.041 \times 10^{14}\ \mathrm{m}^{-3} \tag{2.98}$$

so that

$$eA = \sqrt{\frac{e^2\hbar N_\omega}{2\epsilon_0 n^2\omega\Omega}} = 0.24 \times 10^{-36}\ \mathrm{kg\cdot m/s} \tag{2.99}$$

Here we let $n = 3$ (for common semiconductors). We mentioned at the beginning of this chapter that the electron mobility μ of GaAs is 8500 cm^2/V·s, it is about 1400 cm^2/V·s in Si. Under common device operation conditions, we apply an external bias of 1.0 V across a device with an effective length of ca 1.0 μm so that the speed of the electron is about $v = \mu E \approx 10^7$ cm/s, where E is the electric field ($=1.0\ \mathrm{V}/1.0\ \mu\mathrm{m} = 10^6$ V/m in our estimation). (Note that 10^7 cm/s is actually close to the saturation velocity in GaAs and Si.) Take that the effective mass of the electron is $0.1m_0$, we obtain the amplitude of the electron momentum

$$p = m^*v = 0.1 \times 9.1096 \times 10^{-31}\ \mathrm{kg} \times 10^7\ \mathrm{cm/s} = 9.1 \times 10^{-27}\ \mathrm{kg\cdot m/s} \tag{2.100}$$

thus

$$\frac{eA}{p} = \frac{0.24 \times 10^{-36}}{9.1 \times 10^{-27}} = 2.6 \times 10^{-11}$$

so that we can treat the second term in (2.96) as a perturbation and neglect the fourth term, and the Hamiltonian of the electron in the presence of an electromagnetic field is now $\left(H_0 + H'\right)$, where

$$H_0 = -\frac{\hbar^2\nabla^2}{2m_0} + V_0(\boldsymbol{r}), \quad H' = \frac{ie\hbar}{m_0}\boldsymbol{A}\cdot\nabla \tag{2.101}$$

where H' is known to be the first-order light-matter interaction. If we increase the light power, or when the optical transitions due to this first-order light-matter interaction is forbidden, we shall need to study the term containing $\boldsymbol{A}^2$, which is beyond the scope of this book.

We now have enough information to estimate the scalar potential ϕ of the electromagnetic field under discussion. By (2.83) and (2.99), the amplitude of the electric field of the electromagnetic field is

$$\frac{A}{\omega} = \frac{0.24 \times 10^{-36} \times 2}{1.05459 \times 10^{-34}} = 4.5 \times 10^{-3}\ \text{V/m} \tag{2.102}$$

for $\hbar\omega = 2.0$ eV, while the electric field induced by our externally applied bias is 10^6 V/m (1.0 V across a length of 1.0 μm). Thus, $-e\phi$ is neglected in (2.96).

The steady-state eigen equation of H_0 is straightforward

$$H_0 \Psi_{\boldsymbol{k}}(\boldsymbol{r}) = E_{\boldsymbol{k}} \Psi_{\boldsymbol{k}}(\boldsymbol{r}) \tag{2.103}$$

which we have studied extensively in Sect. 2.1. Here, $\Psi_{\boldsymbol{k}}(\boldsymbol{r})$ is the total wave function of an electron state defined by a set of quantum numbers denoted by $\boldsymbol{k}$. For Bloch states in a bulk semiconductor, $\boldsymbol{k}$ represents both the wave vector $\boldsymbol{k}$ as well as the band index ℓ; While for electron states in a quantum well which is one-dimensionally confined along the z direction, $\boldsymbol{k} = (\boldsymbol{k}_{xy}, i)$, where $\boldsymbol{k}_{xy}$ is the wave vector in the xy plane and i is the energy sublevel index along the z direction. Furthermore, we will also write the wave function $\Psi_{\boldsymbol{k}}$ as $|\boldsymbol{k}\rangle$, i.e., the bra-ket notation, to simplify mathematical expressions.

When studying the light-matter interaction, it is easier to combine the electron and the photon into one composite system with a total Hamiltonian

$$H_0 + H_\omega + H' = -\frac{\hbar^2 \nabla^2}{2m_0} + V_0(\boldsymbol{r}) + \hbar\omega \left(\boldsymbol{b}^+ \boldsymbol{b} + \frac{1}{2} \right) + \frac{ie\hbar}{m_0} \boldsymbol{A} \cdot \nabla \tag{2.104}$$

When $H' = 0$, the electron and the photon are not interacting so that the final state of the composite system, i.e., $(H_0 + H_\omega)$, is simply the product of the electron's eigen state and the photon's eigen state. After introducing a nonzero light-matter interaction H', there will be optical transitions between eigen states of $(H_0 + H_\omega)$. At steady-state, we can list the initial eigen state $|i\rangle$ and final eigen state $|j\rangle$ of $(H_0 + H_\omega)$ for the optical transitions of Fig. 2.11:

- Photon absorption: the electron transits from its initial state $\boldsymbol{k}$ to a final state $\boldsymbol{q}$ by absorbing one photon while the photon number decreases by 1

$$|i\rangle = |\boldsymbol{k}, N_\omega\rangle, \quad |j\rangle = |\boldsymbol{q}, N_\omega - 1\rangle \tag{2.105}$$

- Photon emission: the electron transits from $\boldsymbol{k}$ to $\boldsymbol{q}$ by emitting one photon so that the photon number increases by 1

$$|i\rangle = |\boldsymbol{k}, N_\omega\rangle, \quad |j\rangle = |\boldsymbol{q}, N_\omega + 1\rangle \tag{2.106}$$

After careful mathematical manipulations, the probability of the optical transition (the transition is called optical since a photon is involved) from electron state $|\boldsymbol{k}\rangle$ to $|\boldsymbol{q}\rangle$ is

$$w_{j\leftarrow i}(\omega)=\frac{2\pi}{\hbar}\left|\langle j|H'|i\rangle\right|^{2}\frac{\Gamma_{qk}}{\left(E_{q}-E_{k}\pm\hbar\omega\right)^{2}+\Gamma_{qk}^{2}} \tag{2.107}$$

which is the well-known Fermi's golden rule. "−" represents photon absorption and "+" photon emission. Γ_{qk} is proportional to the transition matrix element $\langle\boldsymbol{q}|H'|\boldsymbol{k}\rangle$. For more details, please see [20]. One critical issue, however, must be specifically pointed out here.

We do not find the wave vector $\boldsymbol{s}$ of the photon in (2.107). The reason is very simple. In most optical spectral studies of semiconductors, the optical range of interest spans from the visible to far infrared range, maybe even to the terahertz range. Refer to Table 2.2, we see that the corresponding wavelengths of these photons span from about 380 nm to 33 μm so that the amplitudes of the photon wave vectors ($|\boldsymbol{s}|=2\pi/\lambda$) range $(0.02\sim1.65)\times10^{-2}\ \mathrm{nm}^{-1}$. The amplitudes of $\boldsymbol{k}$ and $\boldsymbol{q}$ of the electron states in semiconductors can be estimated from (2.100) that

$$\frac{p}{\hbar}=\frac{9.1\times10^{-27}}{1.05459\times10^{-34}}\ \mathrm{m}^{-1}=8.6\times10^{-2}\ \mathrm{nm}^{-1} \tag{2.108}$$

We thus simply neglect the photon wave vector for most photonics applications. This result has a profound effect on the light-matter interaction, namely, the momentum of the electron is conserved during optical transition. We will come back to this shortly.

By (2.93) we can easily calculate the matrix elements of the light-matter interactions:

- Photon absorption

$$\langle j|\boldsymbol{A}\cdot\nabla|i\rangle=\sqrt{\frac{\hbar N_{\omega}}{2\epsilon_{0}n^{2}\omega\Omega}}\langle\boldsymbol{q}|\boldsymbol{e}_{s}\cdot\nabla|\boldsymbol{k}\rangle \tag{2.109}$$

- Photon emission

$$\langle j|\boldsymbol{A}\cdot\nabla|i\rangle=\sqrt{\frac{\hbar(N_{\omega}+1)}{2\epsilon_{0}n^{2}\omega\Omega}}\langle\boldsymbol{q}|\boldsymbol{e}_{s}\cdot\nabla|\boldsymbol{k}\rangle \tag{2.110}$$

Here $\boldsymbol{e}_{\mathrm{s}}$ is the unit vector of the electric field of the photon, see (2.89). The probability of absorbing one photon is

$$w_{\boldsymbol{q}(N_{\omega}-1)\leftarrow\boldsymbol{k}N_{\omega}}^{\mathrm{abs}}(\omega)=\frac{\pi\hbar^{2}e^{2}N_{\omega}}{m_{0}^{2}\epsilon_{0}n^{2}\omega\Omega}|\langle\boldsymbol{q}|\boldsymbol{e}_{\mathrm{s}}\cdot\nabla|\boldsymbol{k}\rangle|^{2}\frac{\Gamma_{qk}}{\left(E_{q}-E_{k}-\hbar\omega\right)^{2}+\Gamma_{qk}^{2}} \tag{2.111}$$

And the probability of emitting one photon is

$$w^{\mathrm{em}}_{\boldsymbol{q}(N_\omega+1)\leftarrow \boldsymbol{k}N_\omega}(\omega) = \frac{\pi \hbar^2 e^2 (N_\omega + 1)}{m_0^2 \epsilon_0 n^2 \omega \Omega} \left|\langle \boldsymbol{q}|\boldsymbol{e}_{\mathrm{s}} \cdot \nabla|\boldsymbol{k}\rangle\right|^2 \frac{\Gamma_{\boldsymbol{qk}}}{\left(E_{\boldsymbol{q}} - E_{\boldsymbol{k}} + \hbar\omega\right)^2 + \Gamma^2_{\boldsymbol{qk}}} \tag{2.112}$$

We see clearly here that the number of photons after the electron absorbs one photon to transit from state $\boldsymbol{k}$ to state $\boldsymbol{q}$ is decreased by 1; It is increased by 1 when the electron emits one photon to transit from $\boldsymbol{k}$ to $\boldsymbol{q}$. In the following we write simply $w^{\mathrm{abs}}_{\boldsymbol{q}\leftarrow\boldsymbol{k}}(\omega)$ and $w^{\mathrm{em}}_{\boldsymbol{q}\leftarrow\boldsymbol{k}}(\omega)$. Reader can understand the modification of the photon number N_ω by superscript "abs" (for "absorption") or "em" ("emission").

We separate the photon emission probability into a stimulated emission term and a spontaneous emission term

$$w^{\mathrm{em}}_{\boldsymbol{q}\leftarrow\boldsymbol{k}}(\omega) = w^{\mathrm{st}}_{\boldsymbol{q}\leftarrow\boldsymbol{k}}(\omega) + w^{\mathrm{spon}}_{\boldsymbol{q}\leftarrow\boldsymbol{k}}(\omega) \tag{2.113}$$

such that

$$w^{\mathrm{st}}_{\boldsymbol{q}\leftarrow\boldsymbol{k}}(\omega) = \frac{\pi \hbar^2 e^2 N_\omega}{m_0^2 \epsilon_0 n^2 \omega \Omega} \left|\langle \boldsymbol{q}|\boldsymbol{e}_{\mathrm{s}} \cdot \nabla|\boldsymbol{k}\rangle\right|^2 \frac{\Gamma_{\boldsymbol{qk}}}{\left(E_{\boldsymbol{q}} - E_{\boldsymbol{k}} + \hbar\omega\right)^2 + \Gamma^2_{\boldsymbol{qk}}} \tag{2.114}$$

$$w^{\mathrm{spon}}_{\boldsymbol{q}\leftarrow\boldsymbol{k}}(\omega) = \frac{\pi \hbar^2 e^2}{m_0^2 \epsilon_0 n^2 \omega \Omega} \left|\langle \boldsymbol{q}|\boldsymbol{e}_{\mathrm{s}} \cdot \nabla|\boldsymbol{k}\rangle\right|^2 \frac{\Gamma_{\boldsymbol{qk}}}{\left(E_{\boldsymbol{q}} - E_{\boldsymbol{k}} + \hbar\omega\right)^2 + \Gamma^2_{\boldsymbol{qk}}} \tag{2.115}$$

Optical transition processes of (2.111), (2.114) and (2.115) are depicted already in Fig. 2.11.

So far we only study the probabilities of transitions between two electron states $\boldsymbol{k}$ and $\boldsymbol{q}$. We need to include the occupations $f(E_{\boldsymbol{k}})$ and $f(E_{\boldsymbol{q}})$ of these electron states since the electron is fermion obeying the Pauli exclusion principle, i.e., an electron occupying state $\boldsymbol{k}$ will transit to an empty state $\boldsymbol{q}$ with a certain transition probability $w_{\boldsymbol{q}\leftarrow\boldsymbol{k}}$. The net transition from state $\boldsymbol{k}$ to $\boldsymbol{q}$ is

$$W_{\boldsymbol{qk}}(\omega) = w_{\boldsymbol{q}\leftarrow\boldsymbol{k}}(\omega) f(E_{\boldsymbol{k}})\Big[1 - f(E_{\boldsymbol{q}})\Big] - w_{\boldsymbol{k}\leftarrow\boldsymbol{q}}(\omega) f(E_{\boldsymbol{q}})\Big[1 - f(E_{\boldsymbol{k}})\Big] \tag{2.116}$$

Normally, N_ω is much larger than 1 so that $w_{\boldsymbol{q}\leftarrow\boldsymbol{k}} = w_{\boldsymbol{k}\leftarrow\boldsymbol{q}}$, and

$$W_{\boldsymbol{qk}}(\omega) = w_{\boldsymbol{q}\leftarrow\boldsymbol{k}}(\omega)\Big[f(E_{\boldsymbol{k}}) - f(E_{\boldsymbol{q}})\Big] \tag{2.117}$$

Summarizing all electron states results in the net number of optical transitions of electrons

$$W(\omega) = \sum_{\boldsymbol{qk}} w_{\boldsymbol{qk}}(\omega) \tag{2.118}$$

Note that in (2.116), (2.117), we write the occupation of electron state $\boldsymbol{k}$ as $f(E_k)$, not $f_{\mathrm{FD}}(E_k)$ which is expressed as (2.94) representing the occupation at equilibrium. Electrons in a semiconductor device under operation are no longer at equilibrium, while in many cases, they are at steady state. See more in Sect. 4.1.

We have learned early that the light propagates along the direction of its wave vector $\boldsymbol{s}_0$, see (2.85). Without loss of generality, let us focus on a light beam which is propagating along the z axis, i.e., $\boldsymbol{s}_0 = \boldsymbol{z}_0$. The net change of the photon number N_ω in an spatial interval between z and $z + \mathrm{d}z$ is

$$\frac{\mathrm{d}N_\omega(z)}{\mathrm{d}t} = \frac{\partial N_\omega(z)}{\partial t} - \frac{\partial N_\omega(z)}{\partial z}\frac{\partial z}{\partial t} \tag{2.119}$$

where the first term on the right side represents either the generation or annihilation of photons in this spatial region, which equals the total number of optical transitions of electrons, i.e., (2.118), such that

$$\frac{\partial N_\omega}{\partial t} = -W(\omega) \tag{2.120}$$

The second term on the right side of (2.119) represents the spatial propagation of the photons at the speed of $\partial z/\partial t = c$, which is the speed of light in the material. $c = c_0/n$, where c_0 is the speed of light in vacuum and n is the refractive index of the material. At steady state, $\mathrm{d}N_\omega/\mathrm{d}t = 0$ so that (2.119) becomes

$$0 = -W(\omega) - c\,\frac{\partial N_\omega}{\partial z} \tag{2.121}$$

with a general solution

$$N_\omega(z) = N_\omega(0)\, e^{g(\omega)z} \tag{2.122}$$

where $g(\omega)$ is commonly referred to as the optical coefficient of the material

$$g(\omega) = \sum_{qk} \frac{\pi\hbar^2 e^2}{m_0^2\epsilon_0 c_0 n\omega\Omega}|\langle \boldsymbol{q}|\boldsymbol{e}_s\cdot\nabla|\boldsymbol{k}\rangle|^2\,\frac{\Gamma_{qk}}{(E_q - E_k \pm \hbar\omega)^2 + \Gamma_{qk}^2}\left[f(E_q) - f(E_k)\right] \tag{2.123}$$

Equation (2.122), the same as (2.87), indicates that if $g(\omega) \geq 0$, the number of photons in the light beam increases during the light propagation through the material, which is called optical gain (such is the function of a light-emitting diode or a laser), while if $g(\omega) < 0$, the number of photons is reduced, meaning that the light becomes absorbed (e.g., a photodetector). Equation (2.123) shows that a population inversion, i.e., $f(E_q) > f(E_k)$ when $E_q > E_k$, is needed to attain the optical gain ($g(\omega) > 0$), while at equilibrium, $f_{\mathrm{FD}}(E_q) < f_{\mathrm{FD}}(E_k)$ for $E_q > E_k$, see (2.94), so the semiconductor at equilibrium will be always lossy for light transmission. Moreover, the absorption coefficient introduced in the Beer–Lambert law, i.e., (2.86),

$$\alpha(\omega) = -g(\omega)$$

Equation (2.123) is the quantitative description about the fundamental optical properties of semiconductor materials and devices. Almost all optical spectra to be closely studied in the forthcoming chapters of this book are originated from this. Here we see that various optical spectra, such as reflectance, transmittance, photoluminescence or modulation spectroscopy, are composed of many Lorentzian peaks

$$\frac{\Gamma_{qk}}{(E_q - E_k \pm \hbar\omega)^2 + \Gamma_{qk}^2}$$

corresponding to optical transitions between two electron states $\boldsymbol{k}$ and $\boldsymbol{q}$. By analyzing the optical spectrum, i.e., decomposing the light beams, both the incident beam (also known as the probing beam) and the output beam (after light-matter interaction), into monochromatic lights, finding peak powers ($\propto |\langle \boldsymbol{q}|\nabla|\boldsymbol{k}|\rangle^2$), peak wavelengths ($|E_q - E_k|$) and peak widths ($\Gamma_{qk}$), we will identify, though indirectly, then understand the electron states in the semiconductor materials and devices, thereafter optimize as well as design the light-matter interaction for better and/or novel materials and devices.

By looking at (2.85) and (2.87) together with (2.122), it is easy to obtain the optical coefficient

$$g(\omega) = -\frac{2\omega\kappa}{c_0} = -\frac{\omega\epsilon''}{nc_0} \tag{2.124}$$

The second equality in the above equation utilizes the second relation in (2.77). Equation (2.124) shows that the extinction coefficient κ shows how strong the light becomes either absorbed (optical loss) or amplified (optical gain) when propagating through the material.

By inserting (2.123) into (2.124) we obtain the expression of $\epsilon''(\omega)$ as a function of microscopic electron transitions

$$\epsilon''(\omega) = \sum_{qk} \frac{\pi\hbar^2 e^2}{m_0^2\omega^2\Omega} |\langle \boldsymbol{q}|\boldsymbol{e}_s \cdot \nabla|\boldsymbol{k}\rangle|^2 \frac{\Gamma_{qk}}{(E_q - E_k \pm \hbar\omega)^2 + \Gamma_{qk}^2} \left[f(E_k) - f(E_q) \right] \tag{2.125}$$

Equation (2.85) shows that the refractive index n modifies the spatial variation (the temporal variation in the time domain remains unchanged, i.e., the same angular frequency), which is due to the polarization of the material induced by the electric field of light. For common semiconductors, the refractive index is in general inversely proportional to the energy bandgap, i.e., a narrow-bandgap material normally has a large refractive index. For example, narrow-bandgap $Hg_{0.67}Cd_{0.33}Te$, which is commonly used as an infrared optoelectronic material, has a refractive index of ca 3.3 in the infrared range, while the wide-bandgap GaN has a refractive index of 2.3. Reader may correlate this macroscopic phenomenon to the microscopic energy band structure of the semiconductor: (2.125) shows that the polarization is proportional to $\langle \boldsymbol{q}|\nabla|\boldsymbol{k}\rangle$, which is proportional to $p_{cv} = \langle c|\nabla|v\rangle$ when discussing optical transition between the conduction band and the valence band, i.e., the interaction between conduction- and valence-bandedge states u_c and u_v in (2.35), see more details in Chap. 4. p_{cv} is large if the energy distance between u_c and u_v is small, which is the energy bandgap. Thus, a narrow energy bandgap is normally associated with a large p_{cv}, resulting in a large ϵ''. A large ϵ'' normally implies a large ϵ' due to the Kramers–Kronig relationship to be introduced in Sect. 3.4, thereafter a large refractive index. This is one example of spectral analysis.

2.4 Polariton and Spectral Analysis

In the above section we learned that the electron transits from state $\boldsymbol{k}$ to state $\boldsymbol{q}$ in the presence of a photon field, see Fig. 2.11. The effect of such an electron transition on the photon field is that the number of photons is either decreased by 1 or increased by 1. But the story does not end here. Consider an intrinsic semiconductor with a completely occupied valence band and a totally empty conduction band. After absorbing one photon, one electron transits from an occupied valence-band state to an empty conduction-band state. More specifically, the initial electron state is composed of an occupied valence-band state and an empty conduction-band state, and the final electron state is composed of one empty valence-band state, which is called a hole in the valence band and is positively charged, and one occupied conduction-band state. In the electron-hole picture, the initial electron state does not compose any occupied hole state in the valence band and any occupied electron state in the conduction band, which is viewed as a vacuum state. The final electron state composes an occupied hole state in the valence band and an occupied electron state in the conduction band. Since the effective masses of electron and hole are different so that their wave function distributions are also different, this results in a polarization. This polarization will certainly affect the photon field via the Maxwell's equations.

Let us elaborate carefully and quantitatively the electron-hole concept we just introduced. For an intrinsic semiconductor, the conduction band is totally empty, thus there is no electron. The valence band is totally occupied, i.e., no empty state, meaning no hole. Thus, in the electron-hole picture, the intrinsic semiconductor is quantitatively described as a vacuum state Ψ_0, i.e., no electron-hole pair. We impinge a light of $\boldsymbol{E}(\boldsymbol{r}, t)$ starting from $t = 0$. Because of the light-matter interaction, one electron originally occupying one valence-band state will transit to an originally empty conduction-band state, leaving the valence-band state empty, i.e., a hole in the valence band. As compared with Ψ_0, we now have a final state composed of one electron in the conduction band and one hole in the valence band. Such an electron-hole pair is known as an exciton. We denote it by wave function $\Psi_n(\boldsymbol{r}_e, \boldsymbol{r}_h, t)$, where $\boldsymbol{r}_e$ and $\boldsymbol{r}_h$ represent the electron in the conduction band and the hole in the valence band, respectively.

We start with $\Psi_n(\boldsymbol{r}_e, \boldsymbol{r}_h)$. By Sect. 2.1, we formulate (2.40) for the conduction-band electron and a similar equation for the valence-band hole then merge them to form the following Hamiltonian and the Schrödinger equation for the envelope function of the electron-hole pair, i.e., exciton:

$$H_0 = -\frac{\hbar^2\nabla_e^2}{2m_e^*} - \frac{\hbar^2\nabla_h^2}{2m_h^*} - \frac{e^2}{4\pi\epsilon_\infty|\boldsymbol{r}_e - \boldsymbol{r}_h|} + V_e(\boldsymbol{r}_e) + V_h(\boldsymbol{r}_h) \tag{2.126}$$

$$H_0\,\psi_n(\boldsymbol{r}_e, \boldsymbol{r}_h) = E_n\psi_n(\boldsymbol{r}_e, \boldsymbol{r}_h) \tag{2.127}$$

where $V_e(\boldsymbol{r}_e)$ and $V_h(\boldsymbol{r}_h)$ are potential energies for the electron and the hole, respectively, ϵ_∞ is the high-frequency dielectric coefficient of the semiconductor material. The total wave function of the exciton is (including the temporal factor)

$$\Psi_n(\boldsymbol{r}_e, \boldsymbol{r}_h, t) = \psi_n(\boldsymbol{r}_e, \boldsymbol{r}_h)u_c(\boldsymbol{r}_e)u_v(\boldsymbol{r}_h)\,e^{-iE_nt/\hbar} \tag{2.128}$$

$u_c(\boldsymbol{r}_e)$ and $u_v(\boldsymbol{r}_h)$ are Bloch functions at the conduction and valence bandedges.

In bulk material, $V_e(\boldsymbol{r}_e)$ and $V_h(\boldsymbol{r}_h)$ are constant in the whole space so that the Hamiltonian of (2.126) is similar to the one of a hydrogen atom. Alike the Bohr model of a hydrogen atom, the ground state of the exciton is characterized by the so-called exciton Bohr radius:

$$a_{Br} = \frac{4\pi\epsilon_\infty\hbar^2}{e^2}\left(\frac{1}{m_e^*} + \frac{1}{m_h^*}\right) = \frac{\epsilon_\infty m_0}{\epsilon_0 m_r^*}\,a_0 \tag{2.129}$$

where $1/m_r^* = 1/m_e^* + 1/m_h^*$ is known to be the reduced effective mass. The derivation of the above equation can be found in quantum mechanics and solid-state theory textbooks. Here

$$a_0 = \frac{4\pi\epsilon_0\hbar^2}{m_0} = 0.529\ \text{Å}$$

is the Bohr radius of the hydrogen atom (while it is called the exciton Bohr radius in (2.129)). In Si, the exciton Bohr radius is 4.5 nm, it is 11.6 nm in GaAs, and 38.1 nm in InAs.

We now try to calculate the impact of the exciton generation on the optical properties of the semiconductor, similar to the previous section where the optical transition of one electron from one state to another is studied. We first list the initial and final exciton states:

$$\begin{aligned}
&H_0\Psi_0(t) = E_0\Psi_0(t) = \hbar\omega_0\Psi_0(t)\\
&H_0\Psi_n(\boldsymbol{r}_e, \boldsymbol{r}_h, t) = E_n\Psi_n(\boldsymbol{r}_e, \boldsymbol{r}_h, t) = \hbar\omega_n\Psi_n(\boldsymbol{r}_e, \boldsymbol{r}_h, t)\\
&\Psi_0(t) = \Psi_0 e^{-i\omega_0 t}\\
&\Psi_n(\boldsymbol{r}_e, \boldsymbol{r}_h, t) = \Psi_n(\boldsymbol{r}_e, \boldsymbol{r}_h)e^{-i\omega_n t}
\end{aligned} \tag{2.130}$$

Note that $\Psi_0(t)$ denotes the vacuum state so that $E_0 = \hbar\omega_0 = 0$.

The electric dipole of the electron-hole pair (the exciton) is

$$\boldsymbol{d}(\boldsymbol{r}) = -e\boldsymbol{r}_e\delta(\boldsymbol{r} - \boldsymbol{r}_e) + e\boldsymbol{r}_h\delta(\boldsymbol{r} - \boldsymbol{r}_h) \tag{2.131}$$

where e is the unit charge.

Initially (defined mathematically as $t < 0$), the semiconductor is at its vacuum state. For $t \geq 0$ we switch on an excitation light defined by its electric field $\boldsymbol{E}(\boldsymbol{r}, t)$. An exciton will be created after the semiconductor absorbs one photon and the electric dipole of the exciton interacts with $\boldsymbol{E}(\boldsymbol{r}, t)$ in the form of

$$H' = \int \boldsymbol{d}(\boldsymbol{r}) \cdot \boldsymbol{E}(\boldsymbol{r}, t)\, \mathrm{d}\boldsymbol{r} \tag{2.132}$$

In the previous section, we expressed the electromagnetic field by its vector potential then obtain the light-matter interaction in the form of (2.101). It is easy to show that the two expressions are equivalent. Here we choose (2.132) just because from here it is easier to calculate the electric dipole and then the polarization.

Denote the wave function of the exciton as $|\boldsymbol{r}_e, \boldsymbol{r}_h, t\rangle$ which obeys the following time-dependent Schrödinger equation

$$i\hbar\frac{\partial}{\partial t}|\boldsymbol{r}_e, \boldsymbol{r}_h, t\rangle = (H_0 + H')|\boldsymbol{r}_e, \boldsymbol{r}_h, t\rangle \tag{2.133}$$

where H_0 is the same Hamiltonian of (2.126).

Assume that the semiconductor is initially in its vacuum state $|\Psi_0\rangle$. Under the excitation of $\boldsymbol{E}(\boldsymbol{r}, t)$, the exciton will be excited to excited state $|\Psi_n\rangle$. By the first-order approximation,

$$|\boldsymbol{r}_e, \boldsymbol{r}_h, t\rangle = |\Psi_0(t)\rangle + c_n(t)|\Psi_n(\boldsymbol{r}_e, \boldsymbol{r}_h, t)\rangle \tag{2.134}$$

where $|c_n(t)| \ll 1$ (we will come back to this assumption shortly).

Insert (2.134) into (2.133)

$$\begin{aligned} &i\hbar\frac{\partial\Psi_0(t)}{\partial t} + i\hbar\frac{\partial c_n(t)}{\partial t}\Psi_n(\boldsymbol{r}_e, \boldsymbol{r}_h, t) + i\hbar c_n(t)\frac{\partial\Psi_n(\boldsymbol{r}_e, \boldsymbol{r}_h, t)}{\partial t} \\ &= H_0\Psi_0(t) + c_n(t)H_0\Psi_n(\boldsymbol{r}_e, \boldsymbol{r}_h, t) + H'\Psi_0(t) + c_n(t)H'\Psi_n(\boldsymbol{r}_e, \boldsymbol{r}_h, t) \end{aligned}$$

And by (2.130)

$$\begin{aligned} &\hbar\omega_0\Psi_0(t) + i\hbar\frac{\partial c_n(t)}{\partial t}\Psi_n(\boldsymbol{r}_e, \boldsymbol{r}_h, t) + \hbar\omega_n\, c_n(t)\Psi_n(\boldsymbol{r}_e, \boldsymbol{r}_h, t) \\ &= E_0\Psi_0(t) + c_n(t)E_n\Psi_n(\boldsymbol{r}_e, \boldsymbol{r}_h, t) + H'\Psi_0(t) + c_n(t)H'\Psi_n(\boldsymbol{r}_e, \boldsymbol{r}_h, t) \end{aligned}$$

Eliminate identical terms from the two sides of the equation

$$i\hbar\frac{\partial c_n(t)}{\partial t}\Psi_n(\boldsymbol{r}_e,\boldsymbol{r}_h,t) = H'\Psi_0(t) + c_n(t)H'\Psi_n(\boldsymbol{r}_e,\boldsymbol{r}_h,t) \tag{2.135}$$

Multiply the above equation by $\langle\Psi_n(\boldsymbol{r}_e,\boldsymbol{r}_h,t)|$ then do the spatial integration

$$\frac{\partial c_n(t)}{\partial t} = \frac{1}{i\hbar}\langle\Psi_n(\boldsymbol{r}_e,\boldsymbol{r}_h,t)|\int \boldsymbol{d}(\boldsymbol{r})\cdot\boldsymbol{E}(\boldsymbol{r},t)\,\mathrm{d}\boldsymbol{r}|\Psi_0(t)\rangle \tag{2.136}$$

Note that there is a term on the right-side of the equation

$$\langle\Psi_n(\boldsymbol{r}_e,\boldsymbol{r}_h,t)|\int \boldsymbol{d}(\boldsymbol{r})\cdot\boldsymbol{E}(\boldsymbol{r},t)\,\mathrm{d}\boldsymbol{r}|\Psi_n(\boldsymbol{r}_e,\boldsymbol{r}_h,t)\rangle$$

which is zero and thus not shown in (2.136) since the parity of the product of $\Psi_n^*(\boldsymbol{r}_e,\boldsymbol{r}_h,t)\Psi_n(\boldsymbol{r}_e,\boldsymbol{r}_h,t)$ is even with respect to $\boldsymbol{r}_e$ and $\boldsymbol{r}_h$, while the parity of $\boldsymbol{d}(\boldsymbol{r})$ is odd.

We express the temporal dependence of the electromagnetic field as

$$\boldsymbol{E}(\boldsymbol{r},t) = \boldsymbol{E}(\boldsymbol{r},\omega)\left(e^{-i\omega t} + e^{i\omega t}\right)$$

where $\boldsymbol{E}(\boldsymbol{r},\omega)$ is real. Again we focus on a monochromatic light. And the above expression simply means that the energy of the photon remains the same while the wavelength, the speed, and spatial distribution (reflection, diffraction etc. as we learned in Chap. 1) of light vary when light travels in different medium. Integrate (2.136) in the time domain to obtain coefficient $c_n(t)$

$$\begin{aligned} c_n(t) = & \frac{e^{i(\omega_n-\omega_0-\omega)t}}{\hbar(\omega_n-\omega_0-\omega)}\langle\Psi_n(\boldsymbol{r}_e,\boldsymbol{r}_h)|\int \boldsymbol{d}(\boldsymbol{r})\cdot\boldsymbol{E}(\boldsymbol{r},\omega)\,\mathrm{d}\boldsymbol{r}|\Psi_0\rangle \\ & + \frac{e^{i(\omega_n-\omega_0+\omega)t}}{\hbar(\omega_n-\omega_0+\omega)}\langle\Psi_n(\boldsymbol{r}_e,\boldsymbol{r}_h)|\int \boldsymbol{d}(\boldsymbol{r})\cdot\boldsymbol{E}(\boldsymbol{r},\omega)\,\mathrm{d}\boldsymbol{r}|\Psi_0\rangle \end{aligned} \tag{2.137}$$

Let us analyze the above expression carefully. For an optical excitation that the semiconductor absorbs one photon then creates an exciton, we expect

$$\hbar\omega_n - \hbar\omega_0 = \hbar\omega$$

due to energy conservation consideration. The denominator of the first term on the right side of (2.137) then approaches zero so that the probability of such an optical transition is very large. Such a situation is known as resonance. At this resonant condition, the first term on the right side of the equation becomes time independent. On the other hand, the denominator of the second term will never be zero physically. And for a light beam with a wavelength 700 nm in vacuum (4.3×10^{14} Hz), this term oscillates extremely fast. Physically what we do when we try to observe (measure) $c_n(t)$ is to do an integration over a finite time interval (called exposure time considering taking a photograph). We thus need to go beyond a time resolution of femtosecond (1 fs $= 10^{-15}$ s) in order to observe this fast oscillating term. Therefore, for practical material and device applications, we can neglect the second term.

Focus now on the first term. We notice that at resonance $\hbar\omega_n - \hbar\omega_0 = \hbar\omega$, the denominator becomes zero so c_n approaches infinite. Physically, the maximal value of c_n is of course only 1, i.e., the exciton state can be maximally occupied by one exciton (in the electron-hole picture, it is one electron in the conduction band and one hole in the valence band). What happens in reality is that long before c_n becomes significantly large, the photon emission process starts that the electron which has been optically excited to the conduction band from the valence band by $\boldsymbol{E}(\boldsymbol{r}, \omega)$ will now transit back to the vacant valence-band state under the excitation of the same $\boldsymbol{E}(\boldsymbol{r}, \omega)$, which is the stimulated emission we learned in the previous section. To include this process, we modify (2.134) to

$$|\boldsymbol{r}_e, \boldsymbol{r}_h, t\rangle = c_0(t)|\Psi_0(t)\rangle + c_n(t)|\Psi_n(\boldsymbol{r}_e, \boldsymbol{r}_h, t)\rangle \tag{2.138}$$

then obtain an equation for $c_0(t)$ similar to (2.136). Careful mathematical manipulation results in the following expression

$$c_n(t) = \frac{e^{i(\omega_n - \omega)t}}{\hbar(\omega_n - \omega - i\gamma_n)} \langle \Psi_n(\boldsymbol{r}_e, \boldsymbol{r}_h)| \int \boldsymbol{d}(\boldsymbol{r}) \cdot \boldsymbol{E}(\boldsymbol{r}, \omega)\, \mathrm{d}\boldsymbol{r} |\Psi_0\rangle \tag{2.139}$$

which turns out to be limited, i.e., $|c_n(t)| \ll 1$ even at resonance $\hbar\omega_n = \hbar\omega$, under most material and device working conditions. Here

$$\hbar\gamma_n \propto \langle \Psi_n(\boldsymbol{r}_e, \boldsymbol{r}_h)| \int \boldsymbol{d}(\boldsymbol{r}) \cdot \boldsymbol{E}(\boldsymbol{r}, \omega)\, \mathrm{d}\boldsymbol{r} |\Psi_0\rangle \propto \langle \mathrm{c}|\nabla|\mathrm{v}\rangle \tag{2.140}$$

which is the same $\langle \mathrm{c}|\nabla|\mathrm{v}\rangle$ in the effective mass expression of (2.35).

Note that we have neglected the second term on the right side of (2.137) which is fast oscillating in the time domain. Moreover, we simply let $\hbar\omega_0 = 0$.

Knowing the wave function of the excited exciton, the polarization of the exciton is simply $\boldsymbol{P}_n(\boldsymbol{r}, \omega) = \langle \boldsymbol{r}_e, \boldsymbol{r}_h, t|\boldsymbol{d}(\boldsymbol{r})|\boldsymbol{r}_e, \boldsymbol{r}_h, t\rangle_t$. After detailed mathematical and quantum mechanical operations (reader is recommended to read more, e.g., [21]), the final result is

$$\boldsymbol{P}_n(\boldsymbol{r}, \omega) = \frac{\epsilon_\infty \omega_{\mathrm{LT}} \pi\, a_{\mathrm{Br}}^3}{\omega_n - \omega - i\gamma_n} \psi_n(\boldsymbol{r}, \boldsymbol{r}) \int \psi_n^*(\boldsymbol{r}', \boldsymbol{r}')\boldsymbol{E}(\boldsymbol{r}', \omega)\mathrm{d}\boldsymbol{r}' \tag{2.141}$$

where ω_{LT} is defined by

$$\epsilon_\infty \omega_{\mathrm{LT}} a_{\mathrm{Br}}^3 = \frac{e^2 \boldsymbol{p}_{cv}^2}{\pi \hbar \omega_n^2 m_0^2} \tag{2.142}$$

and is known as the exciton longitudinal-transverse splitting and a_{Br} is the exciton Bohr radius defined by (2.129). Both are fundamental physical parameters that characterize the semiconductor. Here $\boldsymbol{p}_{cv} = \langle \mathrm{c}|\boldsymbol{p}|\mathrm{v}\rangle$. As we are mostly interested in the visible range where the wavelength is about 500 nm, while the spatial extension of the exciton wave function, i.e., the exciton Bohr radius in a semiconductor is only tens of nanometers, see (2.129), the electric field within the spatial extension of an exciton can be treated as uniform. We can therefore move $\boldsymbol{E}(\boldsymbol{r}, \omega)$ out of the spatial integration

$$\boldsymbol{P}_n(\boldsymbol{r}, \omega) = \boldsymbol{E}(\boldsymbol{r}, \omega) \frac{\epsilon_\infty \omega_{\mathrm{LT}} \pi a_{\mathrm{Br}}^3}{\omega_n - \omega - i\gamma_n} \psi_n(\boldsymbol{r}, \boldsymbol{r}) \int \psi_n^*(\boldsymbol{r}', \boldsymbol{r}') \mathrm{d}\boldsymbol{r}' \tag{2.143}$$

Now, back to the Maxwell's equations where we relate the electric displacement field to the polarization so that we introduce an effective dielectric constant

$$\boldsymbol{D}(\boldsymbol{r}, \omega) = \epsilon_\infty \boldsymbol{E}(\boldsymbol{r}, \omega) + \boldsymbol{P}_n(\boldsymbol{r}, \omega) = \epsilon_n(\boldsymbol{r}, \omega)\boldsymbol{E}(\boldsymbol{r}, \omega) \tag{2.144}$$

where ϵ_∞ is the dielectric constant at high frequency, the same as in (2.126). Insert (2.143) into the above equation

$$\epsilon_n(\boldsymbol{r}, \omega) = \epsilon_\infty \left(1 + \frac{\omega_{\mathrm{LT}n}(\boldsymbol{r})}{\omega_n - \omega - i\gamma_n}\right) \tag{2.145}$$

$$\omega_{\mathrm{LT}n}(\boldsymbol{r}) = \epsilon_\infty \omega_{\mathrm{LT}} \, \psi_n(\boldsymbol{r}, \boldsymbol{r}) \int \psi_n^*(\boldsymbol{r}', \boldsymbol{r}') \mathrm{d}\boldsymbol{r}' \tag{2.146}$$

When there is more than one exciton state,

$$\epsilon(\boldsymbol{r}, \omega) = \epsilon_\infty \left(1 + \sum_n \frac{\omega_{\mathrm{LT}n}(\boldsymbol{r})}{\omega_n - \omega - i\gamma_n}\right) \tag{2.147}$$

We have discussed the spectral resolution of an optical spectrum, which is defined as the minimal interval between the wavelengths of two monochromatic lights that we are able to distinguish, either in space such as by using a prism dispersion or a diffraction grating, or in the time domain by using Fourier transform spectroscopy. Such a spectral resolution is reflected by parameter ω in $\epsilon(\boldsymbol{r}, \omega)$ in the above equation. We also notice here that $\epsilon(\boldsymbol{r}, \omega)$ depends on $\boldsymbol{r}$, which is referred to as the spatial resolution. One example is the well-known microscopic photoluminescencespectrum, or µ-PL. Here we focus a probing light on to a very small area of the sample then detect the luminescence from the focusing area. In practice, we make the probing spot as small as possible and detect the light signal from the proving spot. We move the probing spot to scan the sample. We will study a two-dimensional µ-PL spectrum in Sect. 4.4.

In many cases we encounter physical processes of bulk natures that $\psi_n(\boldsymbol{r}, \boldsymbol{r})$ is a weak function of $\boldsymbol{r}$. In this case, we can neglect the spatial dependence of $\omega_{\mathrm{LT}n}(\boldsymbol{r})$ in (2.146). Another extreme situation is that $\psi_n(\boldsymbol{r}, \boldsymbol{r})$ is strongly localized in a very small spatial region such as an exciton in a quantum dot with a radius of less than 10 nm, which is beyond the resolution (about one wavelength) of any optical measurement setup (which is known to be the diffraction limit). Thus what we observe is an average of the dielectric constant over the probing space. For these two cases,

$$\epsilon(\omega) = \epsilon_{\infty}\left(1 + \sum_{n} \frac{\omega_{\mathrm{LT}n}}{\omega_n - \omega - i\gamma_n}\right) \tag{2.148}$$

Note that the above expression is similar to (2.125), but not exactly the same! It is reminded that (2.148) is derived under the assumption that the wave function of the exciton is much confined in space, i.e., the step from (2.141) to (2.143). Equation (2.125) is thus always used in the rest of the book when we perform spectral analysis at the microscopic level.

As mentioned before, a semiconductor is composed by atoms with a density of approximately 10^{23}/cm^3. The responses from a semiconductor to an external probing are results of various transitions of electrons and/or atoms between their initial states and final states, which can be expressed mathematically similar to the exciton in the form of (2.147). Not being able to go in details within the scope of the book, we simply state the final result that the dielectric coefficient of a material can be expressed as

$$\epsilon(\omega) = \epsilon_0\left(1 + \sum_{n=1}^{\infty} \frac{\omega_{\mathrm{LT}n}}{\omega_n - \omega - i\gamma_n}\right) \tag{2.149}$$

where ω_n denotes various transition energy of the electron and/or atom. And instead of ϵ_∞ in (2.147) which is the case when we focus on the exciton effect out of all other transitions, ϵ_0 appears in (2.149). Moreover, to distinguish the two cases, we specifically write the summation in (2.149) from 1 to ∞.

Note that the dielectric coefficient expressed by (2.149) is complex. Its imaginary part is actually equivalent to the optical coefficient $g(\hbar\omega)$ of (2.123) in Sect. 2.3. We will discuss it further in the next chapter.

Reader may see a different but equivalent form of (2.149) in the context of the damped harmonic oscillator. In the close vicinities of resonance, $\omega \approx \omega_n$,

$$\begin{aligned}\frac{1}{\omega_n - \omega - i\gamma_n} = \frac{(\omega_n + \omega)}{(\omega_n + \omega)(\omega_n - \omega - i\gamma_n)} &= \frac{(\omega_n + \omega)}{\omega_n^2 - \omega^2 - i\gamma_n(\omega_n + \omega)} \\ &\approx \frac{2\omega_n}{\omega_n^2 - \omega^2 - 2i\omega\gamma_n}\end{aligned} \tag{2.150}$$

i.e., we multiply the numerator and the denominator simultaneously by $(\omega_n + \omega)$, approximate it in the numerator as $\omega_n + \omega \approx 2\omega_n$, and the term that multiplies γ_n in the denominator as 2ω (both are valid as long as $\omega \approx \omega_n$). Re-define relevant parameters we obtain the dielectric constant in the picture of the damped harmonic oscillator model

$$\epsilon(\omega) = \epsilon_\infty + \sum_n \frac{f_n\,\omega_n^2}{\omega_n^2 - \omega^2 - i\omega\gamma_n} \tag{2.151}$$

where ω_n is the frequency of the harmonic oscillator, i.e., the experimental peak frequency (easily convertible to peak wavelength), $1/\gamma_n$ is the width of the peak, representing the lifetime of the harmonic oscillator, also known as the damping rate. Without damping, the spectral peak becomes a sharp spectral line. f_n is known as the oscillator strength, representing the strength of the response of an electron and/or atom in the semiconductor to the external probing.

We now know that an optical spectrum is composed by spectral peaks characterized by their wavelengths, strengths, and widths. The peak width, i.e., γ_n in (2.151) is proportional to the light-matter interaction, see (2.140). The geometric shape of the spectral peak in (2.151) is called Lorentzian. In reality, there are many factors that collectively determine the geometric shape of a spectral peak. There are basically two categories.

- Homogeneous broadening:
  Lifetime, i.e., γ_n in (2.151), is the principal factor that shapes the spectral peak into the Lorentzian form

$$\frac{1}{(\omega_n - \omega)^2 + \gamma_n^2} \tag{2.152}$$

- Inhomogeneous broadening:
  Doppler effect is expected when there is a relative motion between the light source and the photon absorber. Let the light frequency be ω_n when there is no relative motion between the light source and the photon absorber. With a relative motion of speed v (move close which is defined as $v > 0$ and move away defined as $v < 0$), the frequency of the light that reaches the photon absorber is

$$\omega_n \left(1 + \frac{v}{c}\right) \tag{2.153}$$

  Electrons move at different speeds in a semiconductor so that photons emitted from these electrons will appear at different frequencies when they reach the spectrometer, resulting in the so-called Doppler broadening. There are other types of broadening such as impurity and defect scatterings. All these factors can shape the spectral peak into the following Gaussian peak

$$e^{(\omega-\omega_n)^2/2\gamma_n^2} \tag{2.154}$$

We normally quantify the peak width by the so-called full width at half maximum (FWHM). For a Lorentzian peak of (2.152), FWHM $= 2\gamma_n$, and for a Gaussian peak of (2.154), FWHM $= 2\sqrt{2\ln 2}\,\gamma_n \approx 2.355\gamma_n$. Figure 2.12 displays the Lorentzian and Gaussian peaks. The former has two broad sides, while the latter decreases much fast when moving away from the peak center. In most cases of our interest, the spectral peaks are Lorentzian. In practices, however, the recorded spectrum can be much complicated, see more discussions in Sect. 4 and various spectral fittings in the coming chapters.

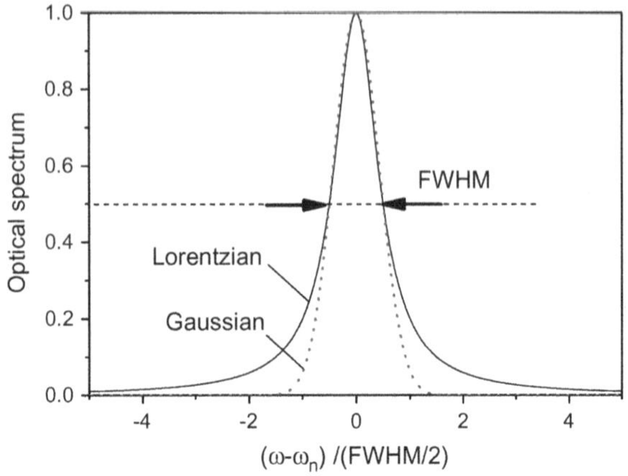

Fig. 2.12 Lorentzian (black solid line) and Gaussian (red dashed line) spectral peaks. ω_n is the frequency of the harmonic oscillator and FWHM is the full width at half maximum

References

1. X. Du, Y. Li, G.S. Herman, A field effect glucose sensor with a nanostructured amorphous In-Ga-Zn-O network. Nanoscale **8**, 18469–18475 (2016)
2. P. Gogl, H.P. Hjalmarson, J.D. Dow, A semi-empirical tight-binding theory of the electronic structure of semiconductors. J. Phys. Chem. Solids **44**, 365–378 (1983)
3. W.-L. Xu, Y. Fu, M. Willander, S.C. Shen, Theory of normal incident absorption for the intersubband transition in n-type indirect-gap semiconductor quantum wells. Phys. Rev. B **49**, 13760–13766 (1994)
4. Y. Fu, *Physical Model of Semiconductor Quantum Devices*, 2nd edn. (Springer, Berlin, 2013), 500 pp
5. I. Vurgaftman, J.R. Meyer, L.R. Ram-Mohan, Band parameters for III-V compound semiconductors and their alloys. J. Appl. Phys. **89**, 5815–5875 (2001)
6. O. Madelung (ed.), *Semiconductors Group IV Elements and III-V Compounds* (Springer, Berlin, 1991)
7. O. Madelung (ed.), *Data in Science and Technology: Semiconductors other than Group IV Elements and III-V Compounds* (Springer, Boston, 1992)
8. Y. Fu, K.A. Chao, Subband structures of GaAs/AlGaAs multiple quantum wells. Phys. Rev. B **40**, 8349–8356 (1989)
9. Y. Fu, K.A. Chao, Effective mass approximation in graded crystals. Int. J. Mod. Phys. B **4**, 2357–2369 (1990)
10. J.C. Hensel, H. Hasegawa, M. Nakayama, Cyclotron resonance in uniaxially stressed silicon. II. Nature of the covalent bond. Phys. Rev. **138**, A225–A238 (1965)
11. E.O. Kane, Band structure of indium antimonide. J. Phys. Chem. Solids **1**, 249–261 (1957)
12. D.C. Reynolds, C.W. Litton, T.C. Collins, Bound-phonon quasiparticle in CdS. Phys. Rev. B **4**, 1868–72 (1971)
13. A.L. Pan, R.B. Liu, B.S. Zou, Phonon-assisted stimulated emission from single CdS nanoribbons at room temperature. Appl. Phys. Lett. **88**, 173102(3) (2006)
14. E.N. Economous, in *Green's Function in Quantum Physics*, ed. by M. Cardona (Springer, Berlin, 1979)
15. R. Haydock, V. Heine, M.J. Kelly, Electronic structure based on the local atomic environment for tight-binding bands. J. Phys. C **5**, 2845–2858 (1972)
16. Y. Fu, Correlations of lattice vibrations. Chin. Phys. Lett. **4**, 309–312 (1987)
17. Y. Fu, W.-L. Xu, Z.B. Zheng, Impurity induced vibrations in light doped silicon. Solid State Commun. **62**, 163–167 (1987)

18. S. Fan, Y. He, B.S. Ung, E. Pickwell-MacPherson, The growth of biomedical terahertz research. J. Phys. D **47**, 374009(11pp) (2014)
19. S.M. Sze, J.C. Irvin, Resistivity, mobility and impurity levels in GaAs, Ge, and Si at 300°K. Solid State Electron. **11**, 599–602 (1968)
20. Y. Fu, Y.-H. Zhou, H. Su, F.Y.C. Boey, H. Ågren, Impact ionization and Auger recombination rates in semiconductor quantum dots. J. Phys. Chem. C **114**, 3743–3747 (2010)
21. Y. Fu, M. Willander, E.L. Ivchenko, A.A. Kiselev, Four-wave mixing in microcavities with embedded quantum wells. Phys. Rev. B **55**, 9872–9879 (1997)

Chapter 3
Reflection and Transmission

Abstract In this chapter we study how a material reflects and transmits a light beam in terms of the macroscopic effective dielectric coefficient of the material. The focus of the chapter is to extract the effective dielectric coefficient of the material using the harmonic oscillator model and the Kramers–Kronig relationship by properly designing the reflection and transmission spectroscopic measurements including the reflection and transmission spectra from and through a thin film and the reflection spectrum from a thin film on a substrate.

An incident plane wave is reflected, refracted and partly transmitted by a sample. The optical power of the reflected light as a function of the wavelength of the light beam is the reflection spectrum, and the optical power of the transmitted light as a function of the wavelength is the transmission spectrum. These two most common, may even be called the simplest, spectra have been key tools in exploring and exploiting semiconductor materials and devices.

Recall Table 2.2 in the previous chapter, we notice words "wave" and "ray" that describe characteristic natures of light in different wavelength ranges. While "wave" emphasizes the wave nature of light in terms of the wave optics, "ray" is in the field of the geometric optics (which is also known as the ray optics). What we do in this chapter is to restrain us to the semiconductor materials and devices whose geometric sizes are much larger than the wavelength of the light beam so that the geometric optics applies when we discuss the propagation of the light beam. This is to be done by studying the reflection and transmission of the light beam at/through boundaries of the semiconductors using the Maxwell's equations.

The reflection spectrum includes not only a single reflected light beam. There can be several reflected beams, such as multiple reflected light beams from a rough or a structured surface, e.g., see Sect. 1.2, or from two surfaces of a thin film (see discussions below). Similar situations occur in the transmission spectrum. The addition of multiple light beams is described by the wave optics.

A monochromatic light beam is described by its electric field

$$\boldsymbol{E}(\boldsymbol{r}, t) = \boldsymbol{E}_0\, e^{i(\boldsymbol{s}\cdot\boldsymbol{r}-\omega t)} + \text{c.c.} \tag{3.1}$$

W. Lu and Y. Fu, *Spectroscopy of Semiconductors*, Springer Series
in Optical Sciences 215, https://doi.org/10.1007/978-3-319-94953-6_3

i.e., the first equation of (2.83), where $\boldsymbol{E}_0$ is the amplitude, $\boldsymbol{s}$ the wave vector, and ω the angular frequency. c.c. represents the complex conjugate of the first term on the right side of the expression. Another light beam is similarly expressed but with some modifications

$$\boldsymbol{E}'(\boldsymbol{r}, t) = \boldsymbol{E}_0' e^{i(\boldsymbol{s}'\cdot\boldsymbol{r}-\omega' t)+i\delta} + \text{c.c.} \tag{3.2}$$

where δ is a phase factor with respect to $\boldsymbol{E}(\boldsymbol{r}, t)$. $\boldsymbol{E}_0$ and $\boldsymbol{E}_0'$ can be $\boldsymbol{r}$-dependent due to, e.g., the light-matter interaction studied in Sect. 2.3. $\boldsymbol{E}(\boldsymbol{r}, t)$ and $\boldsymbol{E}'(\boldsymbol{r}, t)$ can be the two beams in Fig. 1.6 where $\boldsymbol{s} = \boldsymbol{s}'$, $\omega = \omega'$, and δ is the phase factor due to the difference in optical paths of the two beams. The addition of the two beams is

$$\boldsymbol{E}(\boldsymbol{r}, t) + \boldsymbol{E}'(\boldsymbol{r}, t) \tag{3.3}$$

There are many factors that contribute the phase factor δ in (3.2). In a laser operation, δ of the stimulated emitted photon is the same as that of the stimulating photon in resonance amplification, while the spontaneously emitted photon has a random phase. When a light beam impinges on an interface between one spatially uniform region (e.g., air) and another spatially uniform region (e.g., a semiconductor), the momentum of light, $\hbar\boldsymbol{s}$, changes its direction due to the collision of photons with the densely packed atoms in the semiconductor, thus the light beam becomes either reflected or diffracted. The collision is normally instant so does not introduce any phase change (which is also denoted as a phase shift). Note that light-matter interaction is a complex phenomenon described fully by quantum electrodynamics. A proper presentation of the quantum electrodynamics is outside the scope of this book. The phases of the emitted photons mentioned here cannot be studied by the theoretical frame we have described in the previous chapter.

The phrase "spatially uniform" in the previous paragraph is actually defined at the macroscopic level. Consider a light beam composed of many photons propagating in the air at the microscopic level. Initially, all photons have the same phase, such as when they leave the cavity of a laser under resonance amplification operation. Some photons in the light beam will collide with atoms and molecules in air which are randomly distributed in space. The detector will receive photons that have not collided with atoms and molecules which maintain their original optical paths, as well as photons that were collided with atoms and molecules. These photons will have different optical paths when they reach the detector. We eventually will detect the light beam composed of photons with a phase distribution as a function of the densities of atoms and molecules in the air.

People introduce a macroscopic property called the coherence length to describe the propagation of a light beam through a macroscopically uniform medium. The coherence length is the propagation distance within which the light beam largely maintains its sinusoidal character of (3.1). Beyond the coherence length, the phase factor δ in (3.2) is no longer solely determined by the geometric length that the light beam travels. In extreme cases, it becomes simply a random number due to the huge number of collision events at the microscopic level. This will modify significantly the optical spectrum.

Having briefly mentioned a few words about the propagation of the light beam, let us try to quantify the reflection and transmission spectra.

3.1 Fresnel's Equations

In the previous chapter we used the Maxwell's equations to describe the light propagation in a free space and in a semiconductor that is extended theoretically in the whole space. When a light beam impinges on an interface between two media, such as from the free space to a semiconductor, the light beam will become reflected and refracted, diffracted, i.e., the propagation direction of the light beam will be changed. There can also be further polarization changes. All these are well described by the Maxwell's equations.

As we learn from the previous chapter, the propagation of a light beam through a material can be described by the effective dielectric coefficient, i.e., ϵ in (2.67) and (2.149), which is a macroscopic physical parameter summarizing all microscopic light-matter interactions when the light beam propagates through the material. Its measurement is therefore a common means to understand microscopic light-matter interactions and thereafter design and apply these light-matter interactions for various applications.

In Sect. 2.3 we study the light propagation in a bulk material which extends infinitely, i.e., (2.81) describe a decaying or growing plane wave in a three-dimensional space. Now we study a light beam that incidents from one medium described by $\tilde{n}_1$ to another medium of $\tilde{n}_2$, the two media are separated by a flat interface, see Fig. 3.1. The light will be reflected and refracted, resulting in a reflected beam back to $\tilde{n}_1$ and a refracted beam in $\tilde{n}_2$. The refracted beam is also referred to as transmitted (from $\tilde{n}_1$ to $\tilde{n}_2$). It is straightforward by applying Maxwell's equations to obtain reflection and transmission coefficients.

Let θ_i and θ_t be the incident and refraction angle, respectively. The reflected angle is denoted as θ_r which will be shown to equal to θ_i. The wave vectors of the incident and reflected beams define the plane of incidence. With respect to the plane of incidence, there are two polarization situations for the electric field. The electric field of the incident light may lies parallel to the plane of incidence, i.e., the $E_\parallel$ field in Fig. 3.1, which is called p-polarized ("parallel" in German, same in English). It may either sticks out of or into the plane of incidence which is called s-polarized ("senkrecht" in German, "perpendicular" in English), i.e., the $E_\perp$ field in Fig. 3.1.

We would like to compute the fractions of a light wave reflected and a light wave transmitted by the flat interface between two media with different refractive indices, which are the reflection and transmission coefficients mentioned before. Fresnel was the first to do this calculation so the reflection and refraction equations are known to be Fresnel's equations. Let $E_{\parallel\mathrm{i}}$, $E_{\parallel\mathrm{r}}$, and $E_{\parallel\mathrm{t}}$ be the amplitudes of

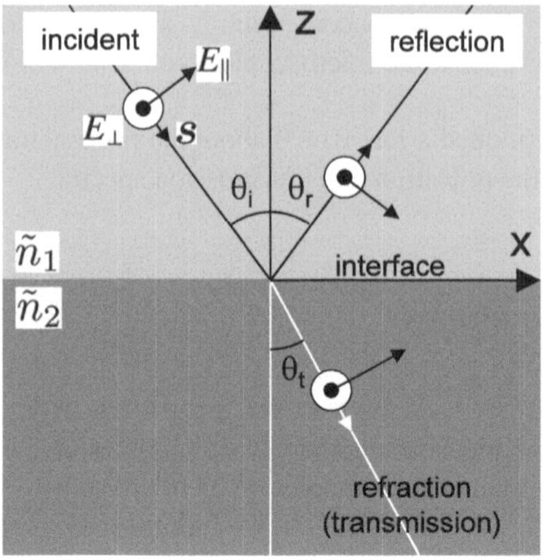

Fig. 3.1 Fresnel's reflection and refraction at the interface between medium $\tilde{n}_1$ and medium $\tilde{n}_2$. The plane of incidence is defined by the wave vectors of the incident and reflected beams. θ_i is the incident angle, θ_r the reflection angle, and θ_t the refraction angle (also known as transmission angle thus subscript "t"). The electric field of p-polarization lies parallel to the plane of incidence, while for s-polarization it sticks out of (in this figure) or into the plane of incidence

p-polarized electric fields of the incident, reflected and transmitted lights, respectively. It is straightforward to obtain the reflection and transmission coefficients

$$r_\parallel = \frac{E_{\parallel r}}{E_{\parallel i}} = \frac{\tilde{n}_1 \cos\theta_t - \tilde{n}_2 \cos\theta_i}{\tilde{n}_1 \cos\theta_t + \tilde{n}_2 \cos\theta_i}, \quad t_\parallel = \frac{E_{\parallel r}}{E_{\parallel i}} = \frac{2\tilde{n}_1 \cos\theta_i}{\tilde{n}_1 \cos\theta_t + \tilde{n}_2 \cos\theta_i} \tag{3.4}$$

The reflection and transmission coefficients of the s-polarized lights $E_{\perp i}$, $E_{\perp r}$, and $E_{\perp t}$

$$r_\perp = \frac{E_{\perp r}}{E_{\perp i}} = \frac{\tilde{n}_1 \cos\theta_i - \tilde{n}_2 \cos\theta_t}{\tilde{n}_1 \cos\theta_i + \tilde{n}_2 \cos\theta_t}, \quad t_\perp = \frac{E_{\perp t}}{E_{\perp i}} = \frac{2\tilde{n}_1 \cos\theta_i}{\tilde{n}_1 \cos\theta_i + \tilde{n}_2 \cos\theta_t} \tag{3.5}$$

Equations (3.4) and (3.5) are the so-called Fresnel's equations of light reflection and refraction.

Along with the derivations of Fresnel's equations come also the Snell's law

$$\theta_i = \theta_r, \quad \tilde{n}_1 \sin\theta_r = \tilde{n}_2 \sin\theta_t \tag{3.6}$$

The first one is called Snell's law of reflection and the second one is Snell's law of refraction.

In practical spectral measurements, the incident light impinges on the interface at a zero incident angle called the normal incidence ($\theta_i = \theta_t = 0$) for which

$$r_\perp = r_\parallel \equiv r = \frac{\tilde{n}_1 - \tilde{n}_2}{\tilde{n}_1 + \tilde{n}_2}, \quad t_\perp = t_\parallel \equiv t = \frac{2\tilde{n}_1}{\tilde{n}_1 + \tilde{n}_2} \tag{3.7}$$

by (3.4) and (3.5).

For an interface between two lossless media, i.e., $\kappa_1 = \kappa_2 = 0$, (3.7) become

$$r = \frac{n_1 - n_2}{n_1 + n_2}, \quad t = \frac{2n_1}{n_1 + n_2} \tag{3.8}$$

We can calculate the reflectance R and transmittance T at this single flat interface

$$R = \frac{|\langle \mathbf{S}_\mathrm{r} \rangle_t|}{|\langle \mathbf{S}_\mathrm{i} \rangle_t|} = \frac{n_1 E_\mathrm{r}^2}{n_1 E_\mathrm{i}^2} = \frac{(n_1 - n_2)^2}{(n_1 + n_2)^2}, \quad T = \frac{|\langle \mathbf{S}_\mathrm{t} \rangle_t|}{|\langle \mathbf{S}_\mathrm{i} \rangle_t|} = \frac{n_2 E_\mathrm{t}^2}{n_1 E_\mathrm{i}^2} = \frac{4n_1 n_2}{(n_1 + n_2)^2} \tag{3.9}$$

And $R + T = 1$, i.e., there is no loss of the optical power when the light beam is being reflected and refracted. In the above equations, $\langle \ldots \rangle_t$ denotes time average, as always.

Fresnel's equations of (3.4), (3.5) and Snell's law of (3.6) are generally valid for both lossless and lossy media. However, the mathematical manipulation and relevant physical interpretation become much more complicated when lossy media are involved. We may retrieve the common textbook expressions by letting $\kappa = 0$ so that the reflection and refraction angles are real, whereas for nonzero κ the angles in the equations will be complex valued and do no longer have the obvious geometrical interpretations. The theme is very interesting, however, it is beyond the scope of this book.

We consider a light beam incident from vacuum or air ($\tilde{n}_1 = 1$) to a medium with a complex refractive index $\tilde{n}_2 = n + i\kappa$, which is the most relevant model of common reflection and transmission measurements (see the next section). It is easy to obtain from (3.7) the reflectance R and transmittance T of the light beam at the single flat interface between air ($\tilde{n}_1 = 1$) and the medium ($\tilde{n}_2 = n + i\kappa$)

$$R = \frac{(n-1)^2 + \kappa^2}{(n+1)^2 + \kappa^2}, \quad T = \frac{4n}{(n+1)^2 + \kappa^2} \tag{3.10}$$

Again $R + T = 1$ indicating energy conservation at the reflection-refraction interface.

Before we move on to analyze experimental reflectance and transmittance spectra, let us take a theoretical/numerical example. Assume that within the optical range of interest, the material of interest has three absorption peaks at photon energies of $\hbar\omega_1 = 150$, $\hbar\omega_2 = 268.5$, and $\hbar\omega_3 = 522.6$ cm^{-1}, respectively, with oscillator strengths of $f_1 = 0.005$, $f_2 = 2.1$ and $f_3 = 0.0005$, and the corresponding damping rates are $\hbar\gamma_1 = 10$, $\hbar\gamma_2 = 2.5$, and $\hbar\gamma_3 = 10$ cm^{-1}. We further assume a high-frequency dielectric coefficientof $\epsilon_\infty = 11.56$. It is easy to calculate the complex refractive index of the material by (2.151), which is repeated below

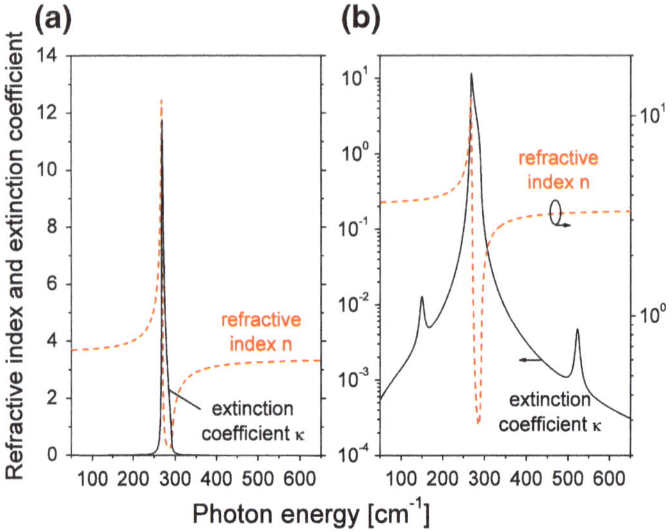

Fig. 3.2 Refractive index n and extinction coefficient κ of a sample material as functions of photon energy $\hbar\omega$. Solid black line is κ and dashed red line is n. Same data in a linear vertical scale (**a**) and a logarithmic vertical scale (**b**)

$$\epsilon(\omega) = \epsilon_\infty + \sum_{\ell=1}^{3} \frac{f_\ell\, \omega_\ell^2}{\omega_\ell^2 - \omega^2 - i\omega\gamma_\ell} \tag{3.11}$$

and the numerical results of n and κ ($n + i\kappa = \sqrt{\epsilon}$) are presented in Fig. 3.2.

In Fig. 3.2a where a linear vertical scale is used, only one peak at $\hbar\omega_2 = 268.5$ cm^{-1} is visible in both n and κ, while the extinction coefficient peaks at $\hbar\omega_1 = 150$ and $\hbar\omega_2 = 522.6$ cm^{-1} are visible only in the logarithmic vertical scale in Fig. 3.2b. In both linear and logarithmic vertical scales, the refractive index n is dominated by a single peak at $\hbar\omega_2 = 268.5$ cm^{-1}.

Insert the complex refractive index of Fig. 3.2 into (3.10) we obtain the reflectance and transmittance spectra presented in Fig. 3.3a at the single flat interface between air $\tilde{n}_1 = 1$ and medium $\tilde{n}_2 = n + i\kappa$. We observe clearly the dominant peak at $\hbar\omega_2 = 268.5$ cm^{-1}, while the two weak peaks at $\hbar\omega_1 = 150$ and $\hbar\omega_3 = 522.6$ cm^{-1} are only observable in the differential spectrum using the logarithmic vertical scale, see Fig. 3.3b where it is shown that the variations are below 10^{-4}, which is beyond the sensitivities of common optical spectrometers.

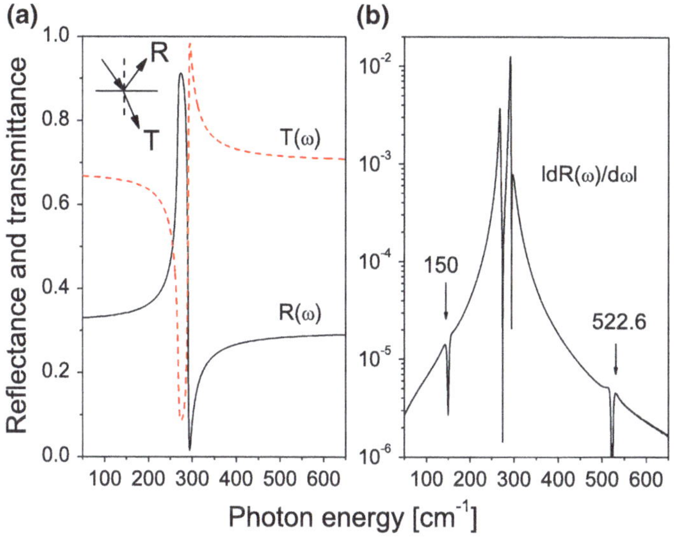

Fig. 3.3 **a** Reflectance $R(\omega)$ (black solid) and transmittance $T(\omega)$ (red dashed line) at the single flat interface between air $\tilde{n}_1 = 1$ and $\tilde{n}_2 = n + i\kappa$ of Fig. 3.2. **b** Variation of the reflectance, $|dR(\omega)/d\omega|$ as a function of the photon energy. Inset in **a** shows the geometry of measuring $R(\omega)$ and $T(\omega)$

3.2 Reflection and Transmission by a Thin Film

Equation (3.10) shows that we are able to extract in principle the microscopic optical properties of the material, i.e., $\tilde{n}_2 = n + i\kappa$, from measuring optical spectra of reflectance R and transmittance T. In reality we cannot measure directly T since it is simply not possible, at least not practical, to put a photodetector in a solid material. A more practical situation is that a light beam incidents to a thin film of the material with a thickness d, see Fig. 3.4. The incident light beam can either be monochromatic or polychromatic. In the latter case, we need a dispersion device (Chap. 1) to separate one light beam of a specific wavelength of interest from the rest of the polychromatic light beam.

For simplicity, let us focus on a monochromatic light with a single angular frequency ω. Assume that the light beam impinges on the upper interface of the thin film at an incident angle θ_i. It will be reflected back to the same side of the light incidence at a reflection angle θ_r; It will also refract into the thin film at a refraction angle θ_t, pass through the thin film then exit the thin film from the lower interface.

In the previous section we learn the reflection and transmission of light at a single interface between two media. In Fig. 3.4 however, the thin film, i.e., medium 2, has two interfaces. Because of the multiple reflections and refractions within the two interfaces, the net reflected and transmitted light beams that reach the two photodetectors S_r and S_t are composed of two different series of light beams with different optical paths. Depending on the wavelength of the incident light beam and

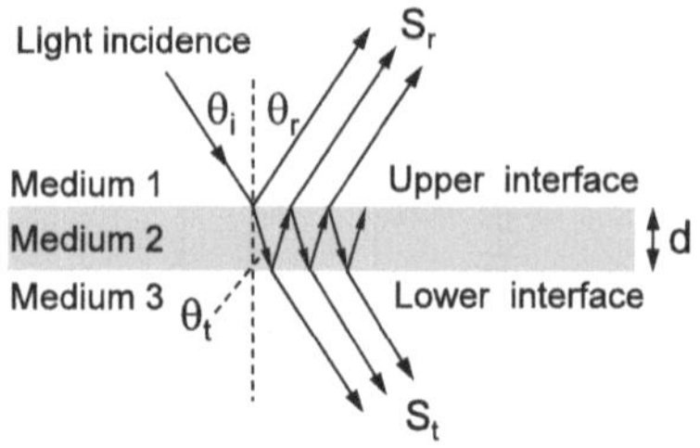

Fig. 3.4 Schematic optical paths of a typical reflection and transmission spectral measurement setup on a thin film (medium 2)

the phase relationships, the series of the reflected and transmitted light beams may interfere constructively or destructively.

Let $\tilde{n} = n + \kappa$ be the complex refractive index of the thin film (denoted as medium 2 in Fig. 3.4), and the measurement is performed in the air so that the refractive indices of the spaces above the upper interface (medium 1) and below the lower interface (medium 3) are 1. Moreover, we assume that the light beam strikes the upper interface at the zero incident angle, i.e., the normal incidence. By (3.7), the reflection and refraction coefficients at the upper and lower interfaces are

$$r_{12} = \frac{1-\tilde{n}}{1+\tilde{n}}, \quad t_{12} = \frac{2}{1+\tilde{n}}, \quad r_{21} = r_{23} = \frac{\tilde{n}-1}{\tilde{n}+1}, \quad t_{21} = t_{23} = \frac{2\tilde{n}}{\tilde{n}+1} \tag{3.12}$$

for a single beam. Here r_{12} is the reflection of the light beam from medium 1 back to medium 1 reflected by the upper interface of the thin film, t_{12} is the refraction from medium 1 into medium 2 at the upper interface. r_{23} and t_{23} are likewise defined but at the lower interface. Note that $r_{12} = -r_{23}$.

The most important aspect about the thin film are the multiple reflections and refractions at the upper (lower) interface for the light beams that are reflected from the lower (upper) interface, denoted as r_{21} and t_{21} (r_{23} and t_{23}).

Let E_i be the electric field of the incident light beam. Since we work on the normal incidence, we do not need to care about the vector nature of the fields (see the previous section that the p- and s-polarized fields are identically reflected and refracted at the normal incidence condition). The direct reflected light beam from the upper interface, called the zeroth-order reflected light beam, is $E_i r_{12}$. The first-order reflected light beam, that comes from one refraction through the upper interface, one reflection from the lower interface, and one refraction through the upper interface, is

$$E_i t_{12} e^{i\delta} r_{23} e^{i\delta} t_{21} \tag{3.13}$$

where $\delta = \omega\tilde{n}d/c_0$ is the extra phase due to the optical path from the upper interface to the lower interface then back to the upper interface. We can similarly construct all other multiple-reflected light beams. The total reflected electric field thus becomes

$$E_r = E_i r_{12} + E_i t_{12} e^{i\delta} r_{23} e^{i\delta} t_{21} + E_i t_{12} e^{i\delta} r_{23} e^{i\delta} r_{21} e^{i\delta} r_{23} e^{i\delta} t_{21} + \cdots \tag{3.14}$$

It is easy to see that the result of the above infinite summation is

$$E_{\mathrm{r}} = E_{\mathrm{i}} \frac{r_{12} + e^{2i\delta} r_{23}}{1 + e^{2i\delta} r_{12} r_{23}} \tag{3.15}$$

Similarly we obtain the total electric field of the series of the transmitted light beams

$$E_{\mathrm{t}} = E_{\mathrm{i}} \frac{e^{i\delta} t_{12} t_{23}}{1 + e^{2i\delta} r_{12} r_{23}} \tag{3.16}$$

Knowing the amplitude E of the electric field $\boldsymbol{E}$ in the air, it is easy to calculate the optical power, i.e., the time-averaged amplitude of the Poynting vector

$$|\langle \boldsymbol{S} \rangle_t| = |\langle \boldsymbol{E} \times \boldsymbol{H} \rangle_t| = 2c_0\epsilon_0 |E|^2 \tag{3.17}$$

by (2.83), (2.84). Note that the light beams under study are now in the air.

$$S_{\mathrm{r}} = 2c_0\epsilon_0 |E_{\mathrm{r}}|^2, \quad S_{\mathrm{t}} = 2c_0\epsilon_0 |E_{\mathrm{t}}|^2 \tag{3.18}$$

are therefore the optical powers of the reflected and transmitted lights measured in Fig. 3.4. And the amount of the absorbed light beam is $\left(S_{\mathrm{i}} - S_{\mathrm{r}} - S_{\mathrm{t}}\right)$, where $S_{\mathrm{i}} = 2c_0\epsilon_0 |E_{\mathrm{i}}|^2$.

Equations (3.15), (3.16) show that the optical powers of reflected and transmitted light beams are direct functions of n and κ, which is exactly the purpose of measuring the reflected light beam S_{r} and the transmitted light beam S_{t} to find information about $\tilde{n} = n + i\kappa$, and thereafter microscopic processes related to $\tilde{n}$.

Now let us take a look at the reflected and transmitted light beams from the thin film of the material with $\tilde{n} = n + i\kappa$ in Fig. 3.2, i.e., the reflection and transmission spectra of Fig. 3.4. The numerical results of (3.15), (3.16), (3.18) are presented in Fig. 3.5 as reflectance $R = S_{\mathrm{r}}/S_{\mathrm{i}}$ and transmittance $T = S_{\mathrm{t}}/S_{\mathrm{i}}$ for three different film thicknesses $d = 0.1$, 1.0 and 10.0 cm, respectively.

We first observe strong oscillations in both the reflection and transmission spectra and the oscillations depend strongly on the photon energy. This is due to the constructive and destructive interferences of the multiple reflected and refracted light beams.

The dominant peak at $\hbar\omega_2 = 268.5\ \mathrm{cm}^{-1}$ in Figs. 3.2 and 3.3 exists always clearly in the reflection spectrum, independent of the film thickness d. And the reflectance of the thin film approaches the one shown in Fig. 3.3a when d becomes large. The reason is mathematical and very simple. For a very thick sample, the imaginary part of the extra phase

$$\delta = \frac{\omega d \tilde{n}}{c_0} = \frac{\omega d(n + i\kappa)}{c_0}$$

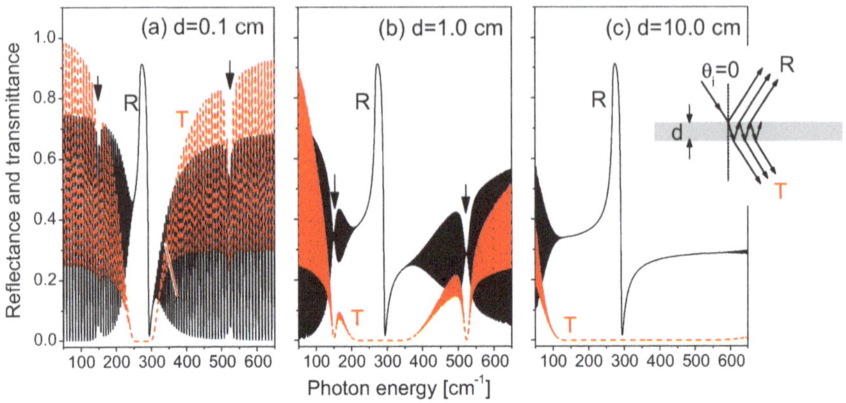

Fig. 3.5 **a** Reflectance (black solid line) and transmittance (red dashed line) of a thin film (inset in **c**) with a complex refractive index shown in Fig. 3.2 obtained under the normal incidence condition ($\theta_i = 0$ while inset in **c** shows a much tilted incidence only for easy visualization). The film thickness is $d = 0.1$ cm. **b** and **c**: same as **a** but for $d = 1.0$ and $d = 10.0$ cm, respectively. The numerical results are obtained by using (3.15), (3.16), (3.18)

becomes very large so that $e^{i\delta} \to 0$, and therefore $E_r = E_i r_{21}$ by (3.15), resulting in $S_r/S_i = |r_{21}|^2 = R(\omega)$. This result is of course limited to the case of a very thick film (in theory) and/or a film with a large extinction coefficient. The first condition, i.e., a thick thin film, may not be sufficient. As mentioned at the very beginning of this chapter, the light transmission through a long distance in a medium depends strongly on the coherence length of the light in the medium. In other words, the coherence length of the light should be long enough in order to see the effect of the light extinction effect. We will discuss shortly the issue more carefully.

All three peaks in $\tilde{n}(\omega)$ are in the form of valleys in the transmission spectrum. And the valley width and depth depend on the values of κ at these photon energies when the film is not too thick. This is understandable since the light beam cannot transmit through a very thick material (here we mean a lossy material, i.e., with a nonzero κ). The absorption at $\hbar\omega_2 = 268.5$ cm^{-1} is already too strong even for a film with a thickness of only $d = 0.1$ cm, which results in an opaque band of 250–300 cm^{-1} in the transmission spectrum, see Fig. 3.5a. The opaque band becomes much wide when the film thickness is increased to 1.0 cm, under which situation, transmission valleys at peaks at $\hbar\omega_1 = 150$ and $\hbar\omega_3 = 522.6$ cm^{-1} are still well defined, i.e., very sharp. For a thin film of 10 cm (very much alike a bulk material now), basically all lights are absorbed and the transmission spectrum measurement is rendered useless.

However, all three peaks can be clearly distinguished in the reflection spectrum when the film is not too thick, i.e., not the film of $d = 10.0$ cm. Even the two weak peaks in the refractive index n and extinction coefficient κ in Fig. 3.2 at $\hbar\omega_1 = 150$ and $\hbar\omega_2 = 522.6$ cm^{-1} are clearly revealed when the film thickness is properly chosen, see Fig. 3.5b where $d = 1.0$ cm. This is the action of the multiple reflections and refractions by the upper and lower interfaces.

By comparing Fig. 3.5a–c one concludes that all peaks can be extracted from the reflection and transmission spectra when the film thickness is properly chosen, even the two very weak peaks at $\hbar\omega_1 = 150$ and $\hbar\omega_3 = 522.6$ cm^{-1} which are only observable in the logarithmic scale in $\tilde{n}$ and in the differential spectrum of $R(\omega)$ in Fig. 3.2.

In reality, a bulk material is nominally defined to have a thickness in the order of 0.1 cm and beyond, which is already the one in Fig. 3.5a. And the typical thickness of the material for the epitaxial substrate is about 0.3–0.5 mm. The coherence length of common light sources cannot cover these thicknesses, mostly because of the high densities of atoms in the solid material. To simulate such a situation, we can reset the real phase of the reflected and refracted lights in (3.15), (3.16). More specifically, for the coherence situation,

$$\delta_{\text{coherence}} = \frac{\omega \tilde{n} d}{c_0} \tag{3.19}$$

while for the incoherence condition,

$$\delta_{\text{incoherence}} = -\frac{\omega \kappa d}{c_0} \tag{3.20}$$

The results of the incoherence condition are presented in Fig. 3.6. As compared with Fig. 3.5, strong oscillations due to constructive and destructive interferences of the multiple reflections and refractions are all gone so that the three peaks in the reflection spectrum and the three valleys in the transmission spectrum are all clearly visible, especially in the thin film case. Again, the light beam at $\hbar\omega_2 = 268.5$ cm^{-1} is strongly absorbed before it can reach the lower interface for further reflections and transmissions. Therefore, it is only observable in the reflection spectrum.

What we have been doing so far is to obtain the reflection and transmission spectra of a thin film as the results of multiple reflections and refractions at the two interfaces of the thin film from the reflectance and transmittance spectra of a single interface, which are the results of the refractive index $n(\omega)$ and extinction coefficient $\kappa(\omega)$ of the material expressed by (3.11), while the ultimate goal of the spectral analysis is actually the other way around.

The procedure of the spectral analysis is thus the following: We first measure the reflection and transmission spectra of the thin film. Look at the spectra to access and estimate the positions of peaks in the spectra. Modify ω_i, f_i and γ_i in the computer code as fitting parameters, run the computer code repeatedly by adjusting the fitting parameters so that the calculated spectra agree best with the experimental data. Here is one example.

Room-temperature reflection R^{exp} and transmission T^{exp} spectra of an intrinsic (001) GaAs substrate film (thickness $d = 0.05$ cm, both the upper and lower interfaces are well polished) are obtained from a Fourier transform spectrometer, which

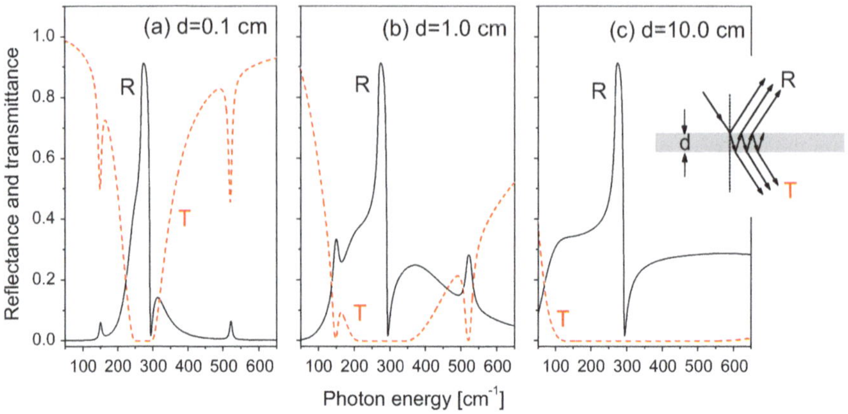

Fig. 3.6 Same as Fig. 3.5 but under the incoherence condition

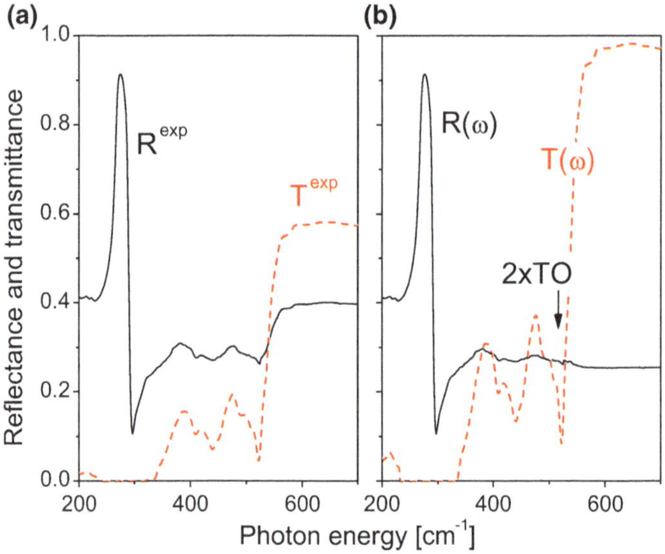

Fig. 3.7 **a** Room-temperature reflection R^{exp} (black solid line) and transmission T^{exp} (red dashed line) spectra of an intrinsic (001) GaAs substrate film (thickness $d = 0.05$ cm). **b** Theoretically extracted reflectance $R(\omega)$ (black solid line) and transmittance $T(\omega)$ (red dashed line) spectra

are presented in Fig. 3.7a, while theoretically extracted reflectance and transmittance spectra are presented in Fig. 3.7b.

First of all, we notice that the photon energy range of the measurement is 200–650 cm^{-1} (24.8–80.6 meV, in the infrared optical range), which is the energy range of the optical phonon dispersion in the bulk GaAs material, see Fig. 2.9.

The general agreements between the experimental data and theoretical fitting are good. However, deviations exist. The most prominent ones are in the range from 350 to 550 cm^{-1}, which is originated from the strong dielectric function modulation induced by the transverse optical (TO) phonons (Fig. 2.9). There are many valleys in R^{exp}, such as 206, 413, 442, 455, 494, 509 and 525 cm^{-1}, corresponding to various two-phonon excitations. Two-phonon excitations (or other multiphonon excitations) are caused by anharmonic effects in the lattice vibrations. Unlike one-phonon excitations, multiphonon excitations occur mostly at energies with high densities of phonon states, they are also temperature dependent [1]. The valley at 525 cm^{-1} in $T(\omega)$, marked by a vertical arrow in Fig. 3.7b, is related to the absorption of two TO phonons at the Γ symmetric point in the $\boldsymbol{k}$ space [2].

3.3 Harmonic Oscillator Model

In the previous section we briefly mentioned the phonon absorptions in the reflection spectrum in the infrared optical range. Infrared reflection spectrum has been widely and effectively used to study phonons, especially optical phonons, in semiconductors. As we have learned from the previous section, the relationship between the reflection spectrum and the dielectric coefficient is rather complicated. Harmonic oscillator model and Kramers–Kronig relationship analysis are commonly used to analyze the infrared reflection spectrum for extracting macroscopic optical parameters (dielectric coefficient) and microscopic physical parameters such as resonance frequencies, oscillator strengths, damping rates of the optical phonons. In this section we introduce the harmonic oscillator model. The Kramers–Kronig relationship analysis will be studied in the next section.

In the harmonic oscillator model, the dielectric coefficient is expressed by

$$\epsilon_{\text{TO}}(\omega) = \epsilon_\infty + \sum_\ell \frac{f_{\text{TO},\ell}\,\omega_{\text{TO},\ell}^2}{\omega_{\text{TO},\ell}^2 - \omega^2 - i\omega\gamma_{\text{TO},\ell}} \tag{3.21}$$

i.e., (2.151). In the above equation, the first term is the high-frequency dielectric coefficient ϵ_∞ which includes contributions of optical transitions of electrons in their energy band structures. The second term sums contributions from various transverse optical (TO) phonon contributions, where $\omega_{\text{TO},\ell}$, $f_{\text{TO},\ell}$ and $\gamma_{\text{TO},\ell}$ denote the energy, the oscillator strength and the damping rate of TO phonon ℓ. Many compound semiconductors have the zincblende lattice structure so that there is normally only one principal reflection peak in the infrared reflection spectrum. In this case, the summation over ℓ in the above expression is limited to just 1.

The solid black line in Fig. 3.8 is a reflection spectrum $R(\omega)$ of a GaAs bulk material. It is to be fitted by using (3.12), (3.15), (3.16), (3.18), (3.21) in order to find the fundamental physical parameters about the TO vibrational mode. The flow chart of the fitting procedure is summarized below

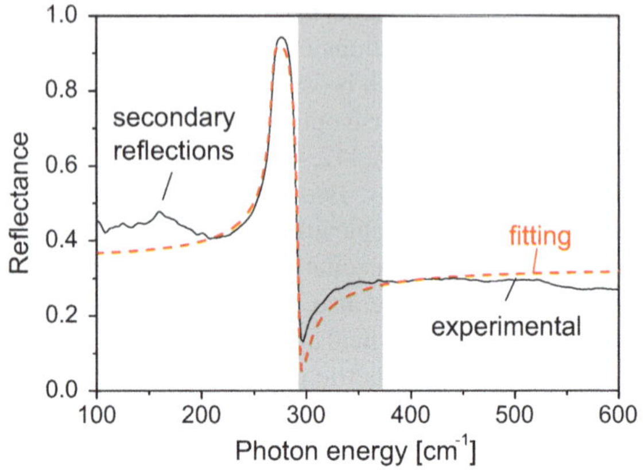

Fig. 3.8 The fitting of the infrared reflection spectrum of bulk GaAs TO phonons by the harmonic oscillator model. Solid black line is the experimental data, and dashed red line is the fitting spectrum

$$
\begin{aligned}
& \epsilon_{\mathrm{TO}}(\omega) = \epsilon_\infty + \sum_\ell \frac{f_{\mathrm{TO},\ell}\,\omega^2_{\mathrm{TO},\ell}}{\omega^2_{\mathrm{TO},\ell} - \omega^2 - i\omega\gamma_{\mathrm{TO},\ell}} \\
& \rightsquigarrow \tilde{n} = \sqrt{\epsilon_{\mathrm{TO}}}, \quad \delta = \frac{\omega\tilde{n}d}{c_0} \\
& \rightsquigarrow r_{12} = \frac{1-\tilde{n}}{1+\tilde{n}}, \quad t_{12} = \frac{2}{1+\tilde{n}}, \quad r_{21} = r_{23} = \frac{\tilde{n}-1}{\tilde{n}+1}, \quad t_{21} = t_{23} = \frac{2\tilde{n}}{\tilde{n}+1} \\
& \rightsquigarrow E_\mathrm{r} = E_\mathrm{i}\,\frac{r_{12} + e^{2i\delta} r_{23}}{1 + e^{2i\delta} r_{12} r_{23}}, \quad E_\mathrm{t} = E_\mathrm{i}\,\frac{e^{i\delta} t_{12} t_{23}}{1 + e^{2i\delta} r_{12} r_{23}} \\
& \rightsquigarrow S_\mathrm{r} = 2c_0\epsilon_0 |E_\mathrm{r}|^2, \quad S_\mathrm{t} = 2c_0\epsilon_0 |E_\mathrm{t}|^2 \\
& \rightsquigarrow R = \frac{S_\mathrm{r}}{S_\mathrm{i}}, \quad T = \frac{S_\mathrm{t}}{S_\mathrm{i}}
\end{aligned}
\tag{3.22}
$$

By following the flow chart of spectral fitting of (3.22) we obtain $f_{\mathrm{TO}} = 2.28$, $\omega_{\mathrm{TO}} = 267.6\ \mathrm{cm}^{-1}$, and $\gamma = 3.4\ \mathrm{cm}^{-1}$. The fitted spectrum is shown in Fig. 3.8 as the red dashed line. The fitting in general is quite good except in the range below $200\ \mathrm{cm}^{-1}$. The higher experimental reflection in this range comes from secondary reflections (mostly, reflections from the lower interface of the sample) which is not described properly in (3.22) (such as the geometric structure of the film, scatterings of photons travelling in the material), resulting in the fitting line below the experimental data.

More critical is the discrepancy in the grey area in Fig. 3.8 that the fitted spectral line is well below the measurement data. This is actually due to the simplification of the harmonic oscillator model. In principle, the damping rate γ in (3.21) is a function of ω. Experimental data show that γ reaches its maximal value when $\omega = \omega_{\mathrm{TO}}$ (resonance), it becomes small when ω deviates from ω_{TO} (off resonance). Reader who is interested in this may refer to (7.15) in Sect. 7.2 where we show both theoretically and experimentally that the energy relaxation of an exciton radiative recombination is a Lorentzian function of the photon energy centered at the exciton energy. The

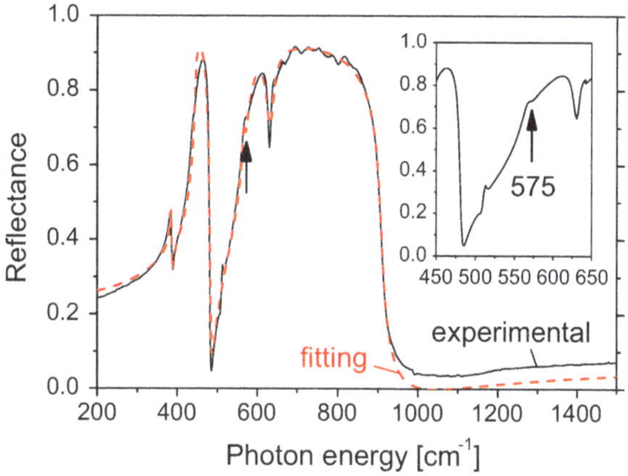

Fig. 3.9 Infrared reflection spectrum of a sapphire substrate. Black solid line is the measured spectrum, and red dashed line is the fitting line. Inset shows the reflection spectrum in the range of 450–650 cm^{-1}

phonon-photon interaction may be studied similarly, which however is beyond the scope of the book.

When there is more than one TO phonon in the materials such as the many ternary and quaternary alloys, there will be many reflection peaks in the infrared reflection spectra. The solid line in Fig. 3.9 is the reflection spectrum of a sapphire substrate with many reflection peaks. The fitting line is red dashed, with extracted TO phonon energies of 384, 444, 564, and 635 cm^{-1} [3].

A closer analysis shows that there is a weak vibrational mode at 575 cm^{-1}, marked by a vertical arrow in Fig. 3.9, see also the inset, very much alike the vibrational mode along the c axis of the hexagonal lattice of the sapphire substrate. Since the c axis is aligned with the normal direction of the sapphire substrate, this vibrational mode is not expected to response to the light of normal incidence, i.e., light incidents to the sapphire substrate along the c axis. In practical reflection measurement setup, we simply cannot align both the light source and the photodetector, which detects the reflected light beam, along the same line normal to the sample surface. Thus, the incidence angle is small, but not zero, commonly around 15°. For many vibrational modes in isotropic materials, this small incident angle can well be approximated as zero. But for anisotropic materials such as the sapphire with an in-plane lattice constant $a = 4.785$ Å and a lattice constant 12.991 Å along the c axis, the setup can no longer be approximated as normal incidence. The weak 575 cm^{-1} peak in Fig. 3.9 (more clearly in the inset) is the result of the interaction between the optical phonon along the c axis with the photon which comes to the sample at a small incident angle.

In this section so far, we apply directly the harmonic oscillator model of only transverse optical (TO) phonons. In Sect. 2.2, lattice vibrations of a semiconductor crystal are much more complicated. They are described by (2.63), (2.64) which are repeated below

$$u_{i,\ell}(\boldsymbol{R}, t) = \frac{\psi_{i,\ell}(\boldsymbol{R})}{\sqrt{M_R}} e^{-i\omega_\ell t} \tag{3.23}$$

$$H|\psi_\ell\rangle = \omega_\ell^2|\psi_\ell\rangle \tag{3.24}$$

where $\boldsymbol{R}$ denotes the spatial position of the lattice site which is occupied by an atom with mass $M_{\boldsymbol{R}}$. $u_{i,\ell}(\boldsymbol{R}, t)$ is the lattice displacement along axis i ($i = x, y, z$). A little different from (2.63), (2.64) is that we add eigen state index ℓ in (3.23), (3.24) for a better narration. Hamiltonian H is a two-dimensional matrix whose elements are

$$H_{ij}(\boldsymbol{R} - \boldsymbol{R}') = \frac{\phi_{ij}(\boldsymbol{R} - \boldsymbol{R}')}{\sqrt{M_{\boldsymbol{R}} M_{\boldsymbol{R}'}}}$$

where $\phi_{ij}(\boldsymbol{R} - \boldsymbol{R}')$ is the interaction between atom $\boldsymbol{R}$ and atom $\boldsymbol{R}'$ defined by (2.49). This model is denoted as the harmonic oscillator model, as introduced in Sect. 2.2. Since this model is derived by neglecting high-order terms in the Taylor expansion of the lattice potential energy, see (2.46), it is also commonly known as the pseudo-harmonic oscillator model. As shown by (3.23), (3.24), the eigen lattice vibration is characterized by its eigen frequency ω_ℓ, i.e., the angular frequency of the phonon. As shown in Fig. 2.8, the phonon spectrum consists of mainly two branches, acoustic-phonon and optical-phonon branches.

Exactly alike what we learned in Sect. 2.3 that an electron initially occupying one electron state can interact with one photon in an incident light beam to transit to another initially empty electron state, the semiconductor crystal lattice can also interact with a photon, either absorbing or emitting one photon, then transits from one vibrational state to another vibrational state.

We can conclude, by simply considering both energy and momentum conservations without going into all mathematical and physical details (reader is encouraged to either do a similar analysis about phonon-photon interaction by following the example of Sect. 2.3, or read relevant literatures), that the phonon-photon interaction occurs most probably for the optical phonon branch, since the photon momentum is very small. Recall that in the case of the light-matter interaction, the small photon momentum results in the requirement of the electron vertical transition in the $\boldsymbol{k}$ space. The same requirement is asked for an significant phonon-photon interaction. Refer to the phonon dispersion relationship of Fig. 2.8a, we understand immediately that only the optical phonons (in the vicinity of ω_3) have very small momenta as well as a significant density of phonon states in Fig. 2.8b. Acoustic phonons have also small momenta close to $\omega = 0$ with however a very low density of phonon states.

Since the optical phonon energies of common semiconductor materials are about 20–50 meV (160–400 cm^{-1}), the corresponding photons are in the infrared optical range, optical-phonon-photon interaction is therefore expected to be reflected mostly in the infrared spectrum.

As briefly mentioned in Sect. 2.2, there are both the transverse optical (TO) phonons whose displacements are perpendicular to the propagation directions of the vibrational waves, and the longitudinal optical (LO) phonons whose displacements are parallel to the propagation directions of the vibrational waves, as schematically presented in Fig. 3.10. Both the TO and LO phonons can interact with photons. However, the electric and magnetic fields of an electromagnetic field are perpendicular

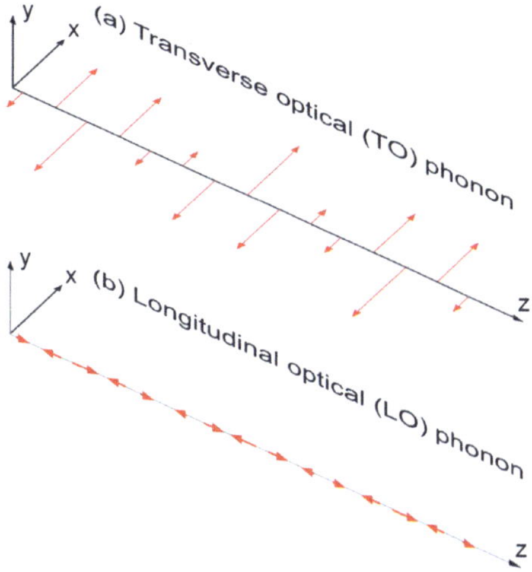

Fig. 3.10 Displacements (red arrows, lengths of the arrows represent the amplitudes of the displacements) of atoms in an optical lattice vibrational mode. **a** Transverse optical (TO) mode, **b** longitudinal optical (LO) mode

to the propagation of the light, see Fig. 2.10 in Sect. 2.3. We can conclude, again without going into mathematical and physical details, that the interaction between a TO phonon and a photon is much stronger than the one between a LO phonon and a photon, simply because the displacements of atoms in a TO mode are aligned with the electric field of the photon. In other words, the oscillator strength f_{TO} is expected to be very large, and thus a strong peak in many spectra such as the refractive index, extinction coefficient, and reflectance, transmittance, reflection, and transmission spectra, based on (3.11). This is the principal reason of the strong absorption in the range of 260–300 cm^{-1} in Fig. 3.7 in the previous section. The same reason applies for (3.22).

In other words, we may introduce the reflections due to harmonic oscillators of LO phonons by letting oscillator strength f_{LO} to be zero. What we can expect then is that when we scan the photon energy from low to high in measuring the reflection spectrum of a semiconductor, the reflection is first very small when the photon energy is well below ω_{TO}. $R(\omega)$ starts to increase and can maximally reach 1.0 when the photon energy approaches ω_{TO} because of the large f_{TO}, i.e., a very large ϵ, see (3.11) and Fig. 3.3. When the energy of the photons reaches ω_{LO}, the photons will easily transmit through the semiconductor since LO phonons do not interact with the photons, resulting a drastic drop in the reflection spectrum. For common semiconductors, the energy differences between $\hbar\omega_{\text{TO}}$ and $\hbar\omega_{\text{LO}}$ are normally less than 10 meV, and there are not many fine structures in the energy range between ω_{TO} and ω_{LO} in their phonon dispersions as well as the densities of phonon states, see, e.g., the density of phonon states of GaAs shown in Fig. 2.9. We therefore expect to see that the reflection increases when the photon energy approaches ω_{TO} from below, it decreases when the photon energy increases further to approach ω_{LO}.

Figure 3.11 is the same reflection spectrum of the bulk GaAs material in the infrared range as Fig. 3.8 except in a narrow photon energy range (100–600 nm in

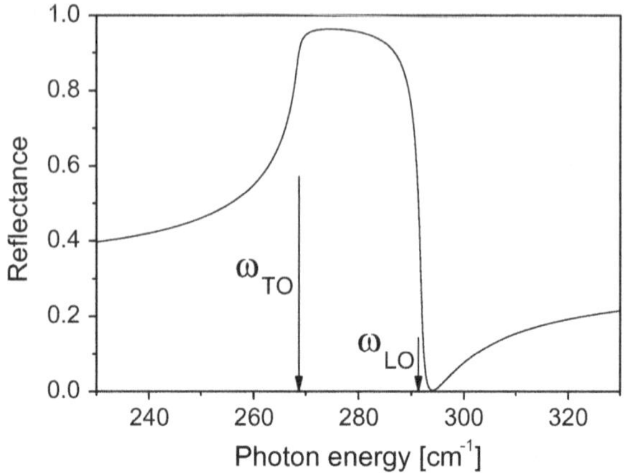

Fig. 3.11 Reflection spectrum in the optical phonon energy range of a bulk GaAs material. Vertical arrows mark the energies of the TO and LO phonons

Fig. 3.8, 230–330 nm in Fig. 3.11). Based on the above analysis, we can determine the energy of the TO phonon by the ascent side and the energy of the LO phonon by the descent side of the reflection peak.

In a brief summary, we apply the harmonic oscillators of TO phonons to fit the experimental reflection spectrum, while the energy of the related LO phonon is obtained by the descent side of the reflection peak (oscillator strength of the LO phonon is zero). In the following we discuss only the TO phonons.

3.4 Kramers–Kronig Relationship

In the previous section we use the harmonic oscillator model to analyze the infrared reflection spectrum of lattice vibrations (mostly optical phonons). However, when there are strong anharmonic effects such that the oscillator strength of the lattice vibration depends strongly on ω, fitting the measured optical spectrum by a set of harmonic-oscillator parameters is no longer valid. In this case, what we have is only the ratio between the optical powers of two light beams, either the reflectance spectrum $R(\omega)$ or the transmittance spectrum $T(\omega)$, see Sect. 3.1. On the other hand, the reflection and transmission coefficients are in general complex, so are the complex refractive index $\tilde{n}$ as well as the dielectric coefficient ϵ. The well-known Kramers–Kronig relationship, commonly abbreviated as the K–K relationship, is widely used to obtain the phase factor of the optical spectrum.

Take $\chi(t)$ as a temporal response of the material to an external field. It is a real function of time t, i.e., $\Im\chi(t) = 0$. Denote $\chi(\omega) = \chi'(\omega) + i\chi''(\omega)$ as its Fourier transform, i.e., $\Re\chi(\omega) = \chi'(\omega)$ and $\Im\chi(\omega) = \chi''(\omega)$. To satisfy causality,

$$\chi'(\omega) = \frac{1}{\pi} \mathrm{P} \int_{-\infty}^{\infty} \frac{\chi''(\omega')}{\omega' - \omega} \mathrm{d}\omega', \quad \chi''(\omega) = \frac{1}{\pi} \mathrm{P} \int_{-\infty}^{\infty} \frac{\chi'(\omega')}{\omega' - \omega} \mathrm{d}\omega' \tag{3.25}$$

These are the most general forms of the Kramers–Kronig relationships. P is the principal value of the singularity at $\omega' = \omega$, i.e.,

$$\mathrm{P}\int_0^{\infty} \mathrm{d}\omega' = \lim_{\delta \to 0}\left[\int_0^{\omega-\delta} \mathrm{d}\omega' + \int_{\omega-\delta}^{\infty} \mathrm{d}\omega'\right] \tag{3.26}$$

The time reversal invariance gives $\chi(-\omega) = \chi^*(\omega)$ resulting in the following integrations over positive ω

$$\chi'(\omega) = \frac{2}{\pi}\,\mathrm{P}\int_0^{\infty} \frac{\omega'\chi''(\omega')}{\omega'^2 - \omega^2}\,\mathrm{d}\omega', \quad \chi''(\omega) = -\frac{2\omega}{\pi}\,\mathrm{P}\int_0^{\infty} \frac{\chi'(\omega')}{\omega'^2 - \omega^2}\,\mathrm{d}\omega' \tag{3.27}$$

The Kramers–Kronig relationships for the relative dielectric function (ϵ_r=1+χ) are

$$\epsilon_r'(\omega) = 1 + \frac{1}{\pi}\,\mathrm{P}\int_{-\infty}^{\infty} \frac{\epsilon_r''(\omega')}{\omega' - \omega}\,\mathrm{d}\omega', \quad \epsilon_r''(\omega) = \frac{1}{\pi}\,\mathrm{P}\int_{-\infty}^{\infty} \frac{\epsilon_r'(\omega')}{\omega' - \omega}\,\mathrm{d}\omega' \tag{3.28}$$

Let us check the expression of the dielectric function (2.148), repeated below

$$\frac{\epsilon(\omega)}{\epsilon_\infty} = 1 + \frac{1}{\omega_n - \omega - i\gamma_n} = \epsilon_r' + i\epsilon_r'' \tag{3.29}$$

Recall the Sokhatsky's formula

$$\lim_{\gamma \to 0^+} \frac{1}{x \mp i\gamma} = \mathrm{P}\frac{1}{x} \pm i\pi\delta(x) \tag{3.30}$$

so that

$$\epsilon_r' = 1 + \mathrm{P}\frac{1}{\omega_n - \omega}, \quad \epsilon_r'' = \pi\delta(\omega_n - \omega) \tag{3.31}$$

which satisfy the Kramers–Kronig relationships of (3.28).

To avoid the evaluation of the principal value at singularity, we can write

$$\epsilon_r' = 1 + \frac{\omega_n - \omega}{(\omega_n - \omega)^2 + \gamma_n^2}, \quad \epsilon_r'' = \frac{\gamma_n}{(\omega_n - \omega)^2 + \gamma_n^2} \tag{3.32}$$

We may compare the above macroscopic expressions with (2.125), which is repeated below

$$\epsilon''(\omega) = \sum_{\boldsymbol{qk}} \frac{\pi\hbar^2 e^2}{m_0^2\omega^2\Omega}|\langle\boldsymbol{q}|\boldsymbol{e}_s \cdot \nabla|\boldsymbol{k}\rangle|^2 \frac{\Gamma_{\boldsymbol{qk}}}{(E_{\boldsymbol{q}} - E_{\boldsymbol{k}} \pm \hbar\omega)^2 + \Gamma_{\boldsymbol{qk}}^2}\left[f(E_{\boldsymbol{k}}) - f(E_{\boldsymbol{q}})\right] \tag{3.33}$$

derived in Sect. 3.1 as a function of microscopic electron transitions.

The reflection coefficient $r(\omega)$, defined as the ratio between optical powers of the reflected electromagnetic field and the incident electromagnetic field, is complex and can be expressed as

$$r = r_0 e^{i\delta} \tag{3.34}$$

where $r_0(\omega)$ is the amplitude of the reflection coefficient, and $\delta(\omega)$ is the phase factor. And $R(\omega) = r_0^2(\omega)$ is the reflectance spectrum. For simple and clear presentation we do not explicitly display the ω-dependence of $r(\omega)$, $r_0(\omega)$ and $\delta(\omega)$ in equations.

We reformulate (3.34)

$$\ln r = \ln r_0 + i\delta = \frac{1}{2} \ln R + i\delta \tag{3.35}$$

so that using the second equation of (3.27) we obtain

$$\delta(\omega) = -\frac{\omega}{\pi} \int_0^{\infty} \frac{\ln R(\omega') - \ln R(\omega)}{\omega'^2 - \omega^2} \mathrm{d}\omega' \tag{3.36}$$

Under the normal incidence condition, $r_0(\omega)$ and $\delta(\omega)$ can be expressed as functions of the complex refractive index by Fresnel's equation (3.7):

$$r = \frac{1 - \tilde{n}}{1 + \tilde{n}} \quad \rightsquigarrow \quad \tilde{n} = \frac{1 - r}{1 + r} = \frac{1 - r_0^2 - 2ir_0 \sin\delta}{1 + r_0^2 + 2r_0 \cos\delta} \tag{3.37}$$

so that by measuring $r_0 = \sqrt{R}$ then calculating δ by (3.36) we obtain the complex refractive index $\tilde{n} = n + i\kappa$, where,

$$n = \frac{1 - r_0^2}{1 + r_0^2 - 2r_0 \cos\delta}, \quad \kappa = \frac{2r_0 \sin\delta}{1 + r_0^2 + 2r_0 \cos\delta} \tag{3.38}$$

And the dielectric coefficient $\epsilon = \tilde{n}^2$ can be further obtained.

In practice we are not able to measure any optical spectrum from $\omega = 0$ to $\omega = \infty$ as required by (3.36) such that the application of the Kramers–Kronig relationships cannot be mathematically rigorous. Equation (3.36) however indicates that the contribution to $\delta(\omega)$ from $R(\omega')$ at ω' is inversely proportional to $\left(\omega'^2 - \omega^2\right)$. Furthermore, for many common semiconductors, $R(\omega)$ is more or less constant and relatively small when ω is very low or very high. Let us define the low and high boundaries of ω such that for $\omega < \omega_1$ or $\omega > \omega_2$, $R(\omega)$ depends weakly on ω. We use the following approximation to calculate δ in the optical range $\omega_1 < \omega < \omega_2$

$$\delta(\omega_1 < \omega < \omega_2) \approx -\frac{\omega}{\pi} \int_{\omega_1}^{\omega_2} \frac{\ln R(\omega') - \ln R(\omega)}{\omega'^2 - \omega^2} \mathrm{d}\omega' \tag{3.39}$$

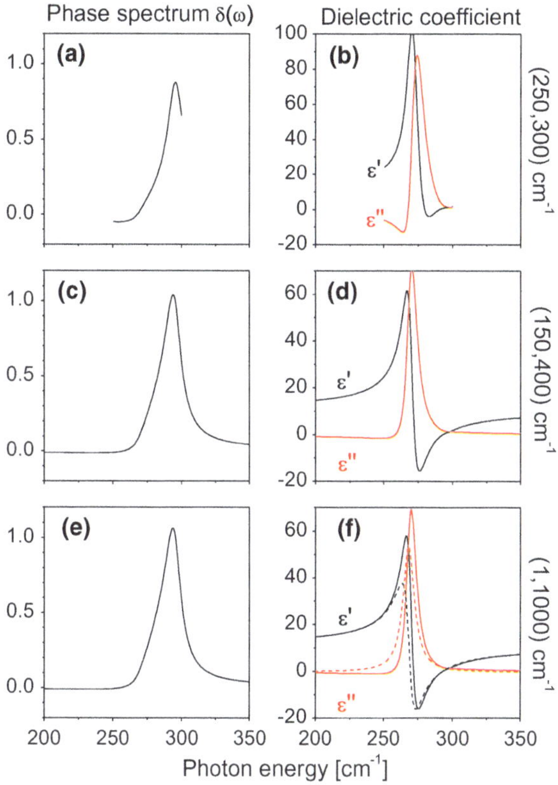

Fig. 3.12 Phase spectrum $\delta(\omega)$ and dielectric coefficient $\epsilon = \epsilon' + i\epsilon''$ obtained from Kramers–Kronig relationships as functions of the optical range of the reflection spectrum. The original $\epsilon = \epsilon' + i\epsilon''$ is shown as dashed lines in **f**

To demonstrate the numerical practice, let us do a simulation

$$\epsilon = \epsilon_2 + \frac{f_2\omega_2^2}{\omega_2^2 - \omega^2 - i\omega\gamma_2}, \quad \tilde{n} = \sqrt{\epsilon}, \quad r_{12} = \frac{1-\tilde{n}}{1+\tilde{n}}, \quad R = |r_{12}|^2 \tag{3.40}$$

where $f_2 = 2.0$, $\omega_2 = 268.7$, $\gamma_2 = 10.0$, $\epsilon_2 = 10.24$. We generate three incomplete reflection spectra in optical ranges of (250, 300), (150, 400), and (1, 1000) cm^{-1} then apply the Kramers–Kronig relationship to retrieve the dielectric coefficient. The effect of the finite optical range of the reflection spectrum is shown in Fig. 3.12. Even the general shape of $\delta(\omega)$ and its magnitude are already well converged when the optical range expands from $\omega \in (250, 300)$ cm^{-1} to $\omega \in (150, 400)$ cm^{-1}, the magnitude of the dielectric coefficients agree with the original ones (dashed lines in Fig. 3.12f) only for $\omega \in (10, 1000)$ cm^{-1}.

Let us work on the reflection spectrum of the bulk GaAs material shown in Fig. 3.8. The phase spectrum from the above Kramers–Kronig relationship is presented as the red dashed line in Fig. 3.13. The major peak in $R(\omega)$ is close to ω_{TO}, as expected, while $\delta(\omega_{\text{TO}})$ is close to zero. Moreover, $\delta(\omega)$ has a peak at ω_{LO}. The physics of these major features is straightforward that the dielectric coefficient at ω_{TO} is the largest

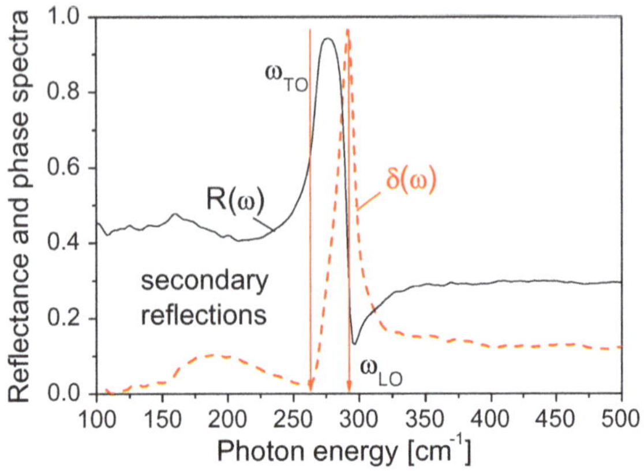

Fig. 3.13 Infrared reflection spectrum (black solid line, same as in Fig. 3.8) of bulk GaAs and the phase spectrum $\delta(\omega)$ (red dashed line) obtained from the Kramers–Kronig relationship of (3.39)

so that the reflection is also the strongest at this photon energy. Since the light at ω_{TO} is directly reflected from the sample without being much refracted in the sample, the phase shift, i.e., $\delta(\omega_{\mathrm{TO}})$, is minimal. At ω_{LO}, the light transmits into the sample easily, resulting in a large phase shift due to multiple reflections and refractions.

We also observe a side peak at the low-frequency side of the major peak, which is due to the secondary reflections, as mentioned before in Fig. 3.8. Similarly, the phase is nonzero on the high-frequency side, largely caused by deviations from ideal measurement setup such as that the light incidents to the sample at a very small (to approach the normal incidence condition) but still finite angle.

Knowing the reflectance $r_0(\omega) = \sqrt{R(\omega)}$ (direct measurement) and the phase factor $\delta(\omega)$ obtained from the Kramers–Kronig relationship of (3.39), we can easily calculate the dielectric coefficient of the material by (3.38), which is shown in Fig. 3.14a corresponding to the infrared reflection spectrum in Fig. 3.13. The dielectric coefficient can be obtained as well by using the harmonic oscillator model introduced in the previous section and the results are presented in Fig. 3.14b. The two sets of spectra are very similar, while they differ numerically by 20–30%.

In general, the dielectric coefficient obtained by the harmonic oscillator model is quite symmetric with respect to ω_{TO}, as shown by Fig. 3.14b compared with the spectra in (a). This is simply the result of the mathematics of the harmonic oscillator model, i.e., (3.21). The Kramers–Kronig relationship, on the other hand, does not have any assumptions (at least physically, see more below) so its results should be more information richer provided that the reflection spectrum is properly measured and pre-assessed. One major issue in assessing the spectrum measurement is about the calibration of the system. As briefly discussed in Sect. 1.5, we normally have a reference spectrum, see (1.34), which may introduce extra signals into the optical spectrum of the sample. In addition, the geometric structure and the size of the sample are also critical. We see the effect of geometric structure as the secondary reflections in Figs. 3.8 and 3.13. The geometric size of the sample is related to the size of the probing light spot that strikes the sample. We may ask a few questions about the

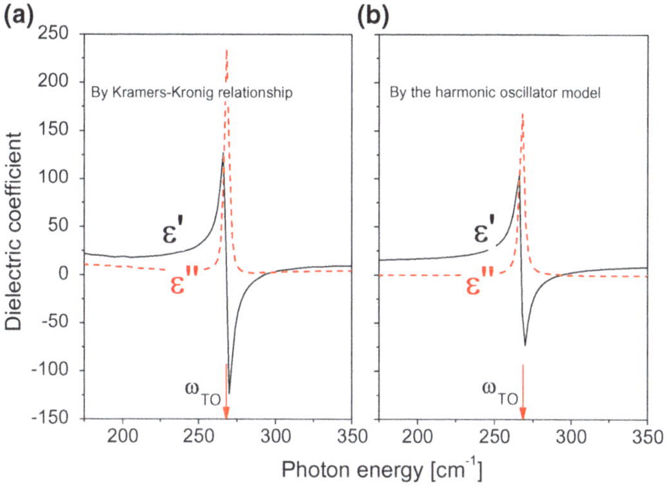

Fig. 3.14 Dielectric coefficient $\epsilon = \epsilon' + i\epsilon''$ of a bulk GaAs material from its infrared reflection spectrum by **a** the Kramers–Kronig relationship and **b** the harmonic oscillator model

measurements such as: Does the light spot cover the whole sample? Is the surface of the sample covered by the light spot flat and smooth?

There is a theoretical aspect about the Kramers–Kronig relationship that it has one mathematical limit that in principle, a reflection spectrum over the entire optical range, i.e., $0 < \omega < \infty$, is needed in order to obtain the phase spectrum. This is impossible in reality, see (3.39), the real spectrum measurement is limited within $\omega_1 < \omega < \omega_2$. This may cause errors which are hardly possible to be properly assessed since we simply do not have the knowledge about the spectrum in $0 \le \omega \le \omega_1$ and $\omega_2 \le \omega \le \infty$.

To estimate the effect of the limited optical range on the results of the Kramers–Kronig relationship, let us do a computer experiment. Assume an optical spectrum composed of two harmonic oscillators described by $\omega_1 = 300\ \text{cm}^{-1}$, $f_1 = 1.0$, $\gamma_1 = 10.0\ \text{cm}^{-1}$, $\omega_2 = 360\ \text{cm}^{-1}$, $f_2 = 4.0$, and $\gamma_2 = 20.0\ \text{cm}^{-1}$. Moreover, $\epsilon_\infty = 10.0$. It is then easy to calculate the dielectric coefficient and eventually the reflection spectrum by (3.22) which is presented as the solid line in Fig. 3.15. It is observed that far away from the frequencies of the two harmonic oscillators such as $\omega \le 100\ \text{cm}^{-1}$ and $\omega \ge 1000\ \text{cm}^{-1}$, the reflection is relatively small and the reflection spectrum is feature-less. So we use (3.39) to calculate the phase spectrum using the reflection spectrum between $\omega_1 = 100\ \text{cm}^{-1}$ and $\omega_2 = 1000\ \text{cm}^{-1}$ and the result is presented as the red dashed line in Fig. 3.15.

Knowing the reflection and phase spectra, we can calculate the dielectric coefficient. Figure 3.16, (a) in a linear scale and (b) logarithmic scale, shows the resulting $\epsilon''(\omega)$, i.e., the imaginary part of the dielectric coefficient. There are a few lines. The black solid line is the calculation result using the reflection and phase spectra ($100 < \omega < 1000\ \text{cm}^{-1}$) from Fig. 3.15, it coincides with the curve obtained

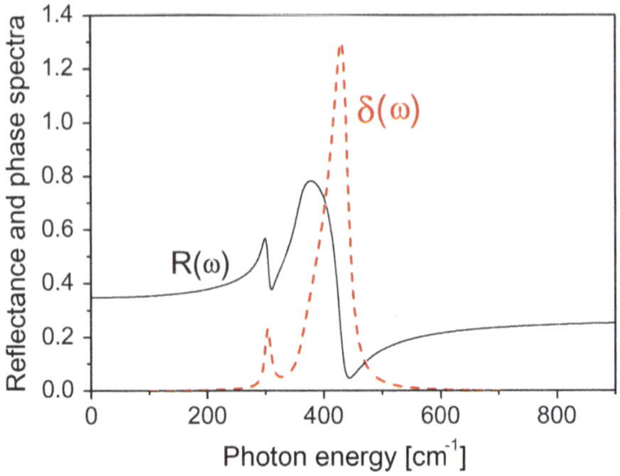

Fig. 3.15 Infrared reflection spectrum (black solid line) of two harmonic oscillators and their phase spectrum (red dashed line) calculated from the Kramers–Kronig relationship using the reflection spectrum between $\omega_1 = 100\ \text{cm}^{-1}$ and $\omega_2 = 1000\ \text{cm}^{-1}$

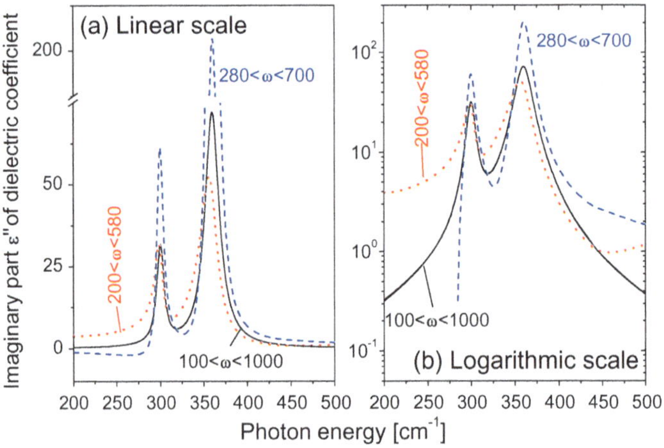

Fig. 3.16 The imaginary part $\epsilon''(\omega)$ of the dielectric coefficient. **a** Linear vertical scale, **b** logarithmic vertical scale. Black solid line is obtained by using the Kramers–Kronig relationship in the optical range of $100 < \omega < 1000\ \text{cm}^{-1}$, red dotted line: $200 < \omega < 580\ \text{cm}^{-1}$, blue dashed line: $280 < \omega < 700\ \text{cm}^{-1}$

from (3.22). Now we reduce the optical range to $200 \leq \omega \leq 580\ \text{cm}^{-1}$, the resulting $\epsilon''(\omega)$ becomes the red dotted line in Fig. 3.16. The peak position remains largely unchanged, while the peak powers and geometric shapes (much clearer in the logarithmic scale) are much modified. If we choose an optical range $280 \leq \omega \leq 700$ cm^{-1} or $0 \leq \omega \leq 460\ \text{cm}^{-1}$ which is not symmetric with respect to ω_1 and ω_2 (this is highly likely in reality since the sensitivity of the spectrometer varies in different optical ranges), the obtained $\epsilon''(\omega)$ becomes asymmetric as well (such as the negative $\epsilon''(\omega)$ below ω_1 for $280 \leq \omega \leq 700\ \mathrm{cm}^{-1}$).

A rule of thumb when assessing the numerical results of the Kramers–Kronig relationship is to check the features of the phase factor in the low and high frequency ranges. If the phase factor approaches zero smoothly in a symmetric manner in these optical ranges, it is fairly safe to say that the numerical results of the Kramers–Kronig relationship are good. Otherwise, more involved analysis about the reflection spectrum is necessary.

3.5 Thin Film on Substrate

Thus far we have studied the reflection and transmission spectra of a light beam through a thin film whose refractive index is characterized by a few absorption peaks. Let us apply the spectral analysis to characterize some as-grown semiconductor structures non-invasively.

First we assess the quality of a multiple-layer structure where the refractive index of each layer is uniform. A good example is the layer-by-layer structure of a quantum well infrared photodetector (QWIP) (whose device function is closely studied in Chap. 6). Figure 3.17a is the schematic structure of the multiple GaAs/$Al_xGa_{1-x}As$ quantum well structure, (b) is the transmission measurement setup along the z axis, and (c) is a typical transmission spectrum.

After measuring the transmission spectrum, we do numerical fitting to obtain the layer thicknesses using the basic formulae we develop in this section. The GaAs/$Al_xGa_{1-x}As$ QWIP structure of Fig. 3.17a contains more than 100 different layers, it is possible to write down all equations then do numerical fitting properly by today's computation facilities. A better way however is to analyze closely to find principal layers that are relevant in terms of the transmission spectrum of Fig. 3.17c.

The back electrode underneath the multiple GaAs quantum wells consists of a Si-doped GaAs layer, a semi-insulating GaAs buffer layer, and a n-type Si-doped GaAs back electrode (the doping level is about $1.5 \times 10^{18}\ \mathrm{cm}^{-3}$). The dielectric coefficients of these three layers do not differ significantly in the optical range of the transmission spectrum in Fig. 3.17c so they can be treated as one composite substrate layer. The designed GaAs quantum well is about 5 nm, and the $Al_xGa_{1-x}As$ barrier is ca 60 nm, i.e., the period of the multiple GaAs/$Al_xGa_{1-x}As$ quantum wells is about 65 nm, which is far smaller than the wavelength, which is about 1 μm, see the horizontal axis of Fig. 3.17c, of the transmission measurement. (This is fundamental in imaging resolution, see, e.g., super-resolution microscopy [4].) The whole multiple GaAs/$Al_xGa_{1-x}As$ quantum wells can therefore be approximated as one composite layer with an effective dielectric coefficient by averaging the ones of GaAs wells and $Al_xGa_{1-x}As$ barriers. The last principal layer is the front electrode layer. The QWIP structure is by now approximated as two composite layers, i.e., the front electrode layer and the composite layer of multiple quantum wells and barriers, on a very

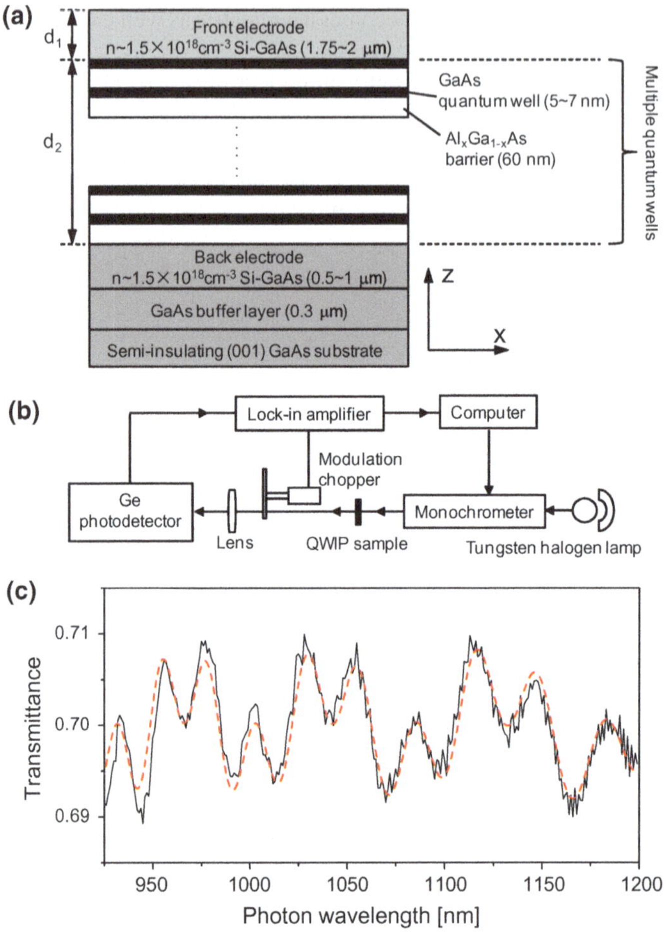

Fig. 3.17 **a** Schematic structure of a GaAs/Al$_x$Ga$_{1-x}$As quantum well infrared photodetector (QWIP), **b** normal-incidence transmission spectrum measurement setup along the z axis, **c** one typical transmission spectrum (solid black line) and its numerically fitted spectrum (dashed red line)

thick substrate layer. And because of the large thickness of the substrate layer, the secondary reflections from the bottom side of the substrate layer can be neglected.

Denote the thickness of the front electrode as d_1, the thickness of the multiple GaAs/Al$_x$Ga$_{1-x}$As quantum wells as d_2 with an effective dielectric coefficient n_{QW} also as a fitting parameter. Because the Al$_x$Ga$_{1-x}$As barrier is much thicker than the GaAs quantum well,

$$n_{\mathrm{QW}} \approx x\, n_{\mathrm{AlAs}} + (1-x) n_{\mathrm{GaAs}} \tag{3.41}$$

It is now easy to obtain the transmittance

$$T_t(n_{1,2}, d_{1,2}, \lambda) = 1 + r_0^2 r_1^2 + r_0^2 r_2^2 + r_1^2 r_2^2$$
$$+2\left(r_0 r_1 + r_0 r_1 r_2^2\right)\cos(2\delta_1) + 2\left(r_1 r_2 + r_0^2 r_1 r_2\right)\cos(2\delta_2)$$
$$+2r_0 r_2 \cos[2(\delta_1 + \delta_2)] + 2r_0 r_1^2 r_2 \cos[2(\delta_1 - \delta_2)] \quad (3.42)$$

where

$$r_0 = \frac{n_0 - n_1}{n_0 + n_1}, \quad r_1 = \frac{n_1 - n_2}{n_1 + n_2}, \quad r_2 = \frac{n_2 - n_3}{n_2 + n_3}, \quad \delta_{1,2} = \frac{2\pi}{\lambda} n_{1,2} d_{1,2} \cos\theta_{1,2}$$

$n_0 = 1$ is the refractive index of the air. $n_{1,2}$ and $d_{1,2}$ are refractive indices and thicknesses of layer 1 (front electrode) and layer 2 (multiple quantum wells/barriers), respectively. $n_1 = n_{\text{GaAs}}$, $n_2 = n_{\text{QW}}$, $n_3 = n_{\text{GaAs}}$. $\theta_{1,2}$ are incident angles, which are zero, as the transmission spectrum of Fig. 3.17c is obtained at the normal incidence.

The numerical fitting starts with inputting the designed values of d_1, d_2 and n_{QW} as initial conditions then modifying the three parameters so that the numerical spectrum $T_t(n_{1,2}, d_{1,2}, \lambda)$ fits best with the experimental spectrum $T_{\text{exp}}(\lambda)$, i.e.,

$$\min\left\{\sum_i \left[T_{\text{exp}}(\lambda_i) - T_t(n_{1,2}, d_{1,2}, \lambda_i)\right]^2\right\} \quad (3.43)$$

where the sum runs over all wavelengths of the transmission measurement. The result of the fitted transmission spectrum is presented in Fig. 3.17c as the dashed red line, and the fitted values of the Al mole fraction x in $\text{Al}_x\text{Ga}_{1-x}\text{As}$ barriers, d_1, and d_2 are listed in Table 3.1 of five QWIP samples.

The purpose of the transmission spectral characterization of the QWIP structure is to assess the basic qualities of the structure directly after its epitaxial growth but before device processing. And the spectral assessment is non-invasive, namely, the as-grown structure is characterized directly. When the assessment shows good

Table 3.1 Comparison of designed and fitted values of a few GaAs/$\text{Al}_x\text{Ga}_{1-x}$As QWIP samples (the unit of thickness d_1 and d_2 is nm)

Sample index	Design			Fitted values		
	d_1	d_2	x	d_1	d_2	x
QWIP-a	1750	3360	0.11	1906.1	3722.8	0.084
QWIP-b	1750	3375	0.11	1863.3	4142.0	0.085
QWIP-c	1750	3390	0.11	2122.4	3724.1	0.059
QWIP-d	1750	3405	0.11	2047.0	4016.1	0.095
QWIP-e	1750	3435	0.11	2048.6	3999.8	0.098

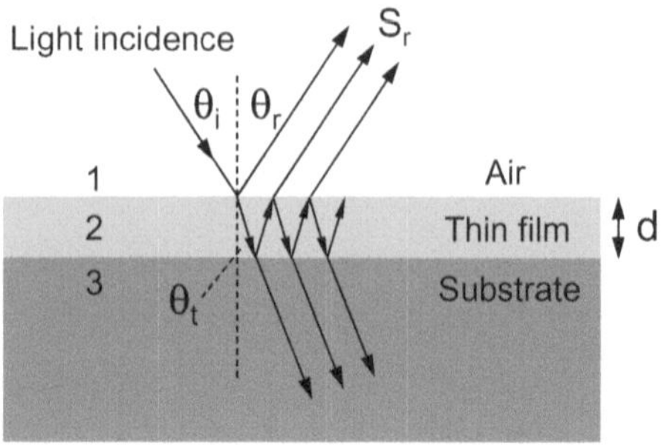

Fig. 3.18 Schematic optical paths of reflected and refracted light beams in a thin-film on a half-infinite-substrate reflection measurement setup

structure quality, device processing will be carried out. (If the assessment has to be invasive, not all as-grown structures can be assessed, some bad structures will be device processed, resulting in a low productivity.) The thickness of the front electrode is not critical for the QWIP performance, while d_2 and x are. Since the QWIP device processing is very expensive, this non-invasive assessment about d_2 and x is very helpful. In Chap. 4 we study the QWIP structure, also non-invasively, by the photoluminescence spectroscopy.

The QWIP structure is a very illustrative example of a thin film structure epitaxially grown on a substrate which is the most common material structure for today's electronics and photonics devices. Let us study this thin film on a substrate in more details using the reflection and transmission spectroscopies. Refer to the schematics in Fig. 3.18, the situation is much alike the thin film we study in Sect. 3.2, see Fig. 3.4, that the light beam incidents from the air to the thin film (complex refractive index $\tilde{n} = n + i\kappa$) except that medium 3 in the structure is now a substrate with an complex refractive index $\tilde{n}_s$.

The calculation of the reflection of the light beam from such a thin film on a substrate is performed similarly to the light reflection from a single thin film in Sect. 3.2. Consider a normal incident light beam with an amplitude E_i, the total amplitude of the reflected light beam is

$$E_\mathrm{r} = E_\mathrm{i} \frac{r_{12} + e^{2i\delta} r_{23}}{1 + e^{2i\delta} r_{12} r_{23}} \tag{3.44}$$

which is exactly the same as (3.15), except that the reflection and refraction coefficients are a bit different:

$$r_{12} = \frac{1 - \tilde{n}}{1 + \tilde{n}}, \quad r_{23} = \frac{\tilde{n} - \tilde{n}_s}{\tilde{n} + \tilde{n}_s}, \quad \delta = \frac{\omega \tilde{n} d}{c_0} \tag{3.45}$$

while they reduce to the ones in (3.12) when $\tilde{n}_\mathrm{s} = 1$ of the substrate, as expected. The amplitude of the reflected electric field is

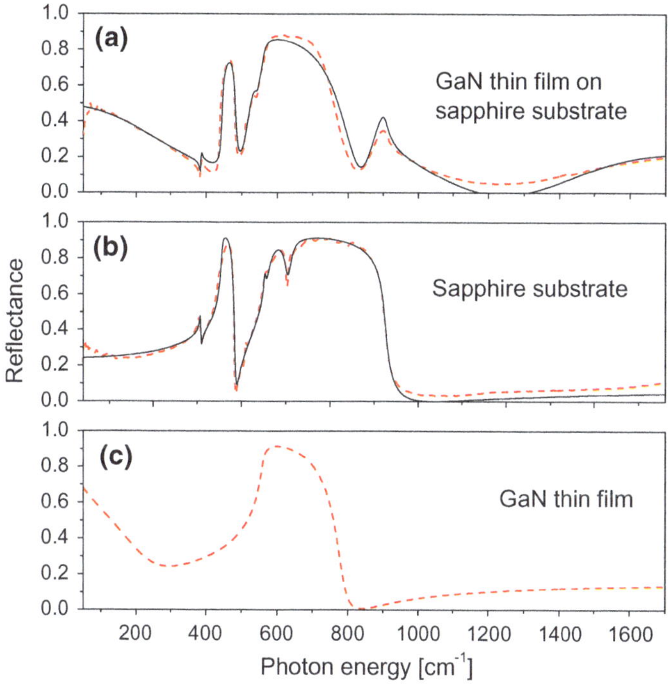

Fig. 3.19 **a** Reflection spectrum of an epitaxial GaN thin film on a sapphire substrate, **b** reflection spectrum of the sapphire substrate. In both **a** and **b**, black solid lines are measured spectra and red dashed lines are fitted spectra. **c** Theoretical reflection spectrum of GaN thin film based on fitting **a** and **b**

$$E_{\mathrm{r}} = E_{\mathrm{i}} \frac{(1-\tilde{n})(\tilde{n}+\tilde{n}_{\mathrm{s}}) + e^{2i\delta}(\tilde{n}+1)(\tilde{n}-\tilde{n}_{\mathrm{s}})}{(1+\tilde{n})(\tilde{n}+\tilde{n}_{\mathrm{s}}) + e^{2i\delta}(1-\tilde{n})(\tilde{n}-\tilde{n}_{\mathrm{s}})} \tag{3.46}$$

and the measured optical power of the reflected light is $S_{\mathrm{r}} = |E_{\mathrm{r}}|^2$. We are not going to measure the transmitted light beam at the other side of the substrate since the substrate is normally rather thick. Thus S_{r} is the only physical parameter to be measured and from which we calculate the reflection spectrum.

We now analyze the reflection spectrum of an epitaxial GaN thin film grown on a sapphire substrate. GaN is a widely applied wide-bandgap material. Typical experimental infrared reflection spectra of the GaN thin film on sapphire and the sapphire substrate are shown in Fig. 3.19a, b [5].

We know from the growth condition that the GaN thin film whose reflection spectra are presented in Fig. 3.19 is Si-doped to a doping level of 10^{17}–10^{19} cm^{-3} so that the density of electrons in the conduction band is very high. In our early studies about the light-matter interaction in Sect. 2.3 and the electromagnetic field propagation in medium in Sect. 3.1 we neglect the electric current and the electron density. Now the conduction-band electron density is so high that we have to take it into account when analyzing the reflection spectra of Fig. 3.19.

The effect of a high electron density on the propagation of the electromagnetic field is mainly characterized by the excitation of the collective motions of the electrons, known as plasmons. Before we do detailed analysis on the GaN reflection spectra in Fig. 3.19, let us introduce the classical Drude model about the plasmons and their effect on the dielectric coefficient.

Refer to Sect. 2.1, the electrons in a crystal structure are classified into core electrons and valence electrons. Here we focus only on the valence electrons. Ga is a group III atom and N group V. In forming GaN, each Ga atom contributes 3 valence electrons and each N atom contributes 5 valence atoms. In an intrinsic GaN crystal, the valence band is completed occupied and the conduction band is completely empty. When GaN is doped with Si atoms, the doped Si atoms substitute Ga atoms then contribute 4 valence electrons per Si atom. The extra valence electron from one doped Si atom has to occupy one of the electron states in the conduction band. These conduction-band electrons make the Si-doped GaN electrically conducting. This is what we learn in Sect. 2.1. A higher Si-doping level means a higher electron density in the conduction band.

In the classical Drude model, these conduction-band electrons are approximated as independent point charges. (The model applies also to the valence-band holes for p-type doped materials.) They collide with each other (electron-electron interaction), as well as with lattice atoms (thus induce lattice vibrations) in the crystal structure. The collisions are random, and the direction of the electron motion directly after each collision is also random. Let the statistical probability of the collision be characterized by a damping rate denoted as γ_p. Let a monochromatic light beam $\boldsymbol{E}(t) = E_0 e^{-i\omega t} z_0$ impinge on the semiconductor material which is polarized along the z direction, where z_0 is the unit vector of the z axis. Note that there is the complex conjugate term in (3.1) to describe the electric field of an electromagnetic field. We adopt this expression here since the same result will be obtained with much simplified mathematical operations. The motion of one conduction-band electron is described by

$$\frac{\mathrm{d}^2 z}{\mathrm{d}t^2} + \gamma_p \frac{\mathrm{d}z}{\mathrm{d}t} = -\frac{e}{m^*} E_0 e^{-i\omega t} \tag{3.47}$$

where m^* is the effective mass of the conduction-band electron, $-e$ is the electric charge carried by the electron. The equation is just Newton's second law of motion where the right side is the electric force the electron experiences in the electric field of the light beam, while the first term on the left side is the acceleration of the electron and the second term the statistical collision effect. It is easy to see that the solution of the above equation is

$$z(t) = z(\omega)\, e^{-i\omega t}, \quad z(\omega) = \frac{eE_0}{m^*(\omega^2 + i\omega\gamma_p)} \tag{3.48}$$

Let ρ be the density of the conduction-band electrons. The polarization P of these electrons per unit volume is

$$P = -\rho e z(\omega) = -\frac{\rho e^2}{m^*\left(\omega^2 + i\omega\gamma_p\right)} E_0 \tag{3.49}$$

and the electric displacement field becomes

$$D = \epsilon_\infty E_0 + P = \epsilon(\omega) E_0 \tag{3.50}$$

Let

$$\omega_p^2 = \frac{\rho e^2}{m^* \epsilon_\infty} \tag{3.51}$$

we then obtain the contribution of the conduction-band electrons to the dielectric coefficient $\epsilon(\omega)$

$$\epsilon(\omega) = \epsilon_\infty \left(1 - \frac{\omega_p^2}{\omega^2 + i\omega\gamma_p}\right) \tag{3.52}$$

Here ω_p is known to be the plasmon frequency. It represents the collective motion of the conduction-band electrons in the crystal. As shall be observed below, ω_p is normally in the infrared optical range, and thus is frequently present in the infrared reflection spectrum.

We are ready to analyze the reflection spectrum of the Si-doped GaN epitaxial film on sapphire. The total dielectric coefficient is now

$$\epsilon(\omega) = \epsilon_\infty + \sum_\ell \frac{f_\ell\, \omega_{\mathrm{TO},\ell}^2}{\omega_{\mathrm{TO},\ell}^2 - \omega^2 - i\omega\gamma_{\mathrm{TO},\ell}} - \frac{\epsilon_\infty \omega_p^2}{\omega^2 + i\omega\gamma_p} \tag{3.53}$$

which includes the transverse optical (TO) phonons and the plasmon of the conduction-band electrons. By modifying various parameters in (3.53) to fit the experimental reflection spectrum we will be able to study the plasmon in the thin film which is further related to other physical properties such as the electron mobility.

The reflection spectrum of the pure GaN thin film on a sapphire substrate is rather simple, see Fig. 3.19c to be discussed in detail below. It is the reflection spectrum of the substrate that is more complicated, see Fig. 3.19b, which is also present in the total reflection spectrum of the GaN thin film on a substrate. Thus, a critical step in analyzing the reflection spectrum of a thin film on a substrate is to find ways to separate/extract the spectrum of the pure thin film from the one of the thin film on the substrate. More specifically, we need to measure the reflection spectrum of the substrate.

For a thin film on a thick substrate sample, the method is very simple that we basically measure the reflection spectra from the thin film side then measure again

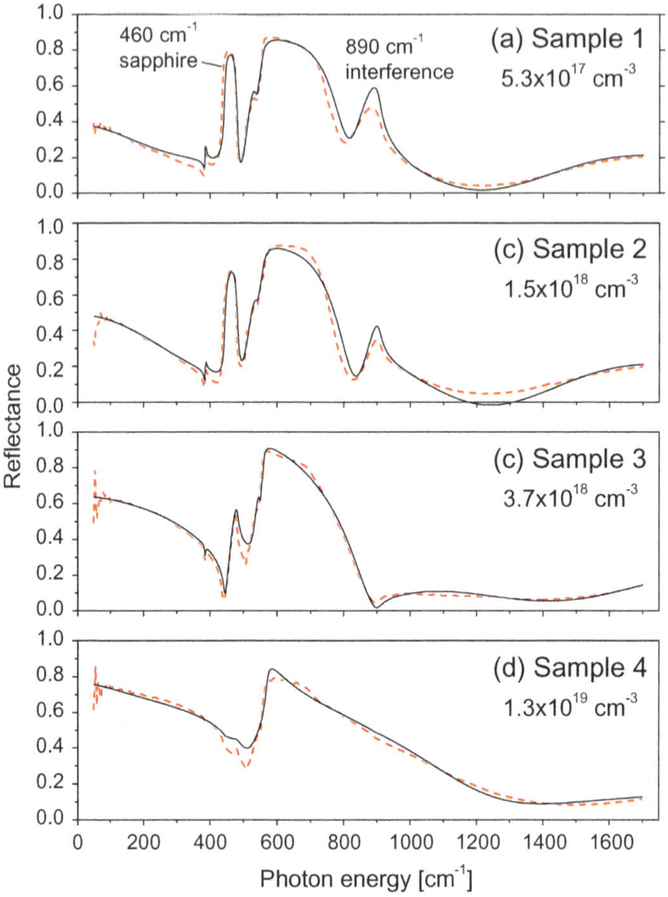

Fig. 3.20 Infrared reflection spectra of GaN thin films on sapphire doped to different Si-doping levels. Black solid lines are experimental data and red dashed lines are theoretical fitting spectra

after flipping the sample so the other side of the substrate now facing the incident light. This was exactly how the two spectra, black solid lines in Fig. 3.19a and b, were obtained [5]. The next step in the spectral analysis is to fit the spectrum of the sapphire substrate, then fit the spectrum of the GaN thin film on the sapphire substrate by modifying only the GaN parameters without modifying the sapphire parameters obtained before. Figure 3.19c is the theoretical spectrum of the pure GaN thin film.

Let us move on to more complicated samples. Figure 3.20 shows the reflection spectra of four samples, both experimental and numerically fitting spectra. The four samples are Si-doped to four different nominal doping levels, which are indicated in their respective spectra. Numerical values of fitting parameters for the transverse optical phonons and the plasmon in the GaN thin films are listed in Table 3.2. The doping level in sample 1 is relatively low so that its reflection spectrum is very similar

Table 3.2 Physical parameters of transverse optical (TO) phonon and plasmon (subscript p) obtained by fitting infrared reflection spectra of GaN thin films. Units of the Si doping level and density ρ of conduction-band electrons are cm^{-3}, units of the frequency and damping rate are cm^{-1}, and the mobility unit is $cm^2/V{\cdot}s$. ρ_{IR} and μ_{IR} denote the electron density and mobility obtained from infrared reflection spectra, while ρ_{Hall} and μ_{Hall} are obtained from Hall measurements

Sample	Doping level	ρ_{IR}	ω_{TO}	γ_{TO}	f_{TO}	ω_p	γ_p	μ_{IR}	ρ_{Hall}	μ_{Hall}	μ_{Hall}/μ_{IR}
1	5.7×10^{17}	8.4×10^{17}	557	11.2	4.43	261	227	205	5.3×10^{17}	370	1.80
2	1.5×10^{18}	1.6×10^{18}	559	13.5	4.36	365	253	184	1.5×10^{18}	320	1.74
3	3.7×10^{18}	4.8×10^{18}	561	14.5	3.84	631	347	134	3.7×10^{18}	234	1.75
4	1.3×10^{19}	1.4×10^{19}	560	10.5	3.56	1062	514	91	1.3×10^{19}	178	1.95

to the one of intrinsic GaN material, see Fig. 3.19a. Increasing the doping level, the reflection below 500 cm^{-1} increases, while the reflection band of 550–800 cm^{-1} is blue-shifted. There is a peak at 890 cm^{-1} originated from the interference of the thin film (multiple reflections between two interfaces), which disappears in the highly-doped sample 3 and 4 because of the doping-induced absorption enhancement. The 460 cm^{-1} peak originates from the sapphire substrate, see Fig. 3.19b.

A little extra work leads us to the relationship between the damping rate and the electron mobility commonly denoted as μ (note that it is also common to use the same Greek character to denote permeability). The electric current is $j = -\rho e v$, where ρ is the electron density and v the average velocity of electrons, which is dz/dt. By (3.48),

$$j = -\rho e \frac{\mathrm{d}z}{\mathrm{d}t} = \frac{i\rho e^2 E\omega}{m^*(\omega^2 + i\omega\gamma_p)} \tag{3.54}$$

where $E = E_0 e^{-i\omega t}$.

Before further analysis, we need to check more carefully the validity of the above expression. When we formulate (3.48), we implicitly assume that the collective motion of the electrons can follow the oscillation of the electric field of the light beam. This is actually only valid when the angular frequency ω of the electric field is much lower than the plasmon frequency ω_p of the electrons. This is also reflected in the above equation that only when $\omega \ll \omega_p$, the resulting j is real

$$j_{\omega \ll \omega_p} = \frac{\rho e^2 E}{m^* \gamma_p} \tag{3.55}$$

We therefore restrain us to the low-frequency condition of $\omega \ll \omega_p$.

On the other hand, the electric current is also expressed as $j = \rho\mu E$, where μ is the electron mobility. We thus obtain the following relationship between the low-frequency mobility $\mu_{\rm IR}$ and the plasmon damping rate γ_p

$$\mu_{\rm IR} = \frac{e^2}{m^*\gamma_p} \tag{3.56}$$

From the frequency and the damping rate of the plasmon in Table 3.2 we can calculate the density of the conduction-band electrons and their mobility by (3.51) and (3.56), respectively. The results are listed in Table 3.2 when $m^* = 0.2m_0$ and $\epsilon_\infty = 5.35$ [6]. We also list electric characterization results of the electron density and the mobility using electric Hall measurements. It is commonly believed that the doped Si atoms in GaN are all ionized so that the density of electrons in the conduction band equals the Si doping level. This is quite correct, as confirmed by both the infrared optical characterization and the electric Hall measurements, however with certain deviations. One important issue is that what we know about the doping level is actually the nominal level, which is not exactly the same as the real doping level.

Table 3.2 shows that the optically characterized $\mu_{\rm IR}$ is about one half of the Hall mobility $\mu_{\rm Hall}$, it is however very close to the drift mobility $\mu_{\rm drift}$ which is about $0.52\mu_{\rm H}$ [7]. Physically, the optically characterized mobility is different from the Hall mobility. Moreover, the electric mobility involves many physical processes which depend on electron energy band structures [8].

References

1. R.H. Stolen, Temperature dependence of far-infrared absorption in GaAs. Phys. Rev. B **11**, 767–770 (1975)
2. C. Patel, T.J. Parker, H. Jamshidi, W.F. Sherman, Phonon frequencies in GaAs. Phys. Stat. Sol. (b), **122**, 461–467 (1984)
3. A.S. Barker Jr., Infrared lattice vibrations and dielectric dispersion in corundum. Phys. Rev. **132**, 1474–1481 (1963)
4. The Nobel Prize in Chemistry 2014 was awarded jointly to Eric Betzig, Stefan W. Hell and William E. Moerner "for the development of super-resolved fluorescence microscopy"; Nobel Prize in Chemistry 2017 was awarded jointly to Jacques Dubochet, Joachim Frank and Richard Henderson "for developing cryo-electron microscopy for the high-resolution structure determination of biomolecules in solution"
5. Z.-F. Li, W. Lu, H.-J. Ye, Z.-Z. Chen, X.-Z. Yuan, H.-F. Dou, S.C. Shen, G. Li, S.J. Chua, Carrier concentration and mobility in GaN epilayers on sapphire substrate studied by infrared reflection spectroscopy. J. Appl. Phys. **86**, 2691(5) (1999)
6. T. Kozawa, T. Kachi, H. Kano, Y. Taga, M. Hashimoto, N. Koide, K. Manabe, Raman scattering from LO phonon-plasmon coupled modes in gallium nitride. J. Appl. Phys. **75**, 1098(4) (1994)
7. S.M. Sze, *Physics of Semiconductor Devices*, 2nd edn. (Willey, New York, 1981), p. 29
8. Y. Fu, K.B. Joelsson, K.J. Grahn, W.-X. Ni, G.V. Hansson, M. Willander, Hall factor in strained p-type doped $Si_{1-x}Ge_x$ alloy. Phy. Rev. B **54**, 11317–11321 (1996)

Chapter 4
Photoluminescence

Abstract Quantum mechanics theory about photoluminescence is first introduced then applied to various case studies to understand the microscopic processes, namely photon excitation, energy relaxation and radiative recombination, involved in the photoluminescence spectroscopy in semiconductor structures ranging from bulk to nanoscale. The study is further deepened by analyzing the photoluminescence spectra of quantum dots under the multiphoton excitation.

In the previous chapter, we have studied the mechanisms of reflection and transmission spectral measurements of how a monochromatic light beam with a particular photon energy after impinging on a material becomes reflected by and transmitted in/through the material. In other words, the energies of the photons in the incident, reflected and transmitted light beams are all the same. In the photoluminescence (PL) spectral measurement, however, we commonly send a probe light beam of relatively high-energy photons to the sample under investigation then collect photons of different energies emitted from the sample. In most common cases, the energies of the emitted photons are lower than the energies of the probing photons. There are extraordinary cases that the energies of the emitted photons are higher than the energies of the probing photons.

The optical path layout of the photoluminescence measurement is quite similar to the reflection and transmission measurement, except that we can collect the emitted photons from any side of the sample, either the front side where the probing light incidents, or the back side, or the edges. We even integrate the emitted photons collected from all different angles if the total number of the emitted photons is relatively low. Under the latter situation, we have to avoid the geometric alignment of the photodetection with the light incidence to avoid the strong signal of the incident light which may saturate or even damage the photodetector, such as at the 90° angle relative to the probing laser in Fig. 1.1. Alternatively we add an optical filter to blockade the probing photons from reaching the photodetector (remember that the energies of the emitted photons are normally different from the energies of the probing photons).

W. Lu and Y. Fu, *Spectroscopy of Semiconductors*, Springer Series
in Optical Sciences 215, https://doi.org/10.1007/978-3-319-94953-6_4

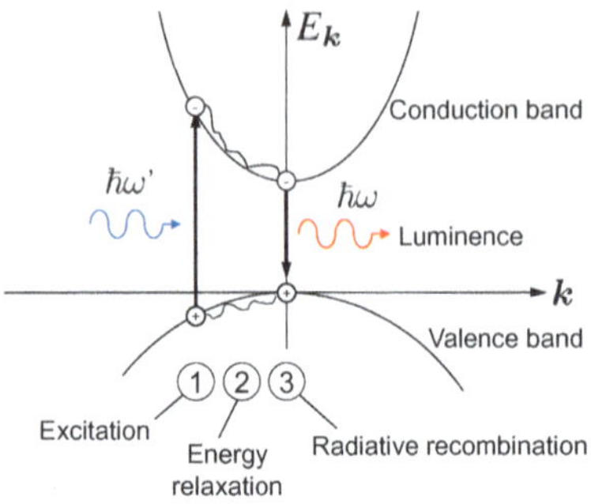

Fig. 4.1 Schematics of photoluminescence process in an intrinsic semiconductor. A probing photon $\hbar\omega'$ incidents to the sample to photo-excite an electron to transit from a valence-band state to a conduction-band state (process 1); The excited electron and hole relax to their respective ground states, i.e., the conduction-bandedge and the valence-bandedge states (process 2); the energy-relaxed electron and hole recombine radiatively to emit a photon $\hbar\omega$ (luminescence, also called florescence)

Refer to Fig. 4.1, the most common microscopic mechanisms of the photoluminescence in an intrinsic semiconductor consists of the following three major physical processes

1. Excitation: an external probing photon with energy $\hbar\omega'$, which is normally higher than the energy bandgap of the semiconductor, comes into the semiconductor to excite an electron to transit from its initially occupied valence-band state to an empty conduction-band state, forming an electron-hole pair (when the electron and hole stay close to each other within the so-called exciton Bohr radius, they form an exciton, see Chap. 2).
2. Energy relaxation: the photogenerated electron and hole from the excitation process are in general in excited states in the conduction and valence bands. They go through various energy relaxation processes such as electron-phonon interactions to relax to their respective ground states, namely, the conduction- and valence-bandedge states.
3. Radiative recombination: the energy-relaxed electron and hole recombine, namely, the electron at the conduction-bandedge state transits to the empty valence-bandedge state, radiatively emits a photon with an energy $\hbar\omega$. This is the same spontaneous emission studied in Sect. 2.3.

In Fig. 4.1, the excitation is induced by a probing photon so that the final luminescence is named photoluminescence. However, the excitation of an electron initially occupying a valence-band state to transit to an initially empty conduction-band state, i.e., process 1, can be induced differently, such as thermally or electrically, resulting in different types of luminescence:

1. Photoluminescence: We have just described the process. In general, the energy of the probing photon is larger than the energy bandgap of the semiconductor. Common lasers used for photoluminescence excitation are the 632.8 nm beam

from He-Ne laser, the 514.5 nm beam from argon ion laser, and lasers in ultraviolet (UV) range. When the optical power of the laser is high enough, there is the so-called multi-photon excitation that even the energy of one probing photon is smaller than the semiconductor energy bandgap, the electron initially occupying a valence-band state can absorb more than one photon in order to reach the available conduction-band state (we will discuss this multiphoton excitation in Sects. 4.6 and 6.4).

2. Electroluminescence, also known as electric-field-induced luminescence. Light-emitting diode (LED) is one example. A LED consists of an n-type doped region and a p-type doped region separated by an intrinsic i-region. When an external bias applies, electrons in the conduction band in the n-type region are driven into the i-region, while the holes in the valence band in the p-type region are driven into the i-region from the other direction. These electrons and holes will radiatively recombine to luminescence when they meet in the i-region. LED is widely used in various displays and has been intensively studied and developed for lighting. The Nobel Prize in Physics 2014 was awarded jointly to Isamu Akasaki, Hiroshi Amano and Shuji Nakamura for the invention of efficient blue light-emitting diodes which has enabled bright and energy-saving white light sources.
3. Thermoluminescence: Heating an intrinsic semiconductor also induces the transitions of electrons from the occupied valence-band states to the empty conduction-band states (recall the Fermi-Dirac distribution of electrons at equilibrium). Such thermally excited electrons in the conduction band can radiatively recombine with holes in the valence band to luminescence, resulting in this thermoluminescence.
4. Cathodoluminescence: High-energy electrons that impinge on the semiconductor from a cathode ray tube will transfer their energies to the valence-band electrons for them to transit to un-occupied conduction-band states. Eventual radiative recombinations of the conduction-band electrons and valence-band holes result in the cathodoluminescence.

One common critical aspect about all these luminescences is of course the final step, namely, the radiative recombination of the electron-hole pair. In reality, the recombination of the electron-hole pair can also be non-radiative, such that the energy from the electron-hole recombination is transferred to the crystal lattice (and heat sink is commonly a must for high-power LED). Therefore, the eventual luminescence power is a measure about the radiative versus non-radiative recombinations of the electron-hole pairs.

In a brief summary, a photoluminescence spectrum can measure the energy difference between the conduction and valence bandedges which is approximately the energy of the photoluminescence peak (for the processes in Fig. 4.1, and we discuss other cases shortly). We can also use the photoluminescence spectrum to study the excitation, the radiative and non-radiative recombination processes, to measure the lifetime of the photoluminescence signal when we use a laser pulse to excite the electrons then measure the so-called time-resolved photoluminescence. There are many different photoluminescence measurement setups, which is the central theme of this chapter.

4.1 Basic Photoluminescence Theory

In Sect. 2.1 we have briefly discussed the photon absorption, stimulated photon emission and spontaneous photon emission shown in Fig. 2.11. In normal photoluminescence processes of Fig. 4.1, the photogenerated electron and hole may radiatively recombine directly after their photogeneration. However, the probability of such a direct radiative recombination is relatively low. The photogenerated electron and hole normally go through various energy relaxation processes within their respective energy bands, such that the electron relaxes in the conduction band, while the hole relaxes in the valence band. The electron and hole recombine after they reach their respective ground states. Thus, photoluminescence is widely used to study low-energy electron states that are close to either the conduction- or the valence-bandedge state. For high-energy states, we may use reflection-transmission spectra (which we studied in the previous chapter) and modulation spectroscopy (to be studied in the next chapter).

In common photoluminescence measurement setup, we use a continuous-wave (CW) laser to excite the electrons in a semiconductor sample and the luminescence signal is recorded continuously as well. All processes involved reach a so-called steady state in close correlations with each other: An electron in a valence-band state absorbs one exciting photon then transits to an empty conduction-band state, leaving the valence-band state unoccupied (which is a hole). This is process 1. The conduction-band electron and the valence-band hole relax to conduction- and valence-bandedge states (which are initially empty), leaving the high-energy conduction-band and the valence-band states empty. This is process 2. The electron at the conduction bandedge and the hole at the valence bandegde recombine to emit a photon. This is process 3. We then return to process 1, followed by step 2 and so on and so forth. We may also have an interlaced scenario such that as soon as process 2 is finished, process 1 may start in parallel with process 3. All depend on the probabilities of the involved processes, as well as the fundamental requirement that each electron state can be maximally occupied by one electron at one time (Pauli exclusion principle). The general mathematical analysis of this steady state consists of formulating then solving rate equations (see Sect. 7.2).

In the following we study the photoluminescence stepwise. Refer to (2.16), the total wave functions of a valence-band state denoted by eigen energy $E_{\mathrm{v}\boldsymbol{k}}$ and a conduction-band state $E_{\mathrm{c}\boldsymbol{q}}$ are

$$\Psi_{\mathrm{v}\boldsymbol{k}}(\boldsymbol{r}) = \frac{1}{\sqrt{N}} e^{-i\boldsymbol{k}\cdot\boldsymbol{r}} u_{\mathrm{v}}(\boldsymbol{r}), \quad \Psi_{\mathrm{c}\boldsymbol{q}}(\boldsymbol{r}) = \frac{1}{\sqrt{N}} e^{-i\boldsymbol{q}\cdot\boldsymbol{r}} u_{\mathrm{c}}(\boldsymbol{r}) \tag{4.1}$$

where N is the number of unit cells. We have neglected the dependence of $u_{\mathrm{c}}(\boldsymbol{r})$ and $u_{\mathrm{v}}(\boldsymbol{r})$ on $\boldsymbol{k}$ since we focus on only low-energy electron and hole states (thus small amplitude of $\boldsymbol{k}$). We have studied this condition early in Chap. 2 and will return to examine this condition again shortly.

By Sect. 2.3, the probability of the spontaneous emission is expressed as

$$w_{q\leftarrow k}^{\text{spon}}(\omega) = \frac{\pi\hbar^2 e^2}{m_0^2\epsilon_0 n^2\omega\Omega}|\langle\Psi_{cq}|e_s\cdot\nabla|\Psi_{vk}\rangle|^2 \frac{\Gamma_{qk}}{\left(E_{cq}-E_{vk}-\hbar\omega\right)^2+\Gamma_{qk}^2} \tag{4.2}$$

i.e., (2.115). As we will see shortly that the photoluminescence is dominant by the radiative recombination of the electrons and holes occupying states close to conduction- and valence bandedges, i.e., $\hbar\omega$ is close to E_g, while the energy of the exciting photon $\hbar\omega'$ is commonly much higher than E_g, the stimulated emission probability at $\hbar\omega'$ is negligible when the above spontaneous emission probability of $\hbar\omega$ around E_g is significant. And initially there is no light with photon energy $\hbar\omega$ so $N_\omega = 0$ so we focus on only the spontaneous emission of (4.2) for a while.

Let us just focus on the optical transition matrix element

$$\langle\Psi_{cq}|e_s\cdot\nabla|\Psi_{vk}\rangle = \frac{1}{N}e_s\cdot\left[\int_\Omega e^{i(q-k)\cdot r}u_c^*(r)\nabla u_v(r)\mathrm{d}r - ik\int_\Omega e^{i(q-k)\cdot r}u_c^*(r)u_v(r)\mathrm{d}r\right] \tag{4.3}$$

by (4.1). The second term on the right side of the above equation is zero due to the wave function orthogonality. u_c and u_v are periodic functions of unit cells in the crystal structure, while $e^{i(q-k)\cdot r}$ varies slowly in space since both q and k are small (remember that this is the condition of our studies), so that we substitute r by the sum of unit cell position R and its spatial coordinate r' with respect to R, i.e., $r \equiv R + r'$. The integration of the first term on the right side of the above equation becomes the summation over R over the whole volume Ω of the semiconductor sample under investigation and the spatial integration over r' within unit cell R

$$\int_\Omega e^{i(q-k)\cdot r}u_c^*(r)\nabla u_v(r)\mathrm{d}r = \sum_{R\in\Omega} e^{i(q-k)\cdot R}\int_{\text{cell}} u_c^*(r')\nabla u_v(r')\mathrm{d}r' \tag{4.4}$$

Since the integration of the Bloch functions within one unit cell is commonly designated as

$$\int_{\text{cell}} u_c^*(r')\nabla u_v(r')\mathrm{d}r' = \frac{p_{cv}}{\hbar} \tag{4.5}$$

where $|p_{cv}| = \sqrt{2m_0E_p}$ is a material parameter known as the optical dipole moment whose values of different semiconductors have been well-documented [1],

$$\int_\Omega e^{i(q-k)\cdot r}u_c^*(r)\nabla u_v(r)\mathrm{d}r = \frac{p_{cv}}{\hbar}\sum_{R\in\Omega} e^{i(q-k)\cdot R} \tag{4.6}$$

Let Ω_{cell} be the volume of the unit cell. Since $\Omega_{\text{cell}} \ll \Omega$, the summation over R over Ω can be evaluated in the following way:

$$\sum_{R\in\Omega} e^{i(q-k)\cdot R} = \frac{1}{\Omega_{\text{cell}}} \sum_{R\in\Omega} e^{i(q-k)\cdot R}\,\Omega_{\text{cell}} \approx \frac{1}{\Omega_{\text{cell}}} \int_{\Omega} e^{i(q-k)\cdot R}\,\mathrm{d}R$$
$$= \delta_{qk}\frac{1}{\Omega_{\text{cell}}}\int_{\Omega}\mathrm{d}R = \delta_{qk}\frac{\Omega}{\Omega_{\text{cell}}} = \delta_{qk}N \tag{4.7}$$

Here $\delta_{qk} = 1$ when $q = k$, it is otherwise zero. Since N is the number of unit cells in Ω, $N\Omega_{\text{cell}} = \Omega$.

With all these we obtain the optical transition matrix element

$$\left\langle \Psi_{cq}(r)\,|e_s \cdot \nabla|\, \Psi_{vk}(r)\right\rangle = \delta_{qk}\frac{e_s \cdot p_{\text{cv}}}{\hbar} \tag{4.8}$$

Insert it into (4.2)

$$w_k^{\text{spon}}(\omega) = \frac{\pi e^2 |e_s \cdot p_{\text{cv}}|^2}{m_0^2 \epsilon_0 n^2 \omega \Omega} \frac{\Gamma_k}{\left(E_{ck} - E_{vk} - \hbar\omega\right)^2 + \Gamma_k^2} \tag{4.9}$$

Here we include δ_{qk} by putting the wave vector of the destination conduction-band state as the one of the initial valence-band state. This is the very reason that the optical excitation and the radiative recombination (process 1 and 3) in Fig. 4.1 are presented vertical in the k space.

The total optical transition rate is obtained by summing (4.9) over all occupied electron states available for optical transitions

$$W(\omega) = \int w_k^{\text{spon}}(\omega)\left[1 - f(E_{vk})\right] f(E_{ck}) \frac{2\mathrm{d}k}{(2\pi)^3/\Omega}$$
$$= \frac{\pi e^2 |e_s \cdot p_{\text{cv}}|^2}{m_0^2 \epsilon_0 n^2 \omega} \int \frac{\Gamma_k\left[1 - f(E_{vk})\right] f(E_{ck})}{\left(E_{ck} - E_{vk} + \hbar\omega\right)^2 + \Gamma_k^2} \frac{2\mathrm{d}k}{(2\pi)^3} \tag{4.10}$$

Here $2\mathrm{d}k/[(2\pi)^3/\Omega]$ is the density of electron states in the k space. $f(E_{vk})$ and $f(E_{cq})$ are probabilities of electrons occupying state E_{vk} and E_{cq}, respectively. Note that $f(E_{vk})$ normally refers to the electron occupation of state E_{vk} so that $\left[1 - f(E_{vk})\right]$ means the probability that E_{vk} is un-occupied by one electron, it is then the probability that E_{vk} is occupied by one hole.

We know that the occupation of one electron state E by one electron is described by Fermi-Dirac distribution function when electrons are at equilibrium

$$f_{\text{FD}}(E) = \frac{1}{\exp\left(\frac{E - E_f}{k_B T}\right) + 1} \tag{4.11}$$

where E_f is the well-defined Fermi level, and T is the temperature of the system. When $E \gg E_f$, $f_{\mathrm{FD}} = 0$; $E \ll E_f$, $f_{\mathrm{FD}} = 1$. And most importantly, when $E = E_f$, $f_{\mathrm{FD}} = 1/2$. In other words, the probability of the Fermi level being occupied by one electron is 1/2.

Not going too deep into details, let us simplify the steady state of photoluminescence measurement by introducing the concept of a quasi-equilibrium state. For most semiconductor samples of interest, we can use different lasers of different wavelengths for measuring their photoluminescence spectra and the resulting photoluminescence spectra are not very sensitive to the wavelength of the excitation laser (as long as the photon energy is quite large as compared with the semiconductor energy bandgaps). Photoluminescence lifetime measurement shows that the luminescence persists for quite a long time after switching off the excitation laser (see more in Sects. 4.4 and 7.2). Refer back to Fig. 4.1 we can conclude that the excitation (process 1) and energy relaxation (process 2) must be much faster than the radiative recombination (process 3). And the photoexcited electrons and holes accumulate at their respective ground states (conduction and valence bandedges). In the first-order approximation we may therefore assume that the accumulations of the electrons and holes in the conduction and valence bands are described by two different quasi Fermi levels, one for the electrons in the conduction band and one for the holes in the valence band. This is the quasi-equilibrium-state approximation:

$$f(E_{\ell k}) = \frac{1}{\exp\left(\dfrac{E_{\ell k} - E_f^{\ell}}{k_B T}\right) + 1} \tag{4.12}$$

where E_f^{ℓ} is the quasi Fermi level of band ℓ (either conduction band "c" or valence band "v"). Note that we denote the occupation probability at quasi-equilibrium-state by f and f_{FD} for equilibrium state. Without any external field such as the excitation laser so the electrons in the semiconductor are at equilibrium, $E_f^{\mathrm{c}} = E_f^{\mathrm{v}} = E_f$ and $f = f_{\mathrm{FD}}$, see Fig. 4.2a.

In an n-type semiconductor there are more electrons in the conduction band than holes in the valence band and the equilibrium-state Fermi level E_f is close to the conduction bandedge (the situation of Fig. 4.2a). Similarly, a p-type material has less electrons in the conduction band than holes in the valence band so that its E_f is close to the valence bandedge. Under the excitation of an external photon field $\hbar\omega'$, electrons that initially occupy the valence band are now excited to the conduction band, leaving holes in the valence band. Together with the initial electrons in the conduction band and holes in the valence band, the photogenerated electrons and holes will reach their respective quasi equilibrium states described by two quasi Fermi levels E_f^{c} and E_f^{v}, see Fig. 4.2b.

In an intrinsic or lightly-doped semiconductor, the numbers of electrons in the conduction band and holes in the valence band are all small. E_f is close to the middle point of the energy bandgap, it is also far away from the conduction and valence bandedges. Under the optical excitations of many common photoluminescence

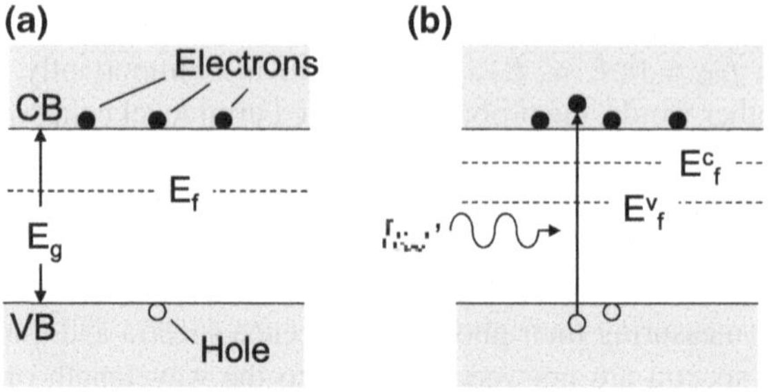

Fig. 4.2 The equilibrium-state Fermi level and steady-state quasi Fermi levels of electrons and holes in a semiconductor. CB denotes the conduction band and VB for the valence band. **a** Without any external field, $E_f^c = E_f^v = E_f$. **b** Quasi equilibrium state under a continuous-wave $\hbar\omega'$ excitation

measurement setups, the numbers of photogenerated electrons and holes are small so that the energy separation between E_f^c and E_f^v is far less than E_g. In other words, $E_{ck} - E_f^c \gg k_B T$ (implicitly, the temperature is not high, mostly room temperature or below), $E_f^v - E_{vk} \gg k_B T$ so that the Fermi-Dirac distribution function (4.11) can be approximated by Boltzmann distribution function

$$f(E_{ck}) \approx \exp\left(-\frac{E_{ck} - E_f^c}{k_B T}\right)$$

$$1 - f(E_{vk}) = 1 - \frac{1}{\exp\left(-\frac{E_{vk} - E_f^v}{k_B T}\right) + 1} = \frac{1}{\exp\left(-\frac{E_f^v - E_{vk}}{k_B T}\right) + 1}$$

$$\approx \exp\left(-\frac{E_f^v - E_{vk}}{k_B T}\right) \quad (4.13)$$

Moreover, as what we have learned in Chap. 2, the energy band structures of the conduction and valence bandedges of many semiconductors can be well approximated as spherical and parabolic

$$E_{ck} = E_c + \frac{\hbar^2 \boldsymbol{k}^2}{2m_c^*}, \quad E_{vk} = E_v - \frac{\hbar^2 \boldsymbol{k}^2}{2m_v^*} \quad (4.14)$$

where E_ℓ is the energy of the bandedge ($\ell = $ c, v) so that $E_c - E_v = E_g$, m_ℓ^* is the effective mass of electron or hole of band ℓ.

By (4.14),

$$E_{ck} - E_{vk} = E_g + \frac{\hbar^2 \boldsymbol{k}^2}{2m_r^*} \quad (4.15)$$

is the optical transition energy at $\boldsymbol{k}$,

$$\frac{1}{m_r^*} = \frac{1}{m_c^*} + \frac{1}{m_v^*}$$

where m_r^* is called the reduced effective mass, the same as in (2.129). We invoke again the weak excitation condition that the energy separation between E_f^c and E_f^v is much smaller than E_g so that

$$\left[1 - f(E_{vk})\right] f(E_{ck}) \approx \exp\left(-\frac{E_{ck} - E_f^c + E_f^v - E_{vk}}{k_B T}\right) = \exp\left(-\frac{E_g + \frac{\hbar^2 k^2}{2m_r^*}}{k_B T}\right) \tag{4.16}$$

Since the numbers of electrons and holes are small, electrons and holes are expected to stay closely to the bandedges, i.e., the amplitudes of the wave vectors of electrons and holes are small so that we may neglect the $\boldsymbol{k}$-dependences of u_{ck} and u_{vk}, therefore also $\boldsymbol{p}_{cv}$, see (2.31).

Putting all these together into (4.10)

$$W(\omega) = \frac{\pi e^2 |\boldsymbol{e}_s \cdot \boldsymbol{p}_{cv}|^2}{m_0^2 \epsilon_0 n^2 \omega} \int \frac{\Gamma_k}{\left(E_g + \frac{\hbar^2 k^2}{2m_r^*} - \hbar\omega\right)^2 + \Gamma_k^2} \exp\left(-\frac{E_g + \frac{\hbar^2 k^2}{2m_r^*}}{k_B T}\right) \frac{2\mathrm{d}\boldsymbol{k}}{(2\pi)^3} \tag{4.17}$$

By (4.15),

$$\frac{2\mathrm{d}\boldsymbol{k}}{(2\pi)^3} = \frac{1}{4\pi^3} 2\pi k^2 \mathrm{d}k = \frac{1}{2\pi^2}\left(\frac{2m_r^*}{\hbar^2}\right)^{3/2} \sqrt{E - E_g}\, \mathrm{d}E \equiv N_3^a(E - E_g)\mathrm{d}E \tag{4.18}$$

where $N_3^a(E)$ is the three-dimensional density of associated electron states, which is valid for $E \geq E_g$, i.e., $\boldsymbol{k}$ must be real. To avoid possible mistakes, we introduce a step function $\theta(x)$ such that for $x \geq 0$, $\theta = 1$, otherwise $\theta = 0$. By this, the three-dimensional density of associated electron states can be expressed as

$$N_3^a(E - E_g) = \frac{1}{2\pi^2}\left(\frac{2m_r^*}{\hbar^2}\right)^{3/2} \sqrt{E - E_g}\, \theta(E - E_g) \tag{4.19}$$

Note that $N_3^a(E - E_g)$ is named as the three-dimensional (thus the subscript "3") density of associated (superscript "a") electron states. This is not the density of electron states in either the single conduction band or the valence band, rather, it is about both the electron state in the conduction band and the hole state in the valence band which are associated by the light-matter interaction, see (4.15), and E here is optical transition energy, and m_r^* is the reduced effective mass.

Finally, the rate of transitions of electrons from the conduction band to the valence band in an intrinsic or a lightly-doped semiconductor is

$$
\begin{aligned}
W(\omega) &= \frac{\pi e^2 |\boldsymbol{e}_s \cdot \boldsymbol{p}_{\mathrm{cv}}|^2}{m_0^2 \epsilon_0 n^2 \omega} \int \frac{\Gamma}{(E-\hbar\omega)^2+\Gamma^2} \exp\left(-\frac{E-E_{\mathrm{g}}}{k_B T}\right) N_3^{\mathrm{a}}(E-E_{\mathrm{g}}) \mathrm{d}E \\
&= \frac{\pi^2 e^2 |\boldsymbol{e}_s \cdot \boldsymbol{p}_{\mathrm{cv}}|^2}{m_0^2 \epsilon_0 n^2 \omega} \exp\left(-\frac{\hbar\omega-E_{\mathrm{g}}}{k_B T}\right) N_3^{\mathrm{a}}(\hbar\omega-E_{\mathrm{g}})\, \theta(\hbar\omega-E_{\mathrm{g}}) \\
&\propto \sqrt{\hbar\omega-E_{\mathrm{g}}}\, \exp\left(-\frac{\hbar\omega-E_{\mathrm{g}}}{k_B T}\right) \theta(\hbar\omega-E_{\mathrm{g}})
\end{aligned} \tag{4.20}
$$

In the above derivations, we use the following mathematical relationship

$$
\lim_{\Gamma\to 0} \frac{\Gamma}{(E-\hbar\omega)^2+\Gamma^2} = \pi\, \delta(E-\hbar\omega) \tag{4.21}
$$

In Sect. 2.3, we have learned that $W(\omega)$ represents the number of photons absorbed or generated from the optical transitions of electrons from the conduction band to the valence band, see (2.120). Therefore, the photoluminescence signal that is recorded by a photodetector is determined by $W(\omega)$ such

$$
S_{\mathrm{bulk}}(\omega) = A\sqrt{\hbar\omega-E_{\mathrm{g}}}\, \exp\left(-\frac{\hbar\omega-E_{\mathrm{g}}}{k_B T}\right) \theta(\hbar\omega-E_{\mathrm{g}}) \tag{4.22}
$$

where we use a parameter A that includes many factors such as the reduced effective mass m_{r}^* and optical dipole moment $\boldsymbol{p}_{\mathrm{cv}}$ which cannot be determined directly from the photoluminescence spectrum. We denote this spectrum of a bulk semiconductor by the subscript "bulk". Since it is directly related to the optical transition between the conduction band and valence band, it is also commonly known as inter-band photoluminescence spectrum.

Figure 4.3 shows the characteristic features of the photoluminescence spectrum of an intrinsic or lightly-doped bulk semiconductor. Here we see that the photoluminescence directly reflect the energy band structure of the semiconductor under investigation, namely, the energy bandgap E_{g}, the density of associated electron states as well as the occupation of the associated electron states.

One may be a bit cautious when matching a measured photoluminescence spectrum with Fig. 4.3. We can expect that photons are emitted uniformly in space in an optically excited bulk semiconductor when electrons and holes recombine radiatively. The emitted photons need to travel from where they are created to the semiconductor surface then refract into air before they reach the photodetector to be recorded. Since photons whose energies are larger than E_{g} will be absorbed by the semiconductor, we would expect that the measured photoluminescence strength at higher photon energies (such as these above 1.5 in Fig. 4.3) will be weaker than what is expected by (4.20) since the emitted high-energy photons will be partially absorbed during their propagation through the semiconductor before reaching the photodetector, while the photoluminescence strength close to E_{g} will be a bit stronger due to the extra optical excitation by the absorption of the emitted high-energy photons. The situation is depicted in Fig. 4.4.

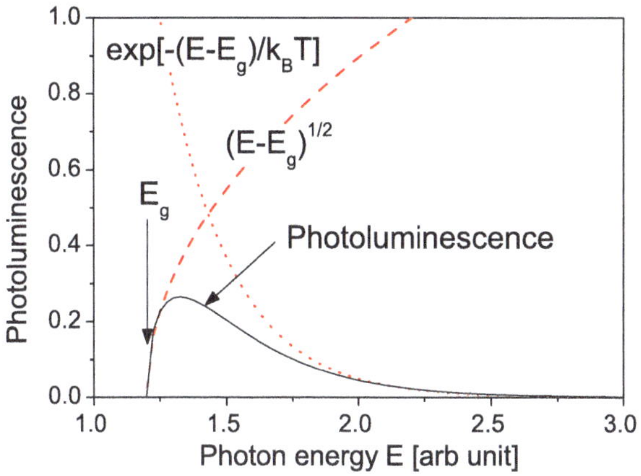

Fig. 4.3 Characteristic feature (black solid line) of the inter-band photoluminescence spectrum of an intrinsic or lightly-doped bulk semiconductor, where $\sqrt{E - E_g}$ (red dashed line) represents the density of associated electron states while $\exp[-(E - E_g)/k_B T]$ (red dotted line) is the occupation probability of the associated electron state at E

A typical experimental photoluminescence spectrum from a II-VI ternary CdZnTe semiconductor is shown as the black solid line in Fig. 4.5, which shows in general the same peak shape as the theoretical spectrum in Fig. 4.3.

As shown by (4.20), there should be no emitted photons whose energies are smaller than E_g, see the step function $\theta(\hbar\omega - E_g)$. The photoluminescence spectrum in Fig. 4.5 shows that in the region close to E_g, the change of the photoluminescence spectrum as a function of $\hbar\omega$ is not the step function. In theory, the step function is the result of the three-dimensional density of associated electron states which says that there is no electron states in the energy bandgap. In reality, the energy bandgap is not totally void of any electron states. There are impurity states, lattice defect states, exciton states etc. We can use the following exponentially decaying function to describe the density of these bandgap states [2]:

$$\frac{N_0}{E_0} \exp\left(-\frac{E_g - E}{E_0}\right) \theta(E_g - E) \tag{4.23}$$

where N_0 is the total number of bandgap states, E_0 represents the diffusion of these states in the energy bandgap. And $E < E_g$, i.e., the bandgap states.

Similar to the derivation of (4.22), we can find that the luminescence from the bandgap states

$$S_{\text{bandgap-states}}(\hbar\omega) = B \exp\left(-\frac{E_g - \hbar\omega}{E_0}\right) \theta(E_g - \hbar\omega) \tag{4.24}$$

where B contains information about the density of bandgap states and optical transitions between bandgap states.

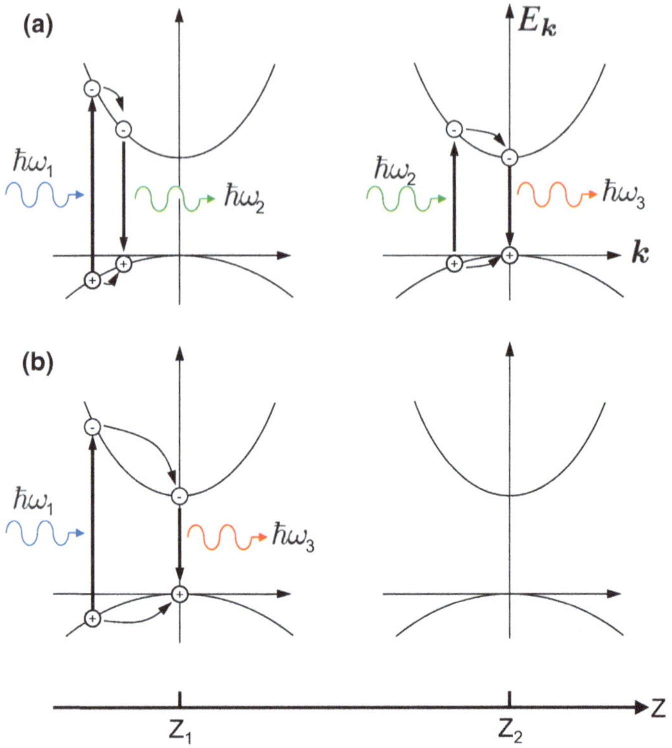

Fig. 4.4 Photoluminescence of a thick sample. A high-energy photon $\hbar\omega_1$ is normally needed to photoexcite the semiconductor while the energy of the emitted photon $\hbar\omega_2$ is low such that $\hbar\omega_1 > \hbar\omega_2$. **a** At z_1, the semiconductor is excited by $\hbar\omega_1$ then emits at $\hbar\omega_2$, which has a certain probability to be re-absorbed at z_2 to emit $\hbar\omega_3$. **b** The energy of $\hbar\omega_3$ emitted at z_1 is too low to excite the semiconductor at z_2

Fig. 4.5 A typical normalized (re-scaled so that the peak optical power becomes 1.0) photoluminescence spectrum of a CdZnTe bulk material. Black solid line: experimental spectrum, while the red dashed line is the fitting spectrum by (4.25) with the fitting E_g marked by the vertical arrow

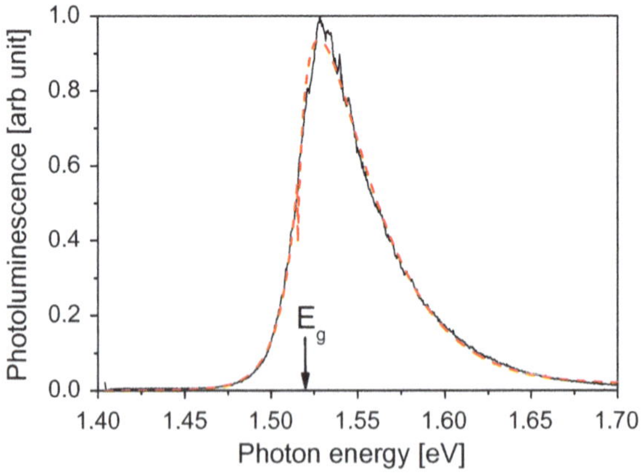

Finally, the total photoluminescence spectrum of an intrinsic or lightly-doped semiconductor can be expressed as

$$S(\hbar\omega) = S_{\text{bulk}}(\hbar\omega) + S_{\text{bandgap-states}}(\hbar\omega) \tag{4.25}$$

We use the above expression to fit the experimental photoluminescence spectrum in Fig. 4.5 with five independent fitting parameters, T, A, B, E_g, and E_0. T, the temperature of the electrons and holes, should be the same as the temperature of the environment where the sample locates. It may need modifications when a strong laser beam is used to excite the photoluminescence spectrum under which conditions the energy relaxations (process 2 in Fig. 4.1) will significantly increase the temperature in the region under the laser irradiation.

The fitted spectrum of the experimental data is shown in Fig. 4.5 as the red dashed line and E_g is marked by a vertical arrow, indicating that our microscopic picture about the photoluminescence spectrum is rather satisfactory.

Note that by the end of Sect. 2.4 we understand that that basic form of a single photon emission process is either Lorentzian or Gaussian, while the photoluminescence peaks described by (4.22), (4.24) and shown in Figs. 4.3 and 4.5 are much more complicated. They are composed of many Lorentzian peaks weighted by the occupation probabilities and the densities of the associated electron states.

The II-VI ternary $Cd_{1-x}Zn_xTe$ semiconductor is widely used as an epitaxial substrate which is lattice-structure matched with the most important infrared-photodetector material $Hg_yCd_{1-y}Te$. The mole fraction x of Zn in $Cd_{1-x}Zn_xTe$ is a critical parameter for the final device performance of $Hg_yCd_{1-y}Te$-based infrared photodetector. Any deviation of the Zn mole fraction in the $Cd_{1-x}Zn_xTe$ substrate from its device design value and non-uniformity of its surface distribution will induce significant lattice distortions and lattice mismatches with the $Hg_yCd_{1-y}Te$ epitaxial layer to be grown on the $Cd_{1-x}Zn_xTe$ substrate.

We know from this section that we can obtain the energy bandgap from measuring the inter-band (between the valence and conduction band) photoluminescence spectrum. And theoretically the energy bandgap $E_g(x)$ of $Cd_{1-x}Zn_xTe$ depends on the Zn mole fraction x [3]

$$E_g(x) = 1.494 + 0.606x + 0.139x^2 \text{ eV} \tag{4.26}$$

at room temperature. Notice the difference in the above equation to that in the reference, where the first term is now 1.494 eV. The reason is that the E_g in the above expression is derived from the physical model fitting from the measurements of Fig. 4.7 (see below), which reflects the cut-off energy of the band-to-band transition, while the commonly cited E_g (as in [3]) is taken as the peak position of the photoluminescence spectrum, and Fig. 4.3 shows that E_g is not the same as the energy of the photoluminescence peak.

A simple effective, also noninvasive, method to evaluate the Zn mole fraction in a $Cd_{1-x}Zn_xTe$ substrate is to measure its inter-band photoluminescence spectrum. In order to find the surface distribution of the Zn mole fraction, we measure the

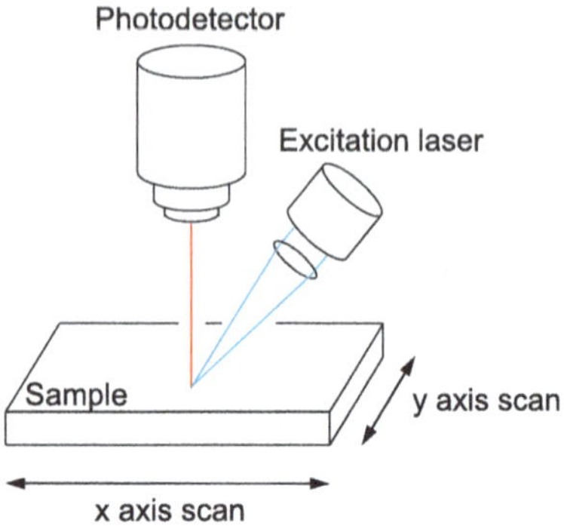

Fig. 4.6 Schematics of the microscopic photoluminescence measurement setup

photoluminescence spectrum as a function of the surface position. This position-dependent photoluminescence measurement is the so-called microscopic photoluminescence spectrum, also known as the μ-photoluminescence, or simply μ-PL.

The basic principle of the μ-photoluminescence is very simple: we impinge the surface on a sample by using a laser beam with a tiny laser spot having a diameter at the μm scale (thus the name μ-photoluminescence) to excite the photoluminescence signal from the tiny region under the laser spot. Moving the laser spot across the whole sample surface or the area of interest will give us the needed surface distribution of the Zn mole fraction. A schematic structure of the μ-photoluminescence measurement setup is shown in Fig. 4.6.

Using the spectral line of 632.8 nm from a He-Ne laser and a 100× microscopic objective to focus the laser beam into a spot with an approximate diameter of 1 μm, we make a μ-photoluminescence scan across the surface of one $Cd_{1-x}Zn_xTe$ epitaxial substrate. The μ-photoluminescence spectra are fitted by using (4.25) to obtain E_g. The resulting two-dimensional maps of E_g are shown in Fig. 4.7, where (a) is the E_g map for the $Cd_{1-x}Zn_xTe$ epitaxial substrate before bromine polishing and (b) is the E_g map after bromine polishing. The E_g map can be easily converted into the map of the Zn mole fraction by using (4.26).

Note that the scales of Fig. 4.7a, b are different, 1.509–1.518 eV (range 9 meV) versus 1.5115–1.5155 eV (range 4 meV). Moreover, the fluctuations of E_g in (b) are much lower than (a), indicating a significant improvement in the distribution of E_g and therefore the x of Zn atoms in the substrate by bromine polishing.

To assess the improvement quantitatively in another way, we do a further data processing that we count the number of pixels within a certain E_g range then present the count as a function of E_g in Fig. 4.8. The visualization about the improvement of the uniformity is even better.

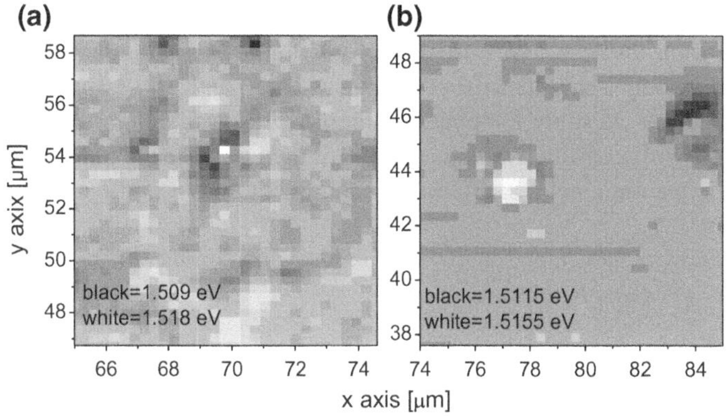

Fig. 4.7 Two-dimensional maps of E_g across the surface of a $Cd_{1-x}Zn_xTe$ epitaxial substrate before (**a**) and after (**b**) bromine polishing

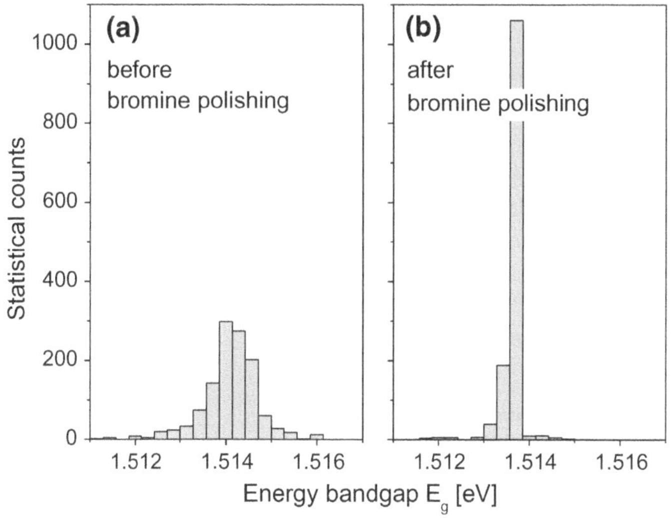

Fig. 4.8 Statistical counts of local energy bandgaps in the $Cd_{1-x}Zn_xTe$ substrate before (**a**) and after bromine polishing (**b**)

4.2 Optical Transitions in Low-Dimensional Structures

The fast development of the microelectronics, nanoelectronics, microphotonics, and nanophotonics relies completely on low-dimensional semiconductor structures. And photoluminescence is one of the best methods to characterize these low-dimensional structures. One obvious reason is that optical characterization does not require physical contacts with the sample under investigation, while electrodes are needed for electric characterizations, which become more and more difficult to be made on tiny low-dimensional structures. This also means that the optical characterization is non-destructive (non-invasive). We only take a look literally (!) at the sample (with a safe dose of the probing light intensity, of course, as not to optically damage the sample).

In this section we study the photoluminescence as a function of the dimensionality of the low-dimensional structure. We will then analyze the photoluminescence spectra of a quantum well, quantum wire and quantum dot in the coming sections, respectively.

As we know from Chap. 2, the wave function of a conduction-band electron in a bulk material is described by

$$\Psi_{ck}(\boldsymbol{r}) = \frac{1}{\sqrt{N}} e^{i\boldsymbol{k}\cdot\boldsymbol{r}} u_{ck}(\boldsymbol{r}), \quad u_{ck}(\boldsymbol{r}) = u_{ck}(\boldsymbol{r}+\boldsymbol{R}) \tag{4.27}$$

where $\boldsymbol{k}$ is the wave vector and $\boldsymbol{R}$ the lattice vector. The wave function of a valence-band hole is similar. The term of $1/\sqrt{N}$ in the above expression is due to the wave function normalization. Without it, the spatial integration of the wave function over the extension of the material which is denoted as Ω is

$$\int_{\Omega} \Psi_{ck}^{*}(\boldsymbol{r})\Psi_{ck}(\boldsymbol{r})\mathrm{d}\boldsymbol{r} = \int_{\Omega} u_{ck}^{*}(\boldsymbol{r})u_{ck}(\boldsymbol{r})\,\mathrm{d}\boldsymbol{r} \tag{4.28}$$

Here we have utilized (2.16). Because $u_{ck}(\boldsymbol{r})$ is a periodic function of unit cells in Ω, we replace $\boldsymbol{r}$ by the lattice vector $\boldsymbol{R}$ of the unit cell and the spatial coordinate $\boldsymbol{r}'$ inside the unit cell, i.e., $\boldsymbol{r} \equiv \boldsymbol{R} + \boldsymbol{r}'$,

$$\sum_{\boldsymbol{R}} \int_{\text{cell}} |u_{ck}(\boldsymbol{r}'+\boldsymbol{R})|^2 \mathrm{d}\boldsymbol{r}' = \sum_{\boldsymbol{R}} \int_{\text{cell}} |u_{ck}(\boldsymbol{r}')|^2 \mathrm{d}\boldsymbol{r}' \tag{4.29}$$

$\int_{\text{cell}}$ represents the integration over $\boldsymbol{r}'$ over the unit cell. Bloch function $u_{ck}(\boldsymbol{r})$ is commonly normalized within the unit cell

$$\int_{\text{cell}} |u_{ck}(\boldsymbol{r})|^2 \mathrm{d}\boldsymbol{r} = 1 \tag{4.30}$$

which indicates that by the SI unit (the Meter-Kilogram-Second MKS unit), the unit of $u_{ck}(\boldsymbol{r})$ is $[\mathrm{m}]^{-3/2}$.

Equation (4.28) now becomes

$$\int_{\Omega} \Psi_{ck}^{*}(\boldsymbol{r})\Psi_{ck}(\boldsymbol{r})\mathrm{d}\boldsymbol{r} = \int_{\Omega} u_{ck}^{*}(\boldsymbol{r})u_{ck}(\boldsymbol{r})\mathrm{d}\boldsymbol{r} = \sum_{\boldsymbol{R}} 1 = N \tag{4.31}$$

where N is the number of unit cells in Ω. And the normalized wave function of the conduction-band electron in the bulk material is

$$\Psi_{ck}(\boldsymbol{r}) = \frac{1}{\sqrt{N}} e^{i\boldsymbol{k}\cdot\boldsymbol{r}} u_{ck}(\boldsymbol{r}) \tag{4.32}$$

This is exactly (2.16) in Chap. 2, the most common expression of the Bloch theorem in solid-state textbooks.

Now we study the electron and hole states in low-dimensional structures. When we reduce the spatial extension along one direction of the three-dimensional Cartesian coordinate system down to tens of nm, or even a few nm, we obtain the so-called quantum well which is confined along the direction of the reduced extension (it is thus also called one-dimensionally confined quantum well, one may read "two-dimensional quantum well" in literature which means two-dimensionally extended quantum well in the context of a three-dimensional bulk material which means a three-dimensionally extended bulk material). Further reduction of the spatial extension in another Cartesian direction results in a two-dimensionally confined quantum wire. When the third Cartesian direction is further reduced, what we have is a three-dimensionally confined quantum dot.

In general, the size of the reduced extension, commonly tens of nm, maybe even a few nm, as mentioned before, is still much larger than the lattice constants of semiconductors (approximately only 0.5 nm). By the envelope function and the effective mass approximation we study in Sect. 2.1, the total wave function of an electron state can be expressed as

$$\Psi_c(\boldsymbol{r}) = \psi_c(\boldsymbol{r})\, u_c(\boldsymbol{r})$$

so that the Schrödinger equation of the effective mass approximation for the envelope function $\psi_c(\boldsymbol{r})$ is

$$\left[\frac{-\hbar^2\nabla^2}{2m_c^*} + V_c(\boldsymbol{r})\right]\psi_c(\boldsymbol{r}) = E_c\psi_c(\boldsymbol{r}) \tag{4.33}$$

i.e., (2.40), where $V_c(\boldsymbol{r})$ is the potential energy for the conduction-band electron.

It is necessary to note that the above equation is, in principle, valid both mathematically and physically only for bulk material, while its application in low-dimensional structures is anyway very successful. One major physical reason is the finite spatial diffusion of atoms so that mathematically sharp interfaces do not exist in the physical world, $V_c(\boldsymbol{r})$ and thereafter $\psi_c(\boldsymbol{r})$ are all slowly varying functions of space. This is especially true for most electronics and photonics applications of low-dimensional structures where only low-energy electron and hole states (thus small wave vectors) close to the conduction- and valence-bandedges are occupied. This is the principal condition for the validity of the envelope function theory of (4.33).

Consider the one-dimensionally confined quantum well in Fig. 2.6. It is confined along the z axis and extended in the xy plane. Because of the extension in the xy plane, the envelope wave function of the electron in this quantum well can be expressed as

$$\psi_i(z)\, e^{i(k_x x + k_y y)} \tag{4.34}$$

$\psi_i(z)$ is the envelope function along the z direction

$$\left[\frac{-\hbar^2}{2m^*}\frac{d^2}{dz^2} + V(z)\right]\psi_i(z) = E_i\psi_i(z) \tag{4.35}$$

where i is the quantum index along the z direction. Here we neglect the subscript "c" that denotes the conduction band (also "v" for the valence band) for simple mathematical expressions. The extension in the xy plane is now represented by wave numbers k_x and k_y. It is easy to see that the removal of $V(z)$ in the z direction results in

$$\psi_i(z) = \psi_{k_z}(z) = e^{ik_z z}$$

so that (4.32) is retrieved.

Envelope function $\psi_i(\boldsymbol{r})$ in (4.33) and (4.35) is commonly normalized in its extension space Ω, i.e.,

$$\int_\Omega |\psi_i(\boldsymbol{r})|^2 d\boldsymbol{r} = 1 \tag{4.36}$$

for which, the unit of $\psi_i(\boldsymbol{r})$ is $[\mathrm{m}]^{-3/2}$, the same as that of Bloch function $u_{\boldsymbol{k}}(\boldsymbol{r})$. $e^{i\boldsymbol{k}\cdot\boldsymbol{r}}$ in (2.16) and (4.32) is unit-less.

For the sake of a unified way to express envelope wave functions in low-dimensional structures, let us take a close look at the wave function normalization issue. We have the Bloch function $u(\boldsymbol{r})$ normalized in one unit cell, i.e., (4.30), and envelope wave function $\psi_i(\boldsymbol{r})$ normalized in Ω, i.e., (4.36). The normalization of the total wave function $\Psi_i(\boldsymbol{r}) = \psi_i(\boldsymbol{r})u(\boldsymbol{r})$ means

$$\int_\Omega |\Psi_i(\boldsymbol{r})|^2 d\boldsymbol{r} = \int_\Omega |\psi_i(\boldsymbol{r})|^2 |u(\boldsymbol{r})|^2 d\boldsymbol{r} = 1 \tag{4.37}$$

As mentioned before, $\psi_i(\boldsymbol{r})$ is a slowly-varying function of space so that replacing $\boldsymbol{r}$ by $\boldsymbol{R} + \boldsymbol{r}'$, the integration in the above equation becomes

$$\sum_{\boldsymbol{R}} |\psi_i(\boldsymbol{R})|^2 \int_{\text{cell}} |u(\boldsymbol{r}')|^2 d\boldsymbol{r}' = \sum_{\boldsymbol{R}} |\psi_i(\boldsymbol{R})|^2 \tag{4.38}$$

Here we utilize the normalization of $u(\boldsymbol{r})$ in the unit cell. Similar to Sect. 2.1, let the lattice constants along the three Cartesian directions x, y, and z be a_x, a_y, and a_z, and the numbers of unit cells N_x, N_y, and N_z, respectively. $\Omega_{\text{cell}} = a_x a_y a_z$ is the volume of the unit cell. With these,

$$\sum_{\boldsymbol{R}} |\psi_i(\boldsymbol{R})|^2 = \frac{1}{\Omega_{\text{cell}}} \sum_{\boldsymbol{R}} |\psi_i(\boldsymbol{R})|^2 \Omega_{\text{cell}} = \frac{1}{\Omega_{\text{cell}}} \int_\Omega |\psi_i(\boldsymbol{r})|^2 d\boldsymbol{r} = \frac{1}{\Omega_{\text{cell}}} \tag{4.39}$$

The last equality comes from the normalization of $\psi_i(\boldsymbol{r})$ in Ω, see (4.36).

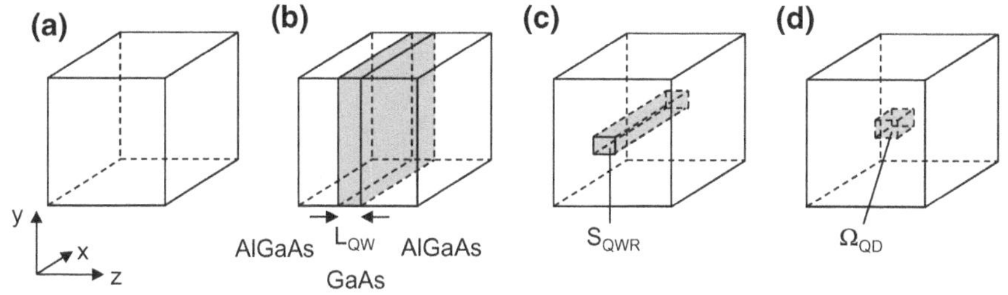

Fig. 4.9 Schematics of various low-dimensional structures. **a** Bulk material, **b** quantum well with well width L_{QW}, **c** quantum wire with cross section S_{QWR}, and **d** quantum dot with confinement volume Ω_{QD}

The above expression indicates that the normalized total wave function must be

$$\Psi_i(\boldsymbol{r}) = \sqrt{\Omega_{\text{cell}}}\, \psi_i(\boldsymbol{r})\, u(\boldsymbol{r}) \tag{4.40}$$

where

$$\int_{\text{cell}} |u(\boldsymbol{r})|^2 \mathrm{d}\boldsymbol{r} = 1, \quad \int_{\Omega} |\psi_i(\boldsymbol{r})|^2 \mathrm{d}\boldsymbol{r} = 1 \tag{4.41}$$

We now can go in details to study the electron states in low-dimensional structures schematically shown in Fig. 4.9.

Bulk material: To model a bulk material, we let $V_c(\boldsymbol{r}) = V_c$ in (4.33), where V_c is the conduction bandedge of the bulk material. The general solution of (4.33) is $\psi_k(\boldsymbol{r}) = C\, e^{i\boldsymbol{k}\cdot\boldsymbol{r}}$, where C is the normalization factor

$$\int_{\Omega} |\psi_k(\boldsymbol{r})|^2 \mathrm{d}\boldsymbol{r} = \int_{\Omega} |C|^2 \mathrm{d}\boldsymbol{r} = |C|^2 \int_{\Omega} \mathrm{d}\boldsymbol{r} = |C|^2 \Omega \tag{4.42}$$

so that $C = 1/\sqrt{\Omega}$ and the normalized envelope function is

$$\psi_k(\boldsymbol{r}) = \frac{1}{\sqrt{\Omega}}\, e^{i\boldsymbol{k}\cdot\boldsymbol{r}} \tag{4.43}$$

Insert this into (4.40) we retrieve (4.32).

As shown by the above equation, the quantum number of the electron states in bulk material is wave vector $\boldsymbol{k}$. Inserting the above equation into (4.33) with $V_c(\boldsymbol{r}) = V_c$ and the energy dispersion relationship is

$$E_k = V_c + \frac{\hbar^2 \boldsymbol{k}^2}{2m^*} \tag{4.44}$$

by which we easily obtain the density of electron states $N_3(E)$

$$\frac{2\mathrm{d}\boldsymbol{k}}{(2\pi)^3} = N_3(E - V_c)\,\mathrm{d}E$$
$$N_3(E - V_c) = \frac{1}{2\pi^2}\left(\frac{2m^*}{\hbar^2}\right)^{3/2}\sqrt{E - V_c}\,\theta(E - V_c) \tag{4.45}$$

Quantum well: For a quantum well which is confined along the z direction by an effective well width L_{QW}, $V_c(\boldsymbol{r}) = V_c(z)$ in (4.33) so that the envelope function can be expressed as $\psi(\boldsymbol{r}) = \psi(z)\psi(x, y)$. Insert it into (4.33)

$$\left[\frac{-\hbar^2\nabla^2}{2m^*} + V_c(z)\right]\psi(z)\psi(x, y) = E\psi(z)\psi(x, y) \tag{4.46}$$

we obtain Schrödinger equations for the z direction and the xy plane

$$\left[\frac{-\hbar^2}{2m^*}\frac{\mathrm{d}^2}{\mathrm{d}z^2} + V_c(z)\right]\psi_i(z) = E_i\psi_i(z)$$
$$\frac{-\hbar^2}{2m^*}\left(\frac{\mathrm{d}^2}{\mathrm{d}x^2} + \frac{\mathrm{d}^2}{\mathrm{d}y^2}\right)\psi_{k_xk_y}(x, y) = E_{k_xk_y}\psi_{k_xk_y}(x, y) \tag{4.47}$$

Here we have already denoted the two envelope functions by their quantum indices. $\psi_i(z)$ and E_i are eigen functions and values of the first Schrödinger equation, respectively, known as the energy sublevels in the quantum well, denoted by quantum index i. By (4.41),

$$\int_{L_{QW}} |\psi_i(z)|^2\mathrm{d}z = 1 \tag{4.48}$$

The general solution of the second Schrödinger equation is $\psi_{k_xk_y}(x, y) = Ce^{i(k_xx+k_yy)}$. And the energy of the total envelope function $\psi_i(z)\psi_{k_xk_y}(x, y)$ is

$$E_{ik_xk_y} = E_i + E_{k_xk_y}, \quad E_{k_xk_y} = \frac{\hbar^2(k_x^2 + k_y^2)}{2m^*} \tag{4.49}$$

The total envelope function is

$$C\,\psi_i(z)\,e^{i(k_xx+k_yy)} \tag{4.50}$$

where C is the normalization factor to be determined by

$$|C|^2\int_{L_{QW}} |\psi_i(z)|^2\mathrm{d}z\int_{xy}\mathrm{d}x\mathrm{d}y = |C|^2S_{QW} = 1 \tag{4.51}$$

where S_{QW} is the extension surface of the quantum well in the xy plane, resulting in $C = 1/\sqrt{S_{QW}}$ so that the normalized envelope function of an electron state in the quantum well is

$$\psi_{ik_xk_y}(\boldsymbol{r}) = \frac{1}{\sqrt{S_{\rm QW}}}\,\psi_i(z)\,e^{i(k_x x + k_y y)} \tag{4.52}$$

and the density of electron states corresponding to the energy dispersion relationship (4.49) is

$$\begin{aligned}\sum_i \frac{2\mathrm{d}k_x\mathrm{d}k_y}{(2\pi)^2} &= \sum_i N_2(E-E_i)\,\mathrm{d}E \\ N_2(E-E_i) &= \frac{m^*}{\pi\hbar^2}\,\theta(E-E_i)\end{aligned} \tag{4.53}$$

Quantum wire: Similarly, we can discuss the electron states in a quantum wire which is extended along the z direction with a length of $L_{\rm QWR}$ and confined in the xy plane with a confinement area $S_{\rm QWR}$

$$\begin{aligned}&\int_{S_{\rm QWR}} |\psi_i(x,y)|^2 \mathrm{d}x\mathrm{d}y = 1 \\ &\psi_{ik_z}(\boldsymbol{r}) = \frac{1}{\sqrt{L_{\rm QWR}}}\,\psi_i(x,y)\,e^{ik_z z} \\ &\sum_i \frac{2\mathrm{d}k_z}{2\pi} = \sum_i N_1(E-E_i)\,\mathrm{d}E \\ &N_1(E-E_i) = \frac{1}{2\pi}\sqrt{\frac{2m^*}{\hbar^2}}\,\frac{1}{\sqrt{E-E_i}}\,\theta(E-E_i)\end{aligned} \tag{4.54}$$

Quantum dot: For a quantum dot confined within $\Omega_{\rm QD}$,

$$\begin{aligned}&\int_{\Omega_{\rm QD}} |\psi_i(\boldsymbol{r})|^2 \mathrm{d}\boldsymbol{r} = 1 \\ &N_0(E) = \sum_i \delta(E-E_i)\end{aligned} \tag{4.55}$$

In other words, the density of electron states in a quantum dot is composed of a series of δ functions.

Let us visualize the energy states in a GaAs-based low-dimensional sample defined by its dimension $L_x \times L_y \times L_z$, which in principle is a quantum dot if L_x, L_y and L_z are in the order of nanometers. When $L_x = L_y = L_z = \infty$, the sample is a bulk material. Reducing significantly L_z results in the quantum well .... We can easily write a computer code about the electron states in the conduction band or valence-band holes. The effective mass of the conduction-band electron in GaAs is $m_{\rm c}^* = 0.067m_0$, where m_0 is the free electron mass. For simplicity, we assume that $V_{\rm c} = 0$ inside the GaAs region and $V_{\rm c} = \infty$ otherwise. The general solution for the Schrödinger equation is

$$\psi_{\ell_x\ell_y\ell_z}(x,y,z) = \frac{1}{\sqrt{L_xL_yL_z}} \sin\left(\frac{\ell_x\pi x}{L_x}\right) \sin\left(\frac{\ell_y\pi y}{L_y}\right) \sin\left(\frac{\ell_z\pi z}{L_z}\right)$$

$$E_{\ell_x\ell_y\ell_z} = \frac{\hbar^2\pi^2}{2m^*}\left(\frac{\ell_x^2}{L_x^2}+\frac{\ell_y^2}{L_y^2}+\frac{\ell_z^2}{L_z^2}\right) \tag{4.56}$$

for $0 \le x \le L_x$, $0 \le y \le L_y$, and $0 \le z \le L_z$. ℓ_x, ℓ_y, and ℓ_z are nonzero integers. They are the three quantum indices in the x, y, and z direction, respectively. In other words, i in (4.55) is now (ℓ_x, ℓ_y, ℓ_z). The numerical results of the densities of electron states are presented in Fig. 4.10. Do we retrieve the cases of quantum wire, quantum well, and bulk when we increase L_x, L_y, and L_z gradually?

Figure 4.10 shows clearly the model sample $L_x \times L_y \times L_z$ works fine to simulate various low-dimensional samples. Note that the δ function in (4.55) is replaced by

$$\delta(E - E_i) \rightarrow \frac{\Gamma}{(E - E_i)^2 + \Gamma^2} \tag{4.57}$$

where Γ is the relaxation energy.

Knowing the wave functions and their corresponding energies that describe the electron states in low-dimensional structures, we now discuss the optical transitions in these structures. Refer to (2.115) in Sect. 2.3, which is valid for structures of any confinement dimensions. We repeat some basic equations here. For a valence-band state $E_{\text{v}\boldsymbol{k}}$ with a total wave function

$$\Psi_{\text{v}\boldsymbol{k}}(\boldsymbol{r}) = \psi_{\text{v}\boldsymbol{k}}(\boldsymbol{r})u_{\text{v}}(\boldsymbol{r})$$

and a conduction-band state $E_{\text{c}\boldsymbol{q}}$ and

$$\Psi_{\text{c}\boldsymbol{q}}(\boldsymbol{r}) = \psi_{\text{c}\boldsymbol{q}}(\boldsymbol{r})u_{\text{c}}(\boldsymbol{r})$$

the spontaneous emission rate is

$$w_{\boldsymbol{q}\leftarrow\boldsymbol{k}}(\omega) = \frac{\pi\hbar^2e^2}{m_0^2\epsilon_0 n^2\omega\Omega}|\langle\Psi_{\text{c}\boldsymbol{q}}|\boldsymbol{e}_s\cdot\nabla|\Psi_{\text{v}\boldsymbol{k}}\rangle|^2\frac{\Gamma_{\boldsymbol{qk}}}{\left(E_{\text{c}\boldsymbol{q}} - E_{\text{v}\boldsymbol{k}} - \hbar\omega\right)^2 + \Gamma_{\boldsymbol{qk}}^2} \tag{4.58}$$

Here we summarize quantum indices as $\boldsymbol{k}$ and $\boldsymbol{q}$. For bulk material, $\boldsymbol{k} = (k_x, k_y, k_z)$; For a quantum well confined along the z direction, $\boldsymbol{k} = (k_x, k_y, i)$. We can write down similar expressions for quantum wire and quantum dot.

For simplicity, we assume that the energy band structures are spherical and parabolic, and $u(\boldsymbol{r})$ is $\boldsymbol{k}$-independent. These two assumptions are valid for the conduction bands of many commonly used semiconductor materials such as Si, GaAs, and GaN, while the valence bands are normally much complicated. However, as well as fortunately, for most electronics and photonics applications, the electron states that are occupied by the active electrons and holes in the devices are very close

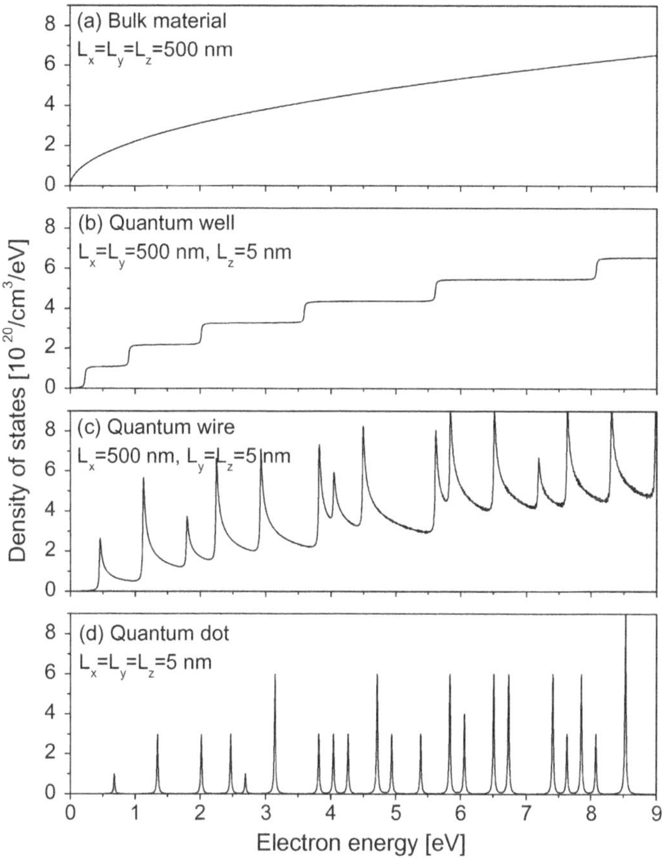

Fig. 4.10 Density of conduction-band electron states of a GaAs-based low-dimensional sample $L_x \times L_y \times L_z$. **a** Bulk material, **b** quantum well, **c** quantum wire, **d** quantum dot. $\Gamma = 10$ meV. The conduction bandedge in GaAs is set to be zero

to the conduction and valence bandedges for which when we make Taylor expansions for these energy bandedge states of very small $\boldsymbol{k}$ values, the above two assumptions are quite alright.

Similar to the discussions in Sect. 4.1 about photoluminescence from inter-band transitions, the photoluminescence spectrum is generally expressed as

$$g(\omega) = \sum_{\boldsymbol{qk}, E_{\mathrm{c}\boldsymbol{q}} > E_{\mathrm{v}\boldsymbol{k}}} \frac{\pi \hbar^2 e^2 \left|\left\langle \Psi_{\mathrm{c}\boldsymbol{q}} \left| \boldsymbol{e}_s \cdot \nabla \right| \Psi_{\mathrm{v}\boldsymbol{k}} \right\rangle\right|^2 \Gamma_{\boldsymbol{kq}} \left[f(E_{\mathrm{c}\boldsymbol{q}}) - f(E_{\mathrm{v}\boldsymbol{k}}) \right]}{m_0^2 \epsilon_0 c_0 n \omega \Omega \left[\left(E_{\mathrm{c}\boldsymbol{q}} - E_{\mathrm{v}\boldsymbol{k}} - \hbar\omega \right)^2 + \Gamma_{\boldsymbol{kq}}^2 \right]} \tag{4.59}$$

i.e., (2.123) in Sect. 2.3, where $f(E_{\mathrm{c}\boldsymbol{q}})$ and $f(E_{\mathrm{v}\boldsymbol{k}})$ are occupations of state $E_{\mathrm{c}\boldsymbol{q}}$ and $E_{\mathrm{v}\boldsymbol{k}}$, respectively.

Bulk material: We study the photoluminescence spectrum of a bulk material in Sect. 4.1 and the results are

$$\Psi_{k}(\boldsymbol{r}) = \sqrt{\Omega_{\text{cell}}}\,\psi_{k}(\boldsymbol{r})\,u_{k}(\boldsymbol{r}), \quad \psi_{k}(\boldsymbol{r}) = \frac{1}{\sqrt{\Omega}}\,e^{i\boldsymbol{k}\cdot\boldsymbol{r}} \tag{4.60}$$

$$\left\langle \Psi_{cq}(\boldsymbol{r})\,|\boldsymbol{e}_s\cdot\nabla|\,\Psi_{vk}(\boldsymbol{r})\right\rangle = \delta_{qk}\frac{\boldsymbol{e}_s\cdot\boldsymbol{p}_{\text{cv}}}{\hbar} \tag{4.61}$$

so that

$$\begin{aligned}
g_{\text{bulk}}(\omega) &= \sum_{qk}\frac{\pi\hbar^2e^2\left|\langle\Psi_{cq}\,|\boldsymbol{e}_s\cdot\nabla|\,\Psi_{vk}\rangle\right|^2\Gamma_{kq}\Big[f(E_{cq})-f(E_{vk})\Big]}{m_0^2\epsilon_0c_0n\omega\Omega\left[\left(E_{cq}-E_{vk}-\hbar\omega\right)^2+\Gamma_{kq}^2\right]}\\
&= \sum_{k}\frac{\pi\hbar^2e^2\,|\langle\Psi_{ck}\,|\boldsymbol{e}_s\cdot\nabla|\,\Psi_{vk}\rangle|^2\,\Gamma_{\text{cv}}\Big[f(E_{ck})-f(E_{vk})\Big]}{m_0^2\epsilon_0c_0n\omega\Omega\left[\left(E_{ck}-E_{vk}-\hbar\omega\right)^2+\Gamma_{\text{cv}}^2\right]}\\
&= \frac{\pi e^2\Gamma_{\text{cv}}|\boldsymbol{e}_s\cdot\boldsymbol{p}_{\text{cv}}|^2}{m_0^2\epsilon_0c_0n\omega\Omega}\int\frac{f(E_{ck})-f(E_{vk})}{\left(E_{ck}-E_{vk}-\hbar\omega\right)^2+\Gamma_{\text{cv}}^2}\,\frac{2\mathrm{d}\boldsymbol{k}}{(2\pi)^3/\Omega}\\
&= \frac{\pi e^2\Gamma_{\text{cv}}|\boldsymbol{e}_s\cdot\boldsymbol{p}_{\text{cv}}|^2}{m_0^2\epsilon_0c_0n\omega}\int\frac{f(E_{ck})-f(E_{vk})}{\left(E_{ck}-E_{vk}-\hbar\omega\right)^2+\Gamma_{\text{cv}}^2}\,\frac{2\mathrm{d}\boldsymbol{k}}{(2\pi)^3}
\end{aligned} \tag{4.62}$$

Quantum well: As discussed before, the quantum well is confined along the z direction with a confinement width L_{QW}. It is extended in the xy plane with an extension surface area S_{QW}. The wave functions of the electron and hole states are

$$\Psi_{ik}(\boldsymbol{r}) = \sqrt{\Omega_{\text{cell}}}\,\psi_{ik}(\boldsymbol{r})\,u(\boldsymbol{r}), \quad \psi_{ik}(\boldsymbol{r}) = \frac{1}{\sqrt{S_{\text{QW}}}}\,\psi_i(z)\,e^{i(k_xx+k_yy)} \tag{4.63}$$

so that the optical transition matrix element is

$$\begin{aligned}
&\left\langle\Psi_{jq}\,|\boldsymbol{e}_s\cdot\nabla|\,\Psi_{ik}\right\rangle\\
&= \frac{\Omega_{\text{cell}}}{S_{\text{QW}}}\,\boldsymbol{e}_s\cdot\int_{\Omega}e^{i\left[(k_x-q_x)x+(k_y-q_y)y\right]}\psi_j^*(z)\psi_i(z)u_{\text{c}}^*(\boldsymbol{r})\nabla u_{\text{v}}(\boldsymbol{r})\mathrm{d}\boldsymbol{r}\\
&-\frac{i(e_xk_x+e_yk_y)\Omega_{\text{cell}}}{S_{\text{QW}}}\int_{\Omega}e^{i\left[(k_x-q_x)x+(k_y-q_y)y\right]}\psi_j^*(z)\psi_i(z)u_{\text{c}}^*(\boldsymbol{r})u_{\text{v}}(\boldsymbol{r})\mathrm{d}\boldsymbol{r}\\
&+\frac{e_z\Omega_{\text{cell}}}{S_{\text{QW}}}\int_{\Omega}e^{i\left[(k_x-q_x)x+(k_y-q_y)y\right]}\psi_j^*(z)\frac{\partial\psi_i(z)}{\partial z}u_{\text{c}}^*(\boldsymbol{r})u_{\text{v}}(\boldsymbol{r})\mathrm{d}\boldsymbol{r}
\end{aligned} \tag{4.64}$$

where $\boldsymbol{e}_s = (e_x, e_y, e_z)$ is the propagation direction of the electromagnetic field. The second term on the right side of the equals sign in above equation is zero due to wave function's orthogonality. The third term can be studied by replacing $\boldsymbol{r}$ with $\boldsymbol{R}+\boldsymbol{r}'$ since the envelope function is a slowly-varying function of space. We have performed similar mathematical manipulations before, such as (4.28), which results in a quantity

$$\int_{\text{cell}} u_c^*(\boldsymbol{r}) u_v(\boldsymbol{r})\, \mathrm{d}\boldsymbol{r}$$

and it is zero since u_c and u_v are eigen functions so are orthogonal. Only the first term on the right side of the equals sign remains, which is called the inter-band transition (between the conduction band and the valence band).

There is a novel phenomenon in low-dimensional structures which does not exist in the bulk material. Let us replace the valence-band state by a new conduction-band state, i.e., replace u_v in (4.64) by u_c. We now see that the first term on the right side of the equals sign in (4.64) becomes zero since the parities of u_c and ∇u_c are opposite so that their spatial integration is zero. The second term is also zero because ψ_i and ψ_j are orthogonal. Only the third term remains, which is known as the intra-band transition (within the same conduction band). The discussion can be extended to the intra-band transition of the valence band. We will discuss it more extensively in Chap. 6 about the photocurrent spectrum of a quantum well infrared photodetector.

Back to the inter-band transition.

$$\begin{aligned}
&\frac{\Omega_{\text{cell}}}{S_{\text{QW}}} \int_{\Omega} e^{i\left[(k_x-q_x)x+(k_y-q_y)y\right]} \psi_j^*(z)\psi_i(z) u_c^*(\boldsymbol{r}) \nabla u_v(\boldsymbol{r}) \mathrm{d}\boldsymbol{r} \\
&= \frac{\Omega_{\text{cell}}}{S_{\text{QW}}} \sum_{\boldsymbol{R}} e^{i\left[(k_x-q_x)R_x+(k_y-q_y)R_y\right]} \psi_j^*(R_z)\psi_i(R_z) \int_{\text{cell}} u_c^*(\boldsymbol{r}) \nabla u_v(\boldsymbol{r}) \mathrm{d}\boldsymbol{r} \\
&= \frac{\Omega_{\text{cell}}}{S_{\text{QW}}} N_{\text{QW}} \delta_{k_x q_x} \delta_{k_y q_y} \frac{\boldsymbol{p}_{\text{cv}}}{\hbar} \frac{1}{a_z} \sum_{R_z} \psi_j^*(R_z)\psi_i(R_z) a_z \\
&= \frac{\Omega_{\text{cell}}}{S_{\text{QW}}} N_{\text{QW}} \delta_{k_x q_x} \delta_{k_y q_y} \frac{\boldsymbol{p}_{\text{cv}}}{\hbar} \frac{1}{a_z} \int_{L_{\text{QW}}} \psi_j^*(R_z)\psi_i(R_z) \mathrm{d}z \\
&= \delta_{k_x q_x} \delta_{k_y q_y} \frac{\boldsymbol{p}_{\text{cv}}}{\hbar} \int_{L_{\text{QW}}} \psi_j^*(z)\psi_i(z) \mathrm{d}z \equiv \delta_{k_x q_x} \delta_{k_y q_y} \frac{\boldsymbol{p}_{\text{cv}}}{\hbar} \langle \psi_j(z) | \psi_i(z) \rangle
\end{aligned} \tag{4.65}$$

where a_z is the lattice constant in the z direction, N_{QW} is the number of unit cells contained in surface area S_{QW}. $\delta_{k_x q_x} = 1$ when $k_x = q_x$, it is zero otherwise.

The inter-band photoluminescence spectrum of the quantum well is thus

$$\begin{aligned}
g_{\text{QW}}(\omega) &= \sum_{jq_xq_y, ik_xk_y} \frac{\pi \hbar^2 e^2 \left|\langle \Psi_{jq_xq_y}(\boldsymbol{r}) | \boldsymbol{e}_s \cdot \nabla | \Psi_{ik_xk_y}(\boldsymbol{r}) \rangle\right|^2 \Gamma_{\text{cv}} \left[f(E_{jq_xq_y}) - f(E_{ik_xk_y})\right]}{m_0^2 \epsilon_0 c_0 n \omega \Omega \left[\left(E_{jq_xq_y} - E_{ik_xk_y} - \hbar\omega\right)^2 + \Gamma_{\text{cv}}^2\right]} \\
&= \frac{\pi e^2 \Gamma_{\text{cv}} |\boldsymbol{e}_s \cdot \boldsymbol{p}_{\text{cv}}|^2}{m_0^2 \epsilon_0 c_0 n \omega \Omega} \sum_{ji} \int\int \frac{\left|\langle \psi_j(z) | \psi_i(z) \rangle\right|^2 \left[f(E_{jk_xk_y}) - f(E_{ik_xk_y})\right]}{\left[\left(E_{jk_xk_y} - E_{ik_xk_y} - \hbar\omega\right)^2 + \Gamma_{\text{cv}}^2\right]} \frac{2\mathrm{d}k_x \mathrm{d}k_y}{(2\pi)^2 / S_{\text{QW}}}
\end{aligned}$$

$$= \frac{\pi e^2 \Gamma_{\text{cv}} |\boldsymbol{e}_s \cdot \boldsymbol{p}_{\text{cv}}|^2}{m_0^2 \epsilon_0 c_0 n \omega L_{\text{QW}}} \sum_{ji} \int\int \frac{|\langle \psi_j(z)|\psi_i(z)\rangle|^2 \left[f(E_{jk_xk_y}) - f(E_{ik_xk_y}) \right]}{\left[\left(E_{jk_xk_y} - E_{ik_xk_y} - \hbar\omega \right)^2 + \Gamma_{\text{cv}}^2 \right]} \frac{2\text{d}k_x \text{d}k_y}{(2\pi)^2} \tag{4.66}$$

where

$$\frac{2\text{d}k_x \text{d}k_y}{(2\pi)^2 / S_{\text{QW}}} = \frac{2 S_{\text{QW}} \text{d}k_x \text{d}k_y}{(2\pi)^2}$$

is known as the sheet density of electron states in the xy plane, $\Omega = S_{\text{QW}} L_{\text{QW}}$.

Quantum wire: We do a similar study about quantum wire

$$\Psi_{ik}(\boldsymbol{r}) = \sqrt{\Omega_{\text{cell}}}\, \psi_{ik}(\boldsymbol{r})\, u_{\boldsymbol{k}}(\boldsymbol{r}), \quad \psi_{ik}(\boldsymbol{r}) = \frac{1}{\sqrt{L_{\text{QWR}}}} \psi_i(x, y)\, e^{ikz} \tag{4.67}$$

$$\begin{aligned}
&\left\langle \Psi_{jq}(\boldsymbol{r}) \left| \boldsymbol{e}_s \cdot \nabla \right| \Psi_{ik}(\boldsymbol{r}) \right\rangle \\
&= \frac{\Omega_{\text{cell}}}{L_{\text{QWR}}} \boldsymbol{e}_s \cdot \int_\Omega e^{i(k-q)z} \psi_j^*(x, y) \psi_i(x, y) u_{\text{c}}^*(\boldsymbol{r}) \nabla u_{\text{v}}(\boldsymbol{r}) \text{d}\boldsymbol{r} \\
&\quad + \frac{i e_z k \Omega_{\text{cell}}}{L_{\text{QWR}}} \int_\Omega e^{i(k-q)z} \psi_j^*(x, y) \psi_i(x, y) u_{\text{c}}^*(\boldsymbol{r}) u_{\text{v}}(\boldsymbol{r}) \text{d}\boldsymbol{r} \\
&\quad + \frac{\Omega_{\text{cell}}}{L_{\text{QWR}}} \int_\Omega e^{i(k-q)z} \psi_j^*(x, y) \left[e_x \frac{\partial \psi_i(x, y)}{\partial x} + e_y \frac{\partial \psi_i(x, y)}{\partial y} \right] u_{\text{c}}^*(\boldsymbol{r}) u_{\text{v}}(\boldsymbol{r}) \text{d}\boldsymbol{r}
\end{aligned} \tag{4.68}$$

Again focus on the inter-band transition,

$$\begin{aligned}
&\frac{\Omega_{\text{cell}}}{L_{\text{QWR}}} \boldsymbol{e}_s \cdot \int_\Omega e^{i(k-q)z} \psi_j^*(x, y) \psi_i(x, y) u_{\text{c}}^*(\boldsymbol{r}) \nabla u_{\text{v}}(\boldsymbol{r}) \text{d}\boldsymbol{r} \\
&= \frac{\Omega_{\text{cell}}}{L_{\text{QWR}}} \boldsymbol{e}_s \cdot \sum_{\boldsymbol{R}} e^{i(k-q)R_z} \psi_j^*(R_x, R_y) \psi_i(R_x, R_y) \int_{\text{cell}} u_{\text{c}}^*(\boldsymbol{r}) \nabla u_{\text{v}}(\boldsymbol{r}) \text{d}\boldsymbol{r} \\
&= \frac{\Omega_{\text{cell}}}{L_{\text{QWR}}} \boldsymbol{e}_s \cdot N_{\text{QWR}} \delta_{kq} \frac{\boldsymbol{p}_{\text{cv}}}{\hbar} \frac{1}{a_x a_y} \sum_{R_x R_y} \psi_j^*(R_x, R_y) \psi_i(R_x, R_y) a_x a_y \\
&= \frac{\Omega_{\text{cell}}}{L_{\text{QWR}}} \boldsymbol{e}_s \cdot N_{\text{QWR}} \delta_{kq} \frac{\boldsymbol{p}_{\text{cv}}}{\hbar} \frac{1}{a_x a_y} \int_{S_{\text{QWR}}} \psi_j^*(x, y) \psi_i(x, y) \text{d}x\text{d}y \\
&= \delta_{kq} \frac{\boldsymbol{e}_s \cdot \boldsymbol{p}_{\text{cv}}}{\hbar} \int_{S_{\text{QWR}}} \psi_j^*(x, y) \psi_i(x, y) \text{d}x\text{d}y \equiv \delta_{kq} \frac{\boldsymbol{e}_s \cdot \boldsymbol{p}_{\text{cv}}}{\hbar} \langle \psi_j(x, y) | \psi_i(x, y) \rangle
\end{aligned} \tag{4.69}$$

$\Omega_{\text{cell}} = a_x a_y a_z$, and $a_z N_{\text{QWR}} = L_{\text{QWR}}$.

$$g_{\rm QWR}(\omega) = \sum_{jq,ik} \frac{\pi\hbar^2 e^2 \left|\langle \Psi_{jq}(\boldsymbol{r})|\boldsymbol{e}_s\cdot\nabla|\Psi_{ik}(\boldsymbol{r})\rangle\right|^2 \Gamma_{\rm cv}\left[f(E_{jq})-f(E_{ik})\right]}{m_0^2\epsilon_0 c_0 n\omega\Omega\left[\left(E_{jq}-E_{ik}-\hbar\omega\right)^2+\Gamma_{\rm cv}^2\right]}$$
$$= \frac{\pi e^2\Gamma_{\rm cv}|\boldsymbol{e}_s\cdot\boldsymbol{p}_{\rm cv}|^2}{m_0^2\epsilon_0 c_0 n\omega\Omega}\sum_{ji}\int \frac{\left|\langle\psi_j(x,y)|\psi_i(x,y)\rangle\right|^2\left[f(E_{jk})-f(E_{ik})\right]}{\left(E_{jk}-E_{ik}-\hbar\omega\right)^2+\Gamma_{\rm cv}^2}\frac{2{\rm d}k}{2\pi/L_{\rm QWR}}$$
$$= \frac{\pi e^2\Gamma_{\rm cv}|\boldsymbol{e}_s\cdot\boldsymbol{p}_{\rm cv}|^2}{m_0^2\epsilon_0 c_0 n\omega S_{\rm QWR}}\sum_{ji}\int \frac{\left|\langle\psi_j(x,y)|\psi_i(x,y)\rangle\right|^2\left[f(E_{jk})-f(E_{ik})\right]}{\left(E_{jk}-E_{ik}-\hbar\omega\right)^2+\Gamma_{\rm cv}^2}\frac{2{\rm d}k}{2\pi} \tag{4.70}$$

Quantum dot: The inter-band transition in the quantum dot is described by

$$\Psi_i(\boldsymbol{r}) = \sqrt{\Omega_{\rm cell}}\,\psi_i(\boldsymbol{r})\,u_{\boldsymbol{k}}(\boldsymbol{r}),\quad \int_{\Omega_{\rm QD}}|\psi_i(\boldsymbol{r})|^2{\rm d}\boldsymbol{r}=1 \tag{4.71}$$

$$\begin{aligned}
&\left\langle\Psi_j(\boldsymbol{r})\left|\boldsymbol{e}_s\cdot\nabla\right|\Psi_i(\boldsymbol{r})\right\rangle\\
&= \Omega_{\rm cell}\,\boldsymbol{e}_s\cdot\int_\Omega\left[\psi_j^*(\boldsymbol{r})\psi_i(\boldsymbol{r})u_{\rm c}^*(\boldsymbol{r})\nabla u_{\rm v}(\boldsymbol{r})+u_{\rm c}^*(\boldsymbol{r})u_{\rm v}(\boldsymbol{r})\psi_j^*(\boldsymbol{r})\nabla\psi_i(\boldsymbol{r})\right]{\rm d}\boldsymbol{r}\\
&= \Omega_{\rm cell}\,\boldsymbol{e}_s\cdot\int_\Omega\psi_j^*(\boldsymbol{r})\psi_i(\boldsymbol{r})u_{\rm c}^*(\boldsymbol{r})\nabla u_{\rm v}(\boldsymbol{r}){\rm d}\boldsymbol{r}\\
&= \Omega_{\rm cell}\,\boldsymbol{e}_s\cdot\sum_{\boldsymbol{R}}\psi_j^*(\boldsymbol{R})\psi_i(\boldsymbol{R})\int_{\rm cell}u_{\rm c}^*(\boldsymbol{r})\nabla u_{\rm v}(\boldsymbol{r}){\rm d}\boldsymbol{r}\\
&= \frac{\boldsymbol{e}_s\cdot\boldsymbol{p}_{\rm cv}}{\hbar}\sum_{\boldsymbol{R}}\psi_j^*(\boldsymbol{R})\psi_i(\boldsymbol{R})\Omega_{\rm cell}\\
&= \frac{\boldsymbol{e}_s\cdot\boldsymbol{p}_{\rm cv}}{\hbar}\int\psi_j^*(\boldsymbol{r})\psi_i(\boldsymbol{r}){\rm d}\boldsymbol{r}\equiv\frac{\boldsymbol{e}_s\cdot\boldsymbol{p}_{\rm cv}}{\hbar}\langle\psi_j(\boldsymbol{r})|\psi_i(\boldsymbol{r})\rangle
\end{aligned} \tag{4.72}$$

$$g_{\rm QD}(\omega) = \sum_{ji}\frac{\pi\hbar^2e^2\left|\langle\Psi_j(\boldsymbol{r})|\boldsymbol{e}_s\cdot\nabla|\Psi_i(\boldsymbol{r})\rangle\right|^2\Gamma_{\rm cv}\left[f(E_j)-f(E_i)\right]}{m_0^2\epsilon_0c_0n\omega\Omega_{\rm QD}\left[\left(E_j-E_i-\hbar\omega\right)^2+\Gamma_{\rm cv}^2\right]}$$
$$= \frac{\pi e^2\Gamma_{\rm cv}|\boldsymbol{e}_s\cdot\boldsymbol{p}_{\rm cv}|^2}{m_0^2\epsilon_0c_0n\omega\Omega_{\rm QD}}\sum_{ji}\frac{\left|\langle\psi_j(\boldsymbol{r})|\psi_i(\boldsymbol{r})\rangle\right|^2\left[f(E_j)-f(E_i)\right]}{\left(E_j-E_i-\hbar\omega\right)^2+\Gamma_{\rm cv}^2} \tag{4.73}$$

Here we include energy relaxation processes so that the δ function is replaced numerically by (4.57).

To understand what we have just learned about the inter-band photoluminescence spectrum as a function of the dimensionality of the sample, let us do a few illustrative

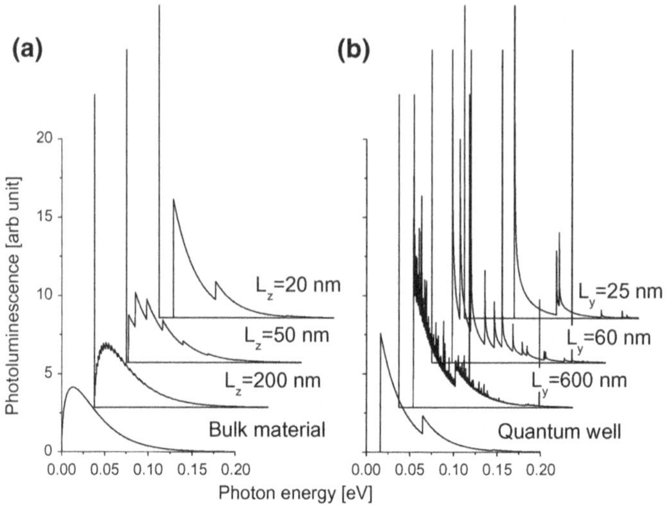

Fig. 4.11 The relationship between the photoluminescence spectrum and the dimensionality of the material. **a** A bulk material evolves into a one-dimensionally confined quantum well. **b** Quantum well evolves into a two-dimensionally confined quantum wire. $L_z = 20$ nm

numerical calculations. Assume a GaAs-based bulk material, for which the electron effective mass and hole effective mass are $m_c^* = 0.067m_0$, and $m_v^* = 0.45m_0$, respectively, where m_0 is the free electron mass. Figure 4.11a presents the evolution of the photoluminescence spectrum when the z dimension of a bulk GaAs is reduced from initially infinite (bulk, which is exactly the same as the one in Fig. 4.3) to $L_z = 200$, 50 and 20 nm, while in Fig. 4.11b the y dimension of the GaAs quantum well ($L_z = 20$ nm, i.e., the last spectrum in Fig. 4.11a) is being reduced ($L_y = 600$, 60 and 25 nm). What we observe here is basically the evolution of the density of electron states following the change of the dimensionality of the sample, see Fig. 4.10, plus of course the occupation of electron states.

A very important result to be closely observed is that the photoluminescence spectrum of a three-dimensionally confined quantum dot consists of a series of individual Lorentzian peaks, while the peaks the photoluminescence spectra of the quantum wire, quantum well and bulk material are not Lorentzian, they are neither Gaussian.

4.3 Photoluminescence of Quantum Well

The potential profiles for conduction-band electrons and valence-band holes in a GaAs/Al_xGa_{1-x}As quantum well are shown schematically in Fig. 2.6 in Sect. 2.1. The solutions of (4.35) depend jointly on the widths of the quantum well and barrier as well as the barrier height. In practical material growth and device fabrication, the

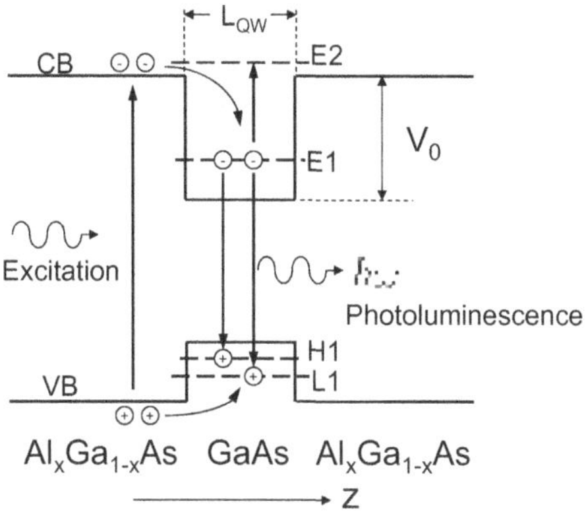

Fig. 4.12 Energy band structure, photoexcitation, carrier transport, and photoluminescence of GaAs/$Al_xGa_{1-x}As$ quantum well infrared photodetector (QWIP)

precise control of the quantum well and barrier widths and the barrier height is very difficult that there are many un-controllable fluctuations in the epitaxial growth as well as to maintain the epitaxial growth conditions for a very long time. In Sect. 3.5, we used the transmission spectroscopy to assess the layer thicknesses and Al mole fractions of four multiple GaAs/$Al_xGa_{1-x}As$ quantum well samples and observed that the deviation of these parameters from their nominal values was quite large. Let us study these quantum well samples using the photoluminescence spectroscopy.

Refer to the energy band diagram of the GaAs/$Al_xGa_{1-x}As$ quantum well in Fig. 4.12, we apply a relatively high-energy excitation light beam to excite electrons from the valence band to the conduction band of the $Al_xGa_{1-x}As$ barrier. Photoexcitation also occurs in the GaAs quantum well. As briefly introduced in Sect. 3.5, the GaAs quantum well is about 5 nm while the $Al_xGa_{1-x}As$ barrier is 60 nm. If the excitation light beam distributes uniformly along the z axis, we expect a similar excitation of electrons in the GaAs region. Since the thickness of the GaAs well is less than one tenth of that of the $Al_xGa_{1-x}As$ barrier, the photoexcitation in the $Al_xGa_{1-x}As$ barrier dominates which is displayed in Fig. 4.12.

The photogenerated electrons and holes will diffuse to the confined energy states in the GaAs quantum well then radiatively recombine to emit photons, i.e., photoluminescence. The photon energy of the photoluminescence peak is $E_{E1} - E_{H1}$, where E_{E1} and E_{H1} denote the ground electron state in the conduction band and the ground heavy hole state in the valence band in the GaAs quantum well. There is also the ground light hole state E_{L1} which is below E_{H1} because of its smaller effective mass.

Interestingly enough, the research and technical development of the quantum well infrared photodetector (QWIP) device has also been accompanied by the demonstration of the basic quantum mechanical processes such as the energy band structure of a one-dimensional square wells (Kronig–Penney model) and the tunneling of an electron through an energy barrier, which are the key models to understand the modern physics.

In order to measure the photoluminescence spectrum of the GaAs multiple quantum wells, the QWIP sample is first to be cleaved. We have to be destructive this

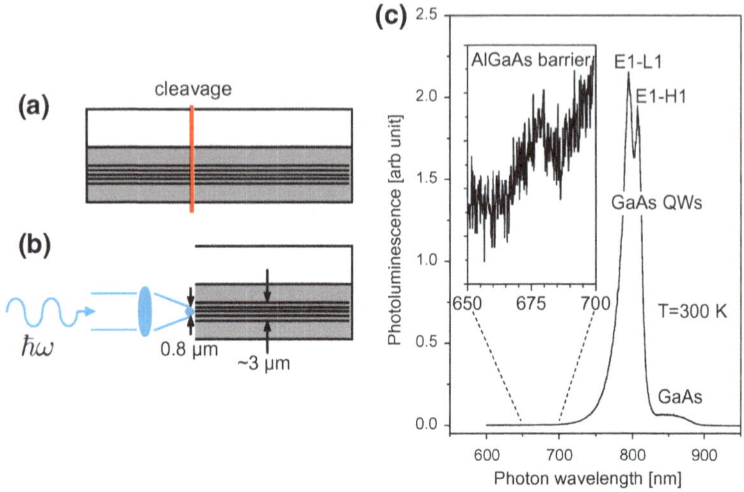

Fig. 4.13 **a** QWIP sample cleavage, **b** photoluminescence measurement setup, **c** one typical room-temperature photoluminescence spectrum of a GaAs/$Al_xGa_{1-x}As$ QWIP. Inset shows the photoluminescence signal at about 680 nm originated from $Al_xGa_{1-x}As$ barriers

time in order to reach the quantum wells cladded in between the substrate and the capping layer. An excitation laser beam is then focused to a light spot with a diameter of ca 0.8 μm on the cleavage plane. Note that for a common QWIP sample, there are 50 GaAs quantum wells (each ca 5 nm thick) sandwiched between $Al_xGa_{1-x}As$ barriers (thickness 60 nm) so that the total thickness of the multiple quantum wells is about $50 \times 0.065 = 3.25$ μm, which is much larger than the size of the laser spot, see Fig. 4.13a, b. Figure 4.13c presents one typical photoluminescence spectrum of a QWIP sample. The main split peak at 800 nm comes from the multiple GaAs quantum wells, and the inset shows a very weak 680 nm peak originated from the $Al_xGa_{1-x}As$ barriers. In addition, we also observe a shoulder on the long-wavelength of the QW peak corresponding to the energy bandgap of GaAs (1.43 eV = 867 nm at room temperature). The appearance of the GaAs bandgap emission is expected since the whole QWIP structure is epitaxially grown on a GaAs substrate, and there are many photon-scattering and carrier-diffusion processes that all contribute to the GaAs substrate's photoluminescence signal.

The splitting of the main peak of ca 20 meV between E1-L1 and E1-H1 in Fig. 4.13, is due to the energy difference between the heavy-hole and light-hole ground states in the valence band. Fitting the photoluminescence signals from GaAs quantum wells and $Al_xGa_{1-x}As$ barriers will give us transition energies $E_{E1} - E_{H1}$ and E_g(AlGaAs). We then obtain the Al mole fraction x and the GaAs quantum well width. The results of 8 QWIP samples are listed in Table 4.1, in comparison with the values of designs. We observe significant deviations which are up to 0.5 nm (note that the lattice constant of GaAs at room temperature is about 0.56 nm [1], this deviation of ca 0.5 nm is thus about one atomic layer) in L_{QW} and 1–2% in Al mole fraction x.

Table 4.1 Comparison between optically characterized structure parameters and their design values. The units of E_g and E_{11H} are all eV

Sample	E_g(AlGaAs)	E_{11H}	Al mole fraction x		Quantum well width L_{QW} (nm)	
			Design	Optical	Design	Optical
QWIP-1	1.817	1.533	0.310	0.315	5.0	4.7
QWIP-2	1.820	1.531	0.310	0.318	5.0	4.8
QWIP-3	1.827	1.532	0.300	0.323	5.0	4.5
QWIP-4	1.809	1.518	0.290	0.309	5.0	5.0
QWIP-5	1.810	1.535	0.300	0.310	5.0	4.5
QWIP-6	1.794	1.533	0.320	0.297	5.0	4.8
QWIP-7	1.828	1.542	0.300	0.324	5.0	4.5
QWIP-8	1.782	1.521	0.300	0.287	5.0	5.0

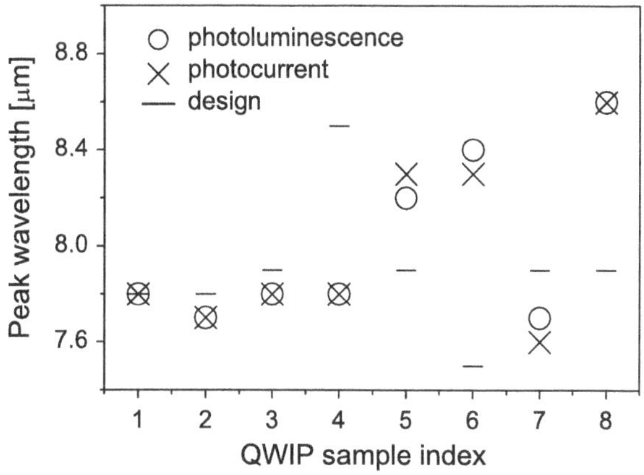

Fig. 4.14 Wavelengths of the photoluminescence peaks, photocurrent peaks, and designed response peaks of 8 QWIP structures/devices

We may speculate that the peak wavelength from the photoluminescence measurement differs from the photocurrent measurement result (to be studied in Chap. 6) since the photocurrent is obtained when the device is biased. Figure 4.14 shows the peak wavelengths of design, photoluminescence, and photocurrent measurements, indicating that the photoluminescence and photocurrent measurement data agree with each other quite well (they differ however quite much from the designs). We therefore conclude that photoluminescence measurement does provide an effective and precise way to characterize the response wavelength of the QWIP device before we actually process the device structure.

There are a few interesting questions that we may ask ourselves. Why the photoluminescence at 680 nm from $Al_xGa_{1-x}As$ barriers is so weak in Fig. 4.13c? The electron states in $Al_xGa_{1-x}As$ barriers are higher than E1 in the GaAs well region so their occupation probability is small. This explains why the peak is weak. The quantitative ratio between this peak and the GaAs photoluminescence peak depends on how fast the electrons in $Al_xGa_{1-x}As$ barriers transfer to the GaAs region, the electrons photo-generated in the middle of an $Al_xGa_{1-x}As$ barrier needs more time to diffuse to the adjacent GaAs well than the electrons photo-generated at the edge of the $Al_xGa_{1-x}As$ barrier. We will study more about the carrier diffusion in the next section.

Another issue about Fig. 4.13 is the sample cleavage. The only reason of this operation is to increase the optical path of the excitation laser beam in the multiple GaAs/$Al_xGa_{1-x}As$ quantum well region in order to obtain a high photoluminescence signal. In principle as well as in reality, the non-invasive measurement setup of Fig. 4.6 will do the job.

4.4 V-Grooved Quantum Wire

In this section we study the photoluminescence spectrum of V-grooved quantum wires to show how much information we can extract about the low-dimensional structures in reality by photoluminescence spectral analysis.

The V-grooved GaAs/$Al_{0.5}Ga_{0.5}As$ quantum wire is prepared by metal organic chemical vapor deposition (MOCVD) method [4]. Refer to Fig. 4.15, V grooves along the y axis (perpendicular to the plane of paper) are first formed on the GaAs (grey) substrate, which are covered by a relatively thick $Al_{0.5}Ga_{0.5}As$ layer. A "thin GaAs layer" (grey) is then deposited followed by another relatively thick $Al_{0.5}Ga_{0.5}As$ cover layer. Majorities of the "thin GaAs layer" are expected to deposit in the bottoms of the V-grooves to form the GaAs quantum wires (QWRs) marked by red circles in Fig. 4.15. However, there are many fine low-dimensional structures in addition to the intentionally deposited GaAs quantum wires, including the side-wall quantum well layer (SQWL), the neck quantum well layer (NQWL), and the top quantum well layer (TQWL). Last but not the least, during the second deposition of the $Al_{0.5}Ga_{0.5}As$ cover layer, Ga atoms do not distribute uniformly through the cover layer, they tend to accumulate at the bottoms of the V-grooves that form the vertical quantum well layers (VQWL) on top of the GaAs quantum wires.

The quantum wire structures are now basically two dimensional in the x and z axes, and they are extended in the y direction (perpendicular to the plane of the page of Fig. 4.15). How can we identify all these details? First of all, we can use the microscopic photoluminescence (μ-PL) setup described in Sect. 4.1 to resolve the geometric structures in the xy plane by the x-axis scan.

An extra feature called a confocal pinhole is added to the μ-PL setup now to resolve the features along the z axis. By tuning properly the z position of the confocal pinhole we can constrain that only those photons originated from the confocal plane in Fig. 4.15 of a limited thickness can reach the photodetector in the optical spectrometer. With this confocal feature we are able to resolve spatial features along the z axis. This is how we identify the photoluminescence signal of the VQWL from the one of the GaAs quantum wire.

Figure 4.16 shows a series of confocal μ-PL spectra obtained from scanning across a single GaAs quantum wire at a x scanning step of 0.1 μm when the pinhole

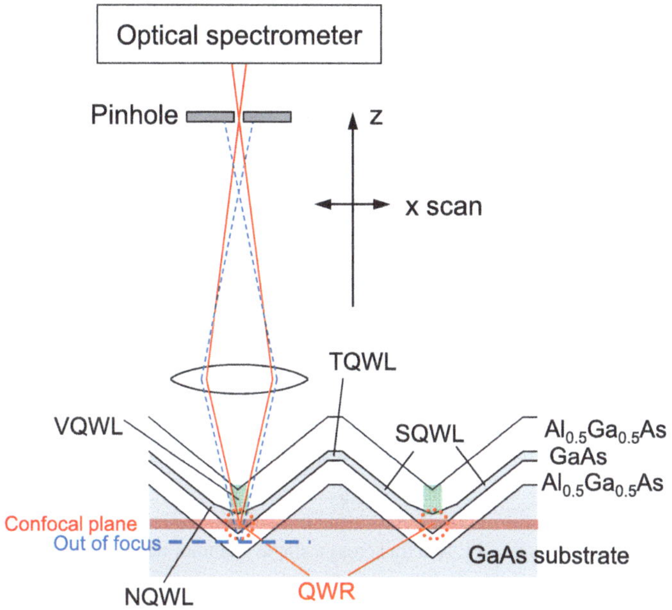

Fig. 4.15 Schematic structures of V-grooved $GaAs/Al_{0.5}Ga_{0.5}As$ quantum wires and confocal μ-PL measurement setup. Features along the z axis are measured by moving the sample vertically. By tuning properly the z position of the confocal pinhole we can constrain that only those photons originated from the confocal plane (red lines) can reach the photodetector in the optical spectrometer, while photons out of focus (blue dotted lines) will be blockaded

is positioned properly that the GaAs quantum wires are within the confocal plane. The spectral features are clearly x dependent.

Let us take a close look at the μ-PL spectra at various spatial locations. We pick up three μ-PL spectra, one at $x = -1.0$ μm, one at $x = 0.0$ and the third at $x = 1.0$ μm, which are presented in Fig. 4.17. The peak at 1.422 eV comes from the GaAs substrate, supported by the shape of the peak compared with the peak in Fig. 4.3. The peak at 2.066 eV comes from the optical transition from the conduction-band X valley to the valence-band Γ point in the $Al_{0.5}Ga_{0.5}As$ cover layer [5]. Between these two energies are various photoluminescence peaks from different low-dimensional structures in the V-grooved quantum wire sample.

By matching the μ-PL scanning spectra of Fig. 4.16 with geometric structures in Fig. 4.15, we identify all principal PL peaks: QWR = 1.743 eV, VQWL = 1.886 eV, NQWL = 1.967 eV, SQWL = 1.939 eV, and TQWL = 1.865 eV.

Diffusion Length of Photoexcited Carrier

There are many other things that can be identified from the μ-PL spectrum. In the room-temperature photoluminescence measurement, the incident radiation is absorbed to generate photoexcited electrons and holes which will relax very quickly to the local conduction- and valence bandedges (in the GaAs substrate and the thick

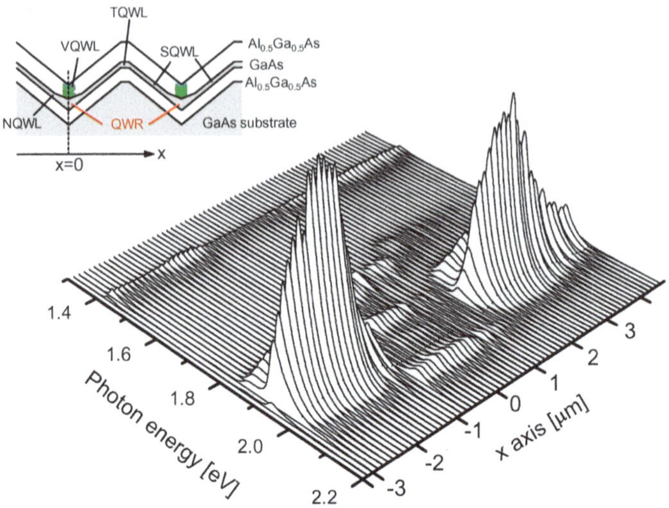

Fig. 4.16 Room-temperature μ-PL spectra across a single quantum wire centered at $x = 0.0$ μm

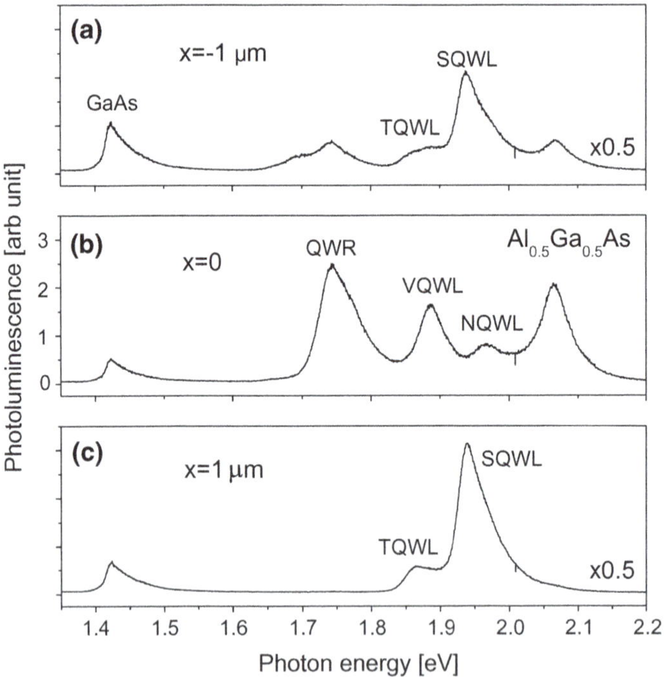

Fig. 4.17 Photoluminescence spectra at **a** $x = -1.0$ μm, **b** $x = 0.0$, and **c** $x = 1.0$ μm from Fig. 4.16. Note that the PL spectra at $x = \pm 1.0$ μm are scaled down by a factor of 0.5

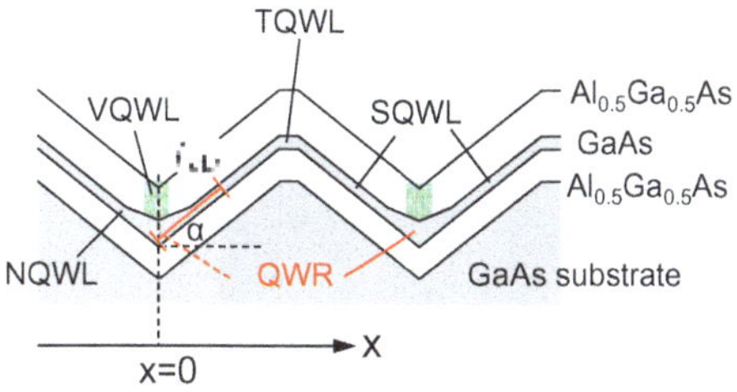

Fig. 4.18 Schematics of the diffusion length ℓ_{cD} for electrons

$Al_{0.5}Ga_{0.5}As$ cover layers), or the ground energy levels in various low-dimensional structures. The conduction and the valence bandedges of the GaAs substrate are the lowest across the whole sample so the photogenerated carriers in the GaAs substrate stay in the GaAs substrate. The total areas of various low-dimensional structures in the xz plane is much smaller than that of the two $Al_{0.5}Ga_{0.5}As$ cover layers, see Fig. 4.15. We can therefore conclude that the photoluminescence signals of various low-dimensional structures originate predominantly from the electrons and holes diffused from the neighboring $Al_{0.5}Ga_{0.5}As$ region. Let us focus on the photogenerated carriers in the two $Al_{0.5}Ga_{0.5}As$ cover layers.

One part of these photogenerated electrons and holes will directly recombine radiatively to emit the photons which is represented by the $Al_{0.5}Ga_{0.5}As$ peak in Fig. 4.17. One part will diffuse to the SQWL and VQWL region where the energy levels are low, another part will diffuse to the GaAs QWR region where the energy levels are the second lowest (remember that it is the GaAs substrate whose energy levels are the lowest). The diffusions of electrons and holes are normally characterized by the so-called diffusion lengths denoted by ℓ_{cD} for electrons and ℓ_{vD} for holes, see Fig. 4.18. Due to the diffusion to the second lowest energy levels in the GaAs QWR region, electrons will be depleted in the part of the SQWL that is within ℓ_{cD} to the GaAs QWR region (similarly, holes will be depleted in the part of the SQWL within ℓ_{cD} to the GaAs QWR region). Thus, there will be no photoluminescence signal from parts of the SQWL regions that are depleted of electrons and/or holes.

Examine more carefully Fig. 4.16 we do find one spatial region approximately between $x = -1.2$ to $x = 1.5$ μm void of the SQWL photoluminescence signal. We therefore can estimate that the longest one of the two diffusion lengths ℓ_{cD} and ℓ_{vD} is about

$$\max(\ell_{cD}, \ell_{vD}) = \frac{1.2 + 1.5}{2\sin\alpha} = 1.8\ \mu\text{m}$$

where $\alpha \approx 45°$ is the angle between the x axis and the SQWL, see Fig. 4.18.

Potential Barrier Between VQWL and QWR

As we can observe in Fig. 4.15, there are a few fine structures around the V-grooved QWR. One neighboring structure is the side quantum well layer (SQWL) which is directly connected to the GaAs QWR region by carriers' diffusions that we just discussed. Another important neighboring structure is the vertical quantum well layer (VQWL). Because of the different mechanisms of the formations of the VQWL

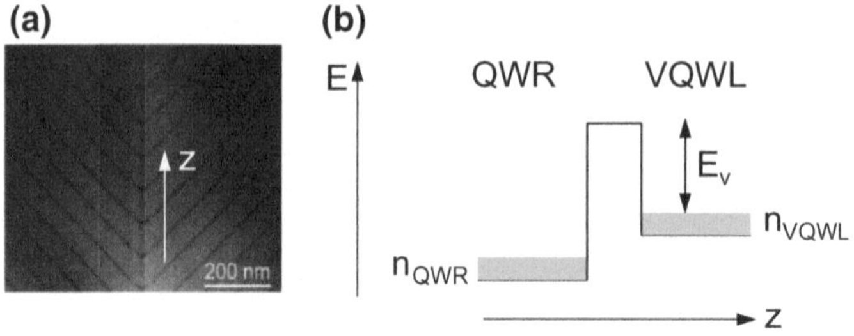

Fig. 4.19 **a** Transmission electron microscopic (TEM) image. **b** Schematics of the energy diagram between VQWL and QWR. n_{VQWL} and n_{QWR} denote the numbers of carriers photoexcited in VQWL and QWR, respectively

(during the deposition of the second $Al_{0.5}Ga_{0.5}As$ cover layer) and the GaAs QWR (thin GaAs film deposition), see Fig. 4.19a, a potential barrier E_v is expected to exist between VQWL and QWR.

Look at the schematic energy diagram Fig. 4.19b under an optical excitation, and assume that there are n_{VQWL} and n_{QWR} carriers photoexcited in VQWL and QWR regions, respectively. The total number of carriers in the QWR region will be

$$N_{\mathrm{QWR}} = n_{\mathrm{QWR}} + \frac{n_{\mathrm{VQWL}}}{1 + e^{E_v/k_BT}} \tag{4.74}$$

after including the thermal diffusion of n_{VQWL} from VQWL to QWR over E_v. And the number of carriers left behind becomes

$$N_{\mathrm{VQWL}} = n_{\mathrm{VQWL}} - \frac{n_{\mathrm{VQWL}}}{1 + e^{E_v/k_BT}} = \frac{n_{\mathrm{VQWL}}}{1 + e^{-E_v/k_BT}} \tag{4.75}$$

At low temperature, nonradiative recombination processes of the carriers are relatively rare. Since the strength of the photoluminescence signal is proportional to the number of carriers, the ratio between the optical powers S_{QWR} and S_{VQWL} of photoluminescence signals from QWR and VQWL is

$$\frac{S_{\mathrm{QWR}}}{S_{\mathrm{VQWL}}} = \frac{n_{\mathrm{QWR}}}{n_{\mathrm{VQWL}}} \left(1 + e^{-E_v/k_BT}\right) + e^{-E_v/k_BT} \tag{4.76}$$

We perform low-temperature photoluminescence measurements on the QWR structure by using the 515.5 nm excitation line from an argon-ion laser in a temperature range from 5 K to 60 K. The photoluminescence spectra are shown in Fig. 4.20. The excitation light spot on the sample has a diameter of ca 50 μm which covers about 10 GaAs QWRs (so this is not μ-PL measurement). Compared with the room-temperature (300 K) photoluminescence spectra shown in Fig. 4.16, all peaks in Fig. 4.20 are blue-shifted by approximately 80 meV, which equals exactly to the change of the GaAs bandgap when the temperature decreases from 300 K to 5 K.

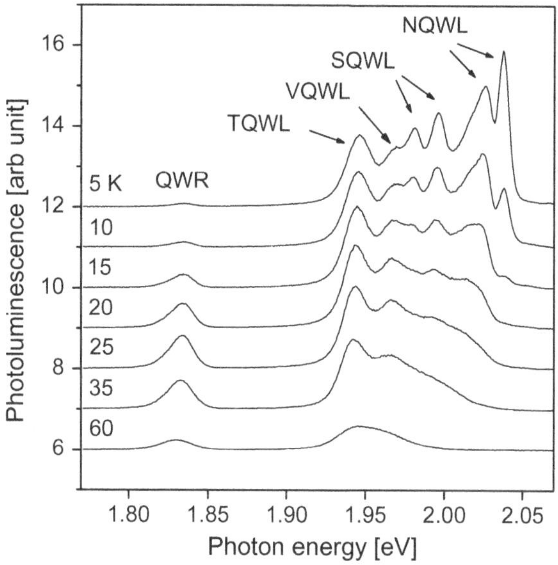

Fig. 4.20 Temperature dependent photoluminescence spectrum of the V-grooved $GaAs/Al_{0.5}Ga_{0.5}As$ quantum wire

Moreover, we now observe double peaks in the side quantum well layer (SQWL) and the neck quantum well layer (NQWL) due to the reduced nonradiative recombination processes.

Back to (4.76) to focus on the low-temperature photoluminescence signals of QWR and VQWL. At 5 K, thermal excitation is basically zero so that

$$\frac{S_{\mathrm{QWR}}}{S_{\mathrm{VQWL}}} = \frac{n_{\mathrm{QWR}}}{n_{\mathrm{VQWL}}} \tag{4.77}$$

Numerical analysis of the 5-K spectrum shows that $S_{\mathrm{QWR}} : S_{\mathrm{VQWL}} \approx 1 : 15$ so that we know $n_{\mathrm{QWR}} : n_{\mathrm{VQWL}} \approx 1 : 15$. Figure 4.20 shows that the optical power of the photoluminescence from QWR increases following the increase of temperature. More specifically, it increases by a factor of 45 when the temperature is increased from 5 K to 35 K. And this increase should be largely due to the thermal diffusion of photoexcited carriers from VQWL to QWR. A direct reflection may show that this is not very likely. If $n_{\mathrm{VQWL}} = 15n_{\mathrm{QWR}}$ as concluded from the 5-K measurement, and we let all n_{VQWL} thermally diffuse to QWR, the photoluminescence strength from the QWR will be $n_{\mathrm{QWR}} + n_{\mathrm{VQWL}} = 16n_{\mathrm{QWR}}$, indicating that the photoluminescence strength from the QWR would only increase by a factor of 16, maximally, not as 45 shown experimentally.

The missing factor of ca 3 is due to the quantum confinement. The brief account of the story is as follow. As discussed in Sect. 2.3, the radiative recombination is described by (2.107) that an electron occupying a conduction-band state $|\boldsymbol{k}\rangle$ transits to an empty valence-band state $|\boldsymbol{q}\rangle$. It is proportional to $\left|\langle \boldsymbol{k}(\boldsymbol{r})|H'|\boldsymbol{q}(\mathrm{r})\rangle\right|^2$. Here the basic requirement for a nonzero radiative recombination probability is a significant spatial overlap between $|\boldsymbol{k}(\boldsymbol{r})\rangle$ and $|\boldsymbol{q}(\boldsymbol{r})\rangle$. In bulk material, both $|\boldsymbol{k}(\boldsymbol{r})\rangle$ and $|\boldsymbol{q}(\boldsymbol{r})\rangle$ are in the form of Bloch functions so are extended in space, meaning that the radiative recombination probability of bulk material is commonly low, which is only perceptible at low temperature (under which condition, the electron and the hole are spatially combined by their Coulombic interaction; At high temperature, the Coulombic interaction between the electron and the hole is easily broken by thermal excitation). In nanostructures the electron and the hole are forced to stay close to each other due to the quantum confinement, resulting in room-temperature photoluminescence [6]. The quantum confinement of a quantum well is one-dimensional, while it is two dimensional in a quantum wire. This is the principal rational behind the observed high radiative recombination probability in the quantum wire.

Knowing $n_{\mathrm{QWR}} : n_{\mathrm{VQWL}}$, we can extract E_{υ} by analyzing the 10-K, 15-K, 20-K etc. photoluminescence spectra in Fig. 4.20. The numerical result is $E_{\upsilon} = 3.5$ meV.

When the temperature becomes higher than 35 K, nonradiative recombination processes, mostly the electron-phonon interactions, start to catch up so that the photoluminescence signals become weak.

Another important aspect that can be observed in Fig. 4.20 is that the photoluminescence signal of SQWL is very strong when the temperature is below 30 K. This indicates that the carriers photoexcited in SQWL remain largely in SQWL. Figure 4.21a shows the time-resolved photoluminescence spectra at different temperature. In time-resolved photoluminescence measurement, a picosecond laser pulse is led into the sample, and the photoluminescence signal of specific wavelength of interest (such as the photoluminescence signal from SQWL) from the sample is detected as a function of time, denoted as $f(t)$. $f(t)$ is commonly described by an exponential decay

$$f(t) = f_0 + f_i\, e^{-(t-t_0)/\beta} \tag{4.78}$$

where β is the radiative lifetime of the carriers (see more in Sect. 7.2). The lifetimes of carriers in SQWL and QWR are presented in Fig. 4.21b as functions of the temperature. Below 10 K, the lifetimes are about 0.32 ns, typical for carriers in quantum wires. By increasing the temperature, the lifetime of carriers in SQWL decreases drastically. At about 60 K, the photoluminescence signal from SQWL disappears, while at the same time, the lifetime of carriers in the QWR reaches its maximal value. A most possible reason of the synchronization of the two phenomena is the transfer of carriers from the SQWL to the QWR region.

There are different means to fabricate quantum wires. InAs nanowires can grow directly, vertically, on n- or p-type Si(111) substrates, by molecular beam epitaxy (MBE). One distinct character of such InAs nanowire is the axial alternation of zinc-blend phase at the nanowire root to wurtzite phase at the middle then zinc-blend phase again at the nanowire top. The embedded wurtzite phase at the middle behaves as a type-II quantum well (quantum well for the conduction-band electron and quantum barrier for the valence-band hole). Reference [7] reports using the conventional photoluminescence to reveal such quantum structures in InAs nanowires.

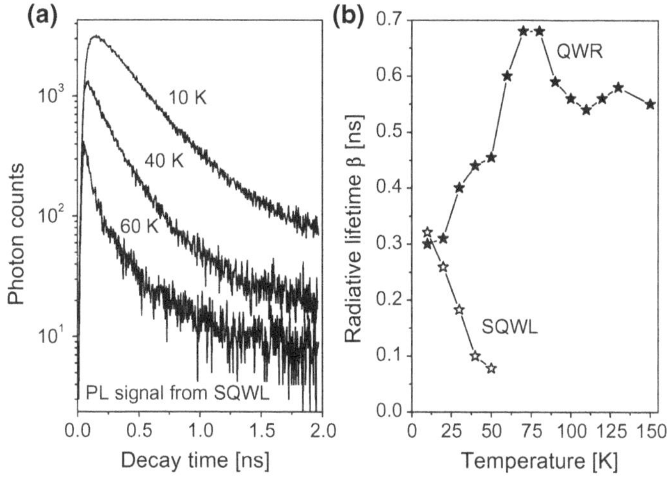

Fig. 4.21 **a** Time-resolved photoluminescence signal from side quantum well layer (SQWL) at different temperatures. **b** Radiative recombination lifetimes of photoexcited carriers in SQWL (hollow stars) and in QWR (solid stars) as functions of temperature

4.5 Quantum Dot

Quantum dot is basically a man-made super atom whose density of electron states is composed of a series of discrete levels, see Fig. 4.10. There are many ways to fabricate quantum dots depending on the applications. One method is called Stranski–Krastanov. Refer to Fig. 4.22, the lattice structures of InAs and GaAs are both zincblende but with different lattice constants. At room temperature, $a_{\mathrm{InAs}} = 6.058$ Å and $a_{\mathrm{GaAs}} = 5.653$ Å [1]. When one InAs monolayer is epitaxially deposited on a GaAs substrate, the lattice of the InAs monolayer has to adjust itself to fit the GaAs lattice, namely, the InAs monolayer has a lattice constant of GaAs on the plane perpendicular to the growth direction (the x direction in Fig. 4.22). In order to reduce the strain thus induced, the y lattice is to be extended, see Fig. 4.22b. Further depositing InAs will initiate the so-called quantum dot nucleation and growth, i.e., Fig. 4.22c, in order to minimize the strain. The InAs quantum dot will eventually be fixed by GaAs capping shown in Fig. 4.22d.

Since the Stranski–Krastanov growth of InAs quantum dots is based on the strain due to lattice mismatch between the epitaxial material InAs and the substrate GaAs, lattice defects exist in the as-grown InAs quantum dots. One way to reduce the density of these lattice defects is the proton implantation [8, 9].

In order to assess the proton implantation, the surface of a sample of InAs quantum dots on GaAs substrate is partitioned into a 8×8 cell array. Of these 64 cells, one cell is un-implanted while the other 63 cells are proton implanted with the implantation doses ranging up to 10^{14} cm^{-2}. The whole sample is then treated by rapid thermal annealing in high pure nitrogen ambient for 60 s at 700 °C. Photoluminescence

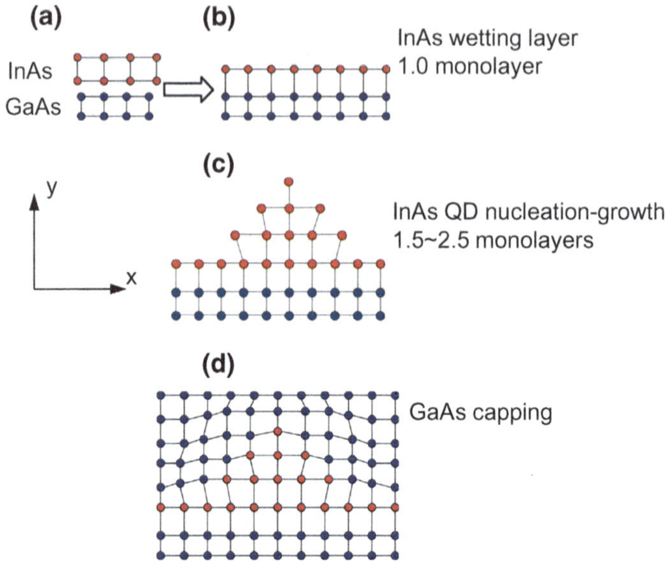

Fig. 4.22 Principal steps of Stranski–Krastanov growth of an InAs quantum dot on a GaAs substrate

measurements are performed at room temperature under excitation of a 514.5 nm line of an Ar^+ laser, and the results are shown in Fig. 4.23. It is observed here that the proton implantation greatly improves the photoluminescence signal of the InAs quantum dots more than 80 times when the proton implantation dose reaches 10^{14} cm^{-2}.

As shown by (4.73), the photoluminescence spectrum of a quantum dot is composed by a series of Lorentzian peaks. At room-temperature (high-temperature therefore fast nonradiative energy relaxations) and common photoluminescence measurement setup (not very high optical excitation), only the low-energy levels of the conduction band and valence band are occupied so that there is normally only one or maximally two photoluminescence peaks. Figure 4.23 shows however two photoluminescence peaks, which is more clearer when only one of photoluminescence spectra in Fig. 4.23 is presented alone in Fig. 4.24a. Using two Lorentzain peaks from (4.73) to fit all the spectra in Fig. 4.23,

$$\frac{p_{\text{ground}}\Gamma_1}{\left(\Delta E_{\text{ground}}-\hbar\omega\right)^2+\Gamma_1^2}+\frac{p_{\text{excited}}\Gamma_2}{\left(\Delta E_{\text{excited}}-\hbar\omega\right)^2+\Gamma_2^2}$$

we obtain the amplitudes p_{ground} and p_{excited}, and the energy relaxations Γ_1 and Γ_2 of the two peaks. A typical fitted spectrum is shown in Fig. 4.24a. The low-energy "ground" peak is associated with the radiative recombination of the ground-state

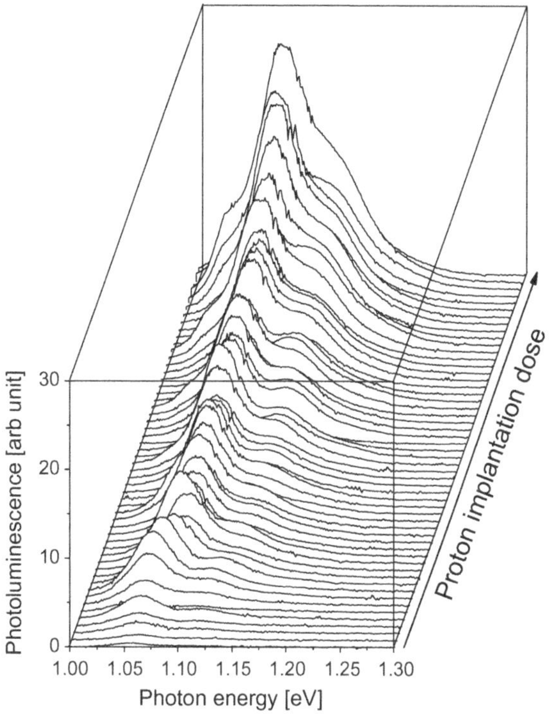

Fig. 4.23 Improvement of InAs quantum dot photoluminescence by proton implantation. Arrow indicates the increase of the proton implantation dose from 0 to 10^{14} cm^{-2}

exciton, and the high-energy "excited" peak with the excited exciton. Their relaxation energies are 29 and 104 meV, respectively. The integrated photoluminescence strengths

$$P_{\text{ground}} = \int \frac{p_{\text{ground}} \Gamma_1}{\left(\Delta E_{\text{ground}} - \hbar\omega\right)^2 + \Gamma_1^2} \mathrm{d}E, \quad P_{\text{excited}} = \int \frac{p_{\text{excited}} \Gamma_2}{\left(\Delta E_{\text{excited}} - \hbar\omega\right)^2 + \Gamma_2^2} \mathrm{d}E$$

vs proton implantation doss are presented in Fig. 4.24b.

The general physical trend reflected in Fig. 4.24 is clear that the proton implantation significantly improves the radiative recombination probability. As schematically shown in Fig. 4.1, the photogenerated electrons and holes relax to the conduction and valence bandedge states, they recombine to emit one photon, resulting in the photoluminescence of photons whose energies are slightly larger than the energy bandgap of the material. In reality, the energy bandgap is not totally void of any electron states so that there are emitted photons whose energies are smaller than the energy bandgap, e.g., see Fig. 4.5.

Figure 4.24 deserve more discussions. Figure 4.24a shows that P_{ground} is narrower than P_{excited}, which is expected. While the peak optical power of P_{ground} is higher than that of P_{excited}, the integrated optical powers of the two peaks are rather similar. Figure 4.24b further demonstrates that the integrated optical power of P_{excited}

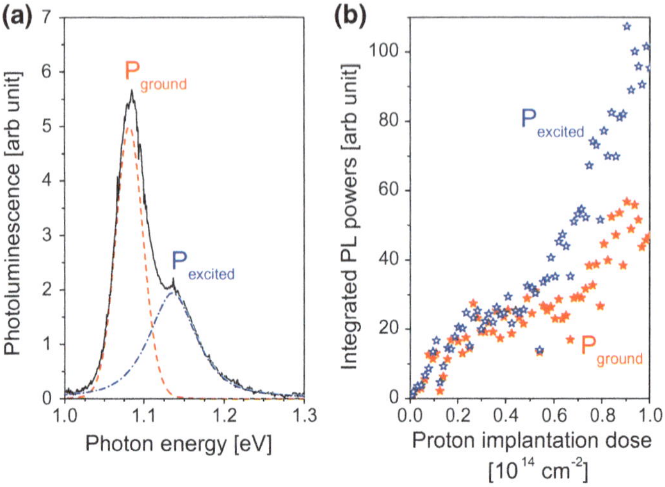

Fig. 4.24 **a** Photoluminescence spectrum (solid line) of InAs quantum dots with proton implanted at 5.0×10^{13} cm^{-2} and rapid thermal annealing. Red dashed and blue dot-dashed lines are fitting peaks. **b** Integrated PL powers of P_{ground} (red solid stars) and P_{excited} (blue hollow stars) vs proton implantation dose

becomes higher than P_{ground}. The principal factor is the density of states that if we approximate the InAs quantum dot as hydrogen-like, the ground state is s-type, i.e., a single state, while the first excited states are p_x, p_y and p_z, which are three-fold degenerate. We therefore observe the persistent increase of P_{excited} following the proton implantation dose, while P_{ground} increases at low proton implantation dose then displays a certain saturation in Fig. 4.24b.

Before we close this section, we like to correlate the photoluminescence with the reflectance spectrum we have studied in the previous chapter by the light-matter interaction theory presented in Chap. 2. In Sect. 2.4 we have learned that the excitation of an excited exciton state, i.e., the simultaneous appearance of the electron and the hole, results in the formation of an electric dipole, an exciton contribution to the polarization, thereafter the modified dielectric coefficient, see (2.148) which is repeated below

$$\epsilon(\omega) = \epsilon_\infty \left(1 + \sum_n \frac{\omega_{\text{LT}n}}{\omega_n - \omega - i\gamma_n}\right) \tag{4.79}$$

By (4.73) we expect to observe a photoluminescence peak at $\hbar\omega = E_j - E_i$ (in this case $\hbar\omega_n = E_j - E_i$ in the above equation, see Sect. 2.4). By the above equation we obtain a peak at in the dielectric constant $\epsilon(\omega_n)$, which should cause a reflection peak (Chap. 3). This so-called exciton polariton effect has been confirmed experimentally. For example, Fig. 4.25 shows a broad reflectance peak at the wavelength of the photoluminescence peak from InAs quantum dots embedded in GaAs matrix

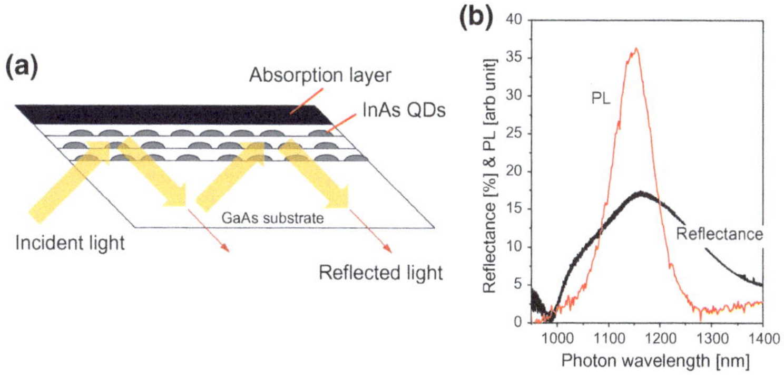

Fig. 4.25 **a** Schematic geometry of the edge polished InAs quantum dot (QD) multiple-layer sample where an absorption layer was added to absorb the transmitted light. **b** Reflectance (black line) and photoluminescence (PL, red line) spectra of the sample

[10]. Notice here that an absorption layer was added on the top GaAs capping layer to absorb the transmitted light so only reflected light was collected. Reflections from the GaAs substrate were negligible since the GaAs substrate was rather thick (Chap. 3).

4.6 Multiphoton Excitation

Thus far, we have studied the optical properties of semiconductor materials and devices under two principal assumptions. First, the materials and devices are always at steady state, and second, the external probing light beams are weak so we use the steady-state perturbation theory. Normal optical characterization setups and device operations fulfill the criterions of these two assumptions, see Sect. 2.3. However, ultra-fast and ultra-intense techniques have been being developed rapidly in the past years due to both the nano-technical development and the new demands. One keyword is the femtosecond laser that can reach a peak optical power of up to 3×10^{10} W/cm^2, i.e., 30 GW/cm^2.

Goeppert-Mayer predicted multiphoton excitation in 1931 [11], which was experimentally confirmed in 1961 when Franken and his colleagues used the laser beam of 694 nm to generate a light beam of 347 nm [12]. In this section, let us study the multiphoton excitation by solving perturbatively the time-dependent Schrödinger equation. One application of the multiphoton excitation is the two-photon microscopy which allows cellular imaging several hundred microns deep in various organs of living animals [13].

When the amplitude of the electric field of the light beam becomes high, we may expect that the $\boldsymbol{A}^2$ in (2.96) can induce optical transitions between two electron states due to a one-step two-photon excitation. Here the keyword is the "one-step", namely, one electron transition from one electron state to another electron state by absorbing simultaneously two photons. For our nano-structure semiconductor system, the energy modulation of the $\boldsymbol{A}^2$ term in the Schrödinger equation is

$$\frac{eA^2}{2m_0}e^{2i(\boldsymbol{s}\cdot\boldsymbol{r}-\omega t)} \tag{4.80}$$

where $\boldsymbol{s}$ is the wave vector of the light. Recall that in Sect. 2.4 we mention that the wavelength of a light beam in common photonics device is about 1 μm, and for a quantum dot with a radius less than 10 nm, the electric field of the light beam can be well approximated as uniform inside the quantum dot so that we obtained (2.148) that describes the dielectric constant of an exciton in the quantum dot. Approximate $\boldsymbol{A}$ as uniform inside the quantum dot, we can see that the optical transition due to (4.80) is zero, i.e.,

$$\langle\psi_i|\boldsymbol{A}^2|\psi_j\rangle \approx \boldsymbol{A}^2\langle\psi_i|\psi_j\rangle = 0$$

for $i \neq j$ due to the orthogonality of the electron wave functions.

Similarly, in a GaAs/$Al_{0.3}Ga_{0.7}As$ based quantum well infrared photodetector (QWIP), the amplitudes of the wave vectors are in the range of 1.03–3.10 $\times$ 10^6 m^{-1} for photons with energies $\hbar\omega$ = 60–180 meV. Because of this long wavelength, the spatial variation of $\boldsymbol{A}^2$ is very small in terms of the thin GaAs quantum wells in the QWIP, the optical transition between electron ground state ψ_1 confined in the GaAs quantum well and continuum states ψ_i above the conduction bandedge of $Al_{0.3}Ga_{0.7}As$ can be approximated as

$$\langle\psi_1|\boldsymbol{A}^2|\psi_i\rangle \approx \boldsymbol{A}^2\langle\psi_1|\psi_i\rangle$$

which again is zero due to the orthogonality of the electron wave functions.

We thus focus on the Hamiltonian of (2.101). Define the total wave function of the electron-photon system

$$\Psi(\boldsymbol{r},t) = \sum_{\ell N_\omega} C_{\ell N_\omega}(t)\psi_\ell(\boldsymbol{r})e^{-iE_\ell t/\hbar}|N_\omega\rangle \tag{4.81}$$

where $|N_\omega\rangle$ describes the photon field, N_ω is the number of photons and $\hbar\omega$ is the photon energy. $\psi_\ell(\boldsymbol{r})$ is the eigen function of the electron Hamiltonian H_0:

$$H_0(\boldsymbol{r})\psi_\ell(\boldsymbol{r}) = E_\ell\psi_\ell(\boldsymbol{r}) \tag{4.82}$$

without the photon field. By (2.93) and (2.101) we solve directly the non-perturbative time-dependent Schrödinger equation

$$i\hbar\frac{\partial\Psi(\boldsymbol{r},t)}{\partial t} = H(\boldsymbol{r},t)\Psi(\boldsymbol{r},t) \tag{4.83}$$

and the equation for coefficient $C_{mN_\omega^*}(t)$ is

$$
\begin{aligned}
i\hbar\frac{\mathrm{d}C_{mN_\omega^*}(t)}{\mathrm{d}t} &= \frac{e}{m_0}\sqrt{\frac{\hbar}{2\epsilon_0 n^2\omega\Omega}}\sum_{\ell N_\omega}\langle\psi_m(\boldsymbol{r})|\boldsymbol{e}_s\cdot\boldsymbol{p}|\psi_\ell(\boldsymbol{r})\rangle C_{\ell N_\omega}(t) \\
&\left\{\sqrt{N_\omega}\exp\left[\frac{i(E_m-E_\ell-\hbar\omega)t}{\hbar}\right]\delta_{N_\omega^*,N_\omega-1}\right. \\
&\left.+\sqrt{N_\omega+1}\exp\left[\frac{i(E_m-E_\ell+\hbar\omega)t}{\hbar}\right]\delta_{N_\omega^*,N_\omega+1}\right\}
\end{aligned}
\tag{4.84}
$$

In the above equation, $\langle\psi_m(\boldsymbol{r})|\boldsymbol{e}_s\cdot\boldsymbol{p}|\psi_\ell(\boldsymbol{r})\rangle$ describes the optical transition between electron occupying state $\psi_m(\boldsymbol{r})$ and $\psi_\ell(\boldsymbol{r})$. The term containing $\sqrt{N_\omega}$ describes the photon absorption and the term with $\sqrt{N_\omega+1}$ describes the photon emission. Note that there are also two δ functions separately attached to the absorption and emission terms. $\delta_{N_\omega^*,N_\omega-1}$ means that the number of photons in the initial electron-photon system ($=\psi_\ell e^{-iE_\ell t/\hbar}|N_\omega\rangle$), i.e., N_ω, is reduced by 1 when the electron absorbs one photon then transits from state $\psi_\ell(\boldsymbol{r})$ to $\psi_m(\boldsymbol{r})$. The number of the photons in the final electron-photon system is $N_\omega^*=N_\omega-1$. $\delta_{N_\omega^*,N_\omega+1}$ means that the number of photons in the initial electron-photon system is increased by 1 when the electron emits one photon and transits from $\psi_\ell(\boldsymbol{r})$ to $\psi_m(\boldsymbol{r})$. The probabilities of other processes are all zero.

Because the peak power of the femtosecond laser can be very high under multiphoton excitation operation, $N_\omega \gg 1$ so that $N_\omega \pm 1 \approx N_\omega$ and we drop off the photon number from coefficient $C_{mN_\omega^*}$ and C_{mN_ω}. By (2.85), we obtain the master equation about the ultra-fast and ultra-intense process

$$
i\hbar\frac{\mathrm{d}C_m(t)}{\mathrm{d}t} = \frac{e}{m_0\omega}\sqrt{\frac{S}{2\epsilon_0 n^2 c}}\sum_{\ell}\langle\psi_m|\boldsymbol{e}_s\cdot\boldsymbol{p}|\psi_\ell\rangle C_\ell(t)\exp\left[\frac{i(E_m-E_\ell)t}{\hbar}\right]2\cos(\omega t)
\tag{4.85}
$$

where S is the optical power of the excitation laser.

In order to discuss the multiphoton excitation, we need to describe the optical field as a beam of multiple photons of different optical powers characterized by h^s, frequencies ω_s and wave vectors $\boldsymbol{s}$ (we can drop off the summation to retrieve the single photon situation)

$$
\sum_s h^s e^{-i\omega_s t}
$$

so that (4.85) becomes

$$
i\hbar\frac{\mathrm{d}C_m(t)}{\mathrm{d}t} = \sum_{s\ell} h^s_{m\ell} C_\ell(t)\exp\left[\frac{i(E_m-E_\ell)t}{\hbar}\right]2\cos(\omega_s t)
\tag{4.86}
$$

where

$$
h^s_{m\ell} = \frac{e}{m_0\omega_s}\sqrt{\frac{S^s}{2\epsilon_0 n^2 c}}\langle\psi_m|\boldsymbol{e}_s\cdot\boldsymbol{p}|\psi_\ell\rangle
\tag{4.87}
$$

Let us first try to simplify the problem in order to obtain a good physical insight. Assume that the electron initially occupies the ground electron state E_k, i.e., $C_k(0) = 1$ and $C_{m(\neq k)}(0) = 0$. And $E_{m(\neq k)} > E_k$. Express the steady-state solution of $C_m(t)$ (note that $m \neq k$) in (4.86) up to the second order in the form of (it is easy to go beyond the second order to study more-than-two-photon processes which is however beyond the scope of the book)

$$C_m(t) = C_m^{(0)} + C_m^{(1)}(t) + C_m^{(2)}(t) \tag{4.88}$$

Note that $C_m^{(0)} \equiv C_m(0) = 0$. When the excitation power is weak so that the ground electron state is always almost occupied, i.e., $C_k(t) \approx C_k(0) \equiv C_k^{(0)}$, the perturbation theory says

$$i\hbar \frac{\mathrm{d}C_m^{(1)}(t)}{\mathrm{d}t} = \sum_s h_{mk}^s C_k^{(0)} \exp\left[\frac{i(E_m - E_k)t}{\hbar}\right] 2\cos(\omega_s t) \tag{4.89}$$

The solution of $C_m^{(1)}$ in the above equation "looks" very simple. A direct mathematical integration over t results in

$$\begin{aligned} C_m^{(1)}(t) = \sum_s & \frac{-h_{mk}^s C_k^{(0)}}{E_m - E_k - \hbar\omega_s} \exp\left[\frac{i(E_m - E_k - \hbar\omega_s)t}{\hbar}\right] \\ & + \frac{-h_{mk}^s C_k^{(0)}}{E_m - E_k + \hbar\omega_s} \exp\left[\frac{i(E_m - E_k + \hbar\omega_s)t}{\hbar}\right] \end{aligned} \tag{4.90}$$

As assumed before, $E_m > E_k$.

Let us consider a semiconductor quantum dot with $E_m > E_k = 100$ meV so that the resonant-response photon energy $\hbar\omega_s$ equals 100 meV, for which $E_m - E_k - \hbar\omega_s = 0$ and $E_m - E_k + \hbar\omega_s = 200$ meV. Under this resonance condition, the oscillation time period of the first exponential function ($= \hbar/(E_m - E_k - \hbar\omega_s) = \hbar/0$) is infinite, while it is ca 3.3 fs ($= \hbar/(E_m - E_k + \hbar\omega_s)$) for the second term of the above equation. Since the physical measurement process runs always in a finite time duration, and in almost all practical measurements including device operations, the time duration is much longer than a few fs (note that the excitation can be short, as in the case of a femtosecond laser pulse), the measurement result of the second term, i.e., its time average,

$$\left\langle \exp\left[\frac{i(E_m - E_k + \hbar\omega_s)t}{\hbar}\right]\right\rangle_t$$

is zero. This is of course physically correct that emitting a photon during an electron transition from a low-energy state to a high-energy state is very unlikely.

The first term in (4.90) is about the photon absorption when the electron transits from state E_k to E_m. The time average of its exponential function becomes 1 at resonance of $E_m - E_k - \hbar\omega_s = 0$. This however results in a zero denominator, which means $C_m^{(1)} \to \infty$. This is not physical. We have done nothing wrong mathematically except that we have missed a physical event accompanying the photon absorption process, which is the photon emission when the electron transits from E_m to E_k. What happens in reality at resonance is that at resonance, $E_m - E_k - \hbar\omega_s = 0$ so that (4.89) becomes

$$i\hbar \frac{\mathrm{d}C_m^{(1)}(t)}{\mathrm{d}t} = \sum_s h_{mk}^s C_k^{(0)} \tag{4.91}$$

in a very short time duration $t \in (0, \delta)$ after switching on the excitation light. (We have neglected the unlikely photon emission process.) At $t = \delta$,

$$C_m^{(1)}(\delta) = \frac{\delta}{i\hbar} \sum_s h_{mk}^s C_k^{(0)} \tag{4.92}$$

Since $C_m^{(1)}(\delta)$ is no longer zero, there will be the probability that the electron, now occupying E_m, transits back to E_k. This is the physical event which we have missed when we perform the direct mathematical integration that results in (4.90).

The proper treatment is the scattering theory which can be found in many quantum mechanical textbooks. It is beyond the scope of this book. However, a common method to circumvent the issue is to introduce a relaxation energy between the high-energy electron state and the low-energy state (between the two electron states the electron transits under the photo-excitation).

Adding the relaxation Γ_{mk} of the excited state then integrating (4.89) from time 0 to t results in

$$C_m^{(1)}(t) = \sum_s \frac{h_{mk}^s C_k^{(0)}}{E_m - E_k - \hbar\omega_s + i\Gamma_{mk}} \left\{1 - \exp\left[\frac{i(E_m - E_k - \hbar\omega_s + i\Gamma_{mk})t}{\hbar}\right]\right\} \tag{4.93}$$

The physical meaning of the above equation is how much the excited electron state E_m will be occupied at time t when the electron, originally occupying E_k has been under photo-excitation from time 0 to time t.

Now the excited state E_m is partially occupied, its occupying electron will further transit to other available states. We change the subscript m in (4.93) to i, also the photon field to $h^v e^{-i\omega_v t}$,

$$C_i^{(1)}(t) = \sum_v \frac{h_{ik}^v C_k^{(0)}}{E_i - E_k - \hbar\omega_v + i\Gamma_{ik}} \left\{1 - \exp\left[\frac{i(E_i - E_k - \hbar\omega_v + i\Gamma_{ik})t}{\hbar}\right]\right\} \tag{4.94}$$

then insert the above expression into (4.86) to obtain $C_m^{(2)}(t)$

$$
\begin{aligned}
i\hbar\frac{\mathrm{d}C_m^{(2)}(t)}{\mathrm{d}t} &= \sum_{si} h_{mi}^{s} C_i^{(1)}(t)\exp\left[\frac{i(E_m-E_i)t}{\hbar}\right]2\cos(\omega_s t) \\
&= \sum_{sv,i}\frac{h_{mi}^{s}h_{ik}^{v}C_k^{(0)}}{E_i-E_k-\hbar\omega_v+i\Gamma_{ik}}\left\{1-\exp\left[\frac{i(E_i-E_k-\hbar\omega_v+i\Gamma_{ik})t}{\hbar}\right]\right\} \\
&\quad\times\exp\left[\frac{i(E_m-E_i)t}{\hbar}\right]2\cos(\omega_s t)
\end{aligned}
\tag{4.95}
$$

The temporal integration of (4.95) is straightforward, while the final expression of $C_m^{(2)}(t)$ is a bit long so we skip it. Remember, however, that it is a two-step process that the electron at E_k transits to E_i under photo-excitation $h^{\boldsymbol{v}}e^{-i\omega_v t}$ followed by a transition from E_i to E_m under $h^{s}e^{-i\omega_s t}$:

$$
E_k \xrightarrow{h^{\boldsymbol{v}}e^{-i\omega_v t}} E_i \xrightarrow{h^{s}e^{-i\omega_s t}} E_m \tag{4.96}
$$

Letting $t\to\infty$, it is easy to obtain the following steady-state perturbation solutions

$$
C_m^{(1)}(\infty) = \sum_{s}\frac{h_{mk}^{s}C_k^{(0)}}{E_m-E_k-\hbar\omega_s+i\Gamma_{mk}} \tag{4.97}
$$

$$
\begin{aligned}
C_m^{(2)}(\infty) &= \sum_{sv,i}\frac{h_{mi}^{s}h_{ik}^{v}C_k^{(0)}}{E_i-E_k-\hbar\omega_v+i\Gamma_{ik}} \\
&\times\left(\frac{1}{E_m-E_i\pm\hbar\omega_s+i\Gamma_{mi}}-\frac{1}{E_m-E_k-\hbar\omega_v\pm\hbar\omega_s+i\Gamma_{mk}}\right)
\end{aligned}
\tag{4.98}
$$

The "±" sign in (4.98) is the result of the occupation of excited state, i.e., $C_i^{(1)}(t)\neq 0$ by (4.94) and (4.97), so that there is a possibility of a photon emission when the electron transits from a high-energy state to a low-energy state.

$C_m^{(1)}(\infty)$ in (4.97) reaches its peak when $E_m-E_k=\hbar\omega_s$, which is known as the one-photon excitation (also called one-photon absorption), see schematically in Fig. 4.26a.

The most interesting result is (4.98). For a single beam excitation, $\boldsymbol{v}=\boldsymbol{s}$ so that it becomes

$$
\begin{aligned}
C_m^{(2)}(\infty) &= \sum_{s,i}\frac{h_{mi}^{s}h_{ik}^{s}C_k^{(0)}}{E_i-E_k-\hbar\omega_s+i\Gamma_{ik}} \\
&\times\left(\frac{1}{E_m-E_i\pm\hbar\omega_s+i\Gamma_{mi}}-\frac{1}{E_m-E_k-\hbar\omega_s\pm\hbar\omega_s+i\Gamma_{mk}}\right)
\end{aligned}
\tag{4.99}
$$

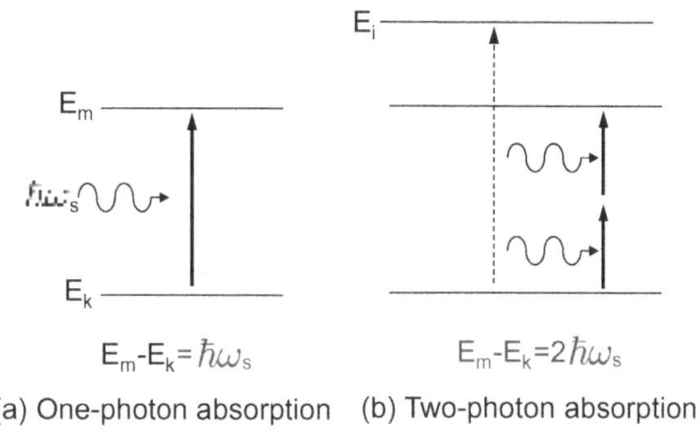

Fig. 4.26 **a** One-photon excitation between state E_m and E_k, **b** two-photon excitation. The dashed arrow shows the off-resonance one-photon excitation

Without loss of generality, we focus on the energy diagram in Fig. 4.26b such that $E_i > E_m > E_k$ in the sense that E_k is the ground state, E_m is the first excited state, and E_i is a higher-energy state. Consider a light beam with photons whose energies $\hbar\omega_s < E_m - E_k$. $E_i - E_k - \hbar\omega_s$ will never be zero so we leave it as it is. There are two terms in the parenthesis which may be resonant so that the above equation reduces to

$$C_m^{(2)}(\infty) = \sum_{s,i} \frac{h^s_{mi} h^s_{ik} C_k^{(0)}}{E_i - E_k - \hbar\omega_s + i\Gamma_{ik}}$$
$$\times \left(\frac{1}{E_m - E_i + \hbar\omega_s + i\Gamma_{mi}} - \frac{1}{E_m - E_k - \hbar\omega_s - \hbar\omega_s + i\Gamma_{mk}} \right) \quad (4.100)$$

The first term in the parenthesis will be significant when $E_m + \hbar\omega_s = E_i$, which is not the case for the moment since we have assumed just a second ago that $\hbar\omega < E_m - E_k$. We however have a chance that the second term becomes significant if $E_m - E_k = 2\hbar\omega_s$. This is the well-known two-photon excitation (also called two-photon absorption) predicted by Goeppert-Mayer [11]:

$$C_m^{*(2)}(\infty) = -\sum_{s,i} \frac{h^s_{mi} h^s_{ik} C_k^{(0)}}{\left(E_i - E_k - \hbar\omega_s + i\Gamma_{ik}\right)\left(E_m - E_k - 2\hbar\omega_s + i\Gamma_{mk}\right)} \quad (4.101)$$

which shows that the electron can transits from E_k to E_m by absorbing two photons whose energies are only one half of $E_m - E_k$. As shown by (4.101), the two-photon excitation (also known as the two-photon absorption) is accompanied by an off-resonance one-photon excitation from E_k to E_i so that the total two-photon excitation probability is very weak, which is to be compensated in experiment by a very high excitation power S_s.

A new type of quantum dots has been vastly studied and exploited for biomedical applications in the past ten years. They are the so-called colloidal quantum dots, see Fig. 1.1 in the introduction of Chap. 1. These colloidal quantum dots are solution-

Table 4.2 Structure and optical properties of three core-multishell CdSe-CdS/$Cd_{0.5}Zn_{0.5}S$/ZnS quantum dots

	QD556	QD600	QD622
CdSe core diameter [nm]	3.6	4.5	5.4
CdS shell [monolayer]	1	2	3
$Cd_{0.5}Zn_{0.5}S$ shell [monolayer]	1	1	1
ZnS shell [monolayer]	1	1	1
Photoluminescence peak wavelength [nm]	556	600	622
FWHM [nm/eV]	32/121	29/104	24/77

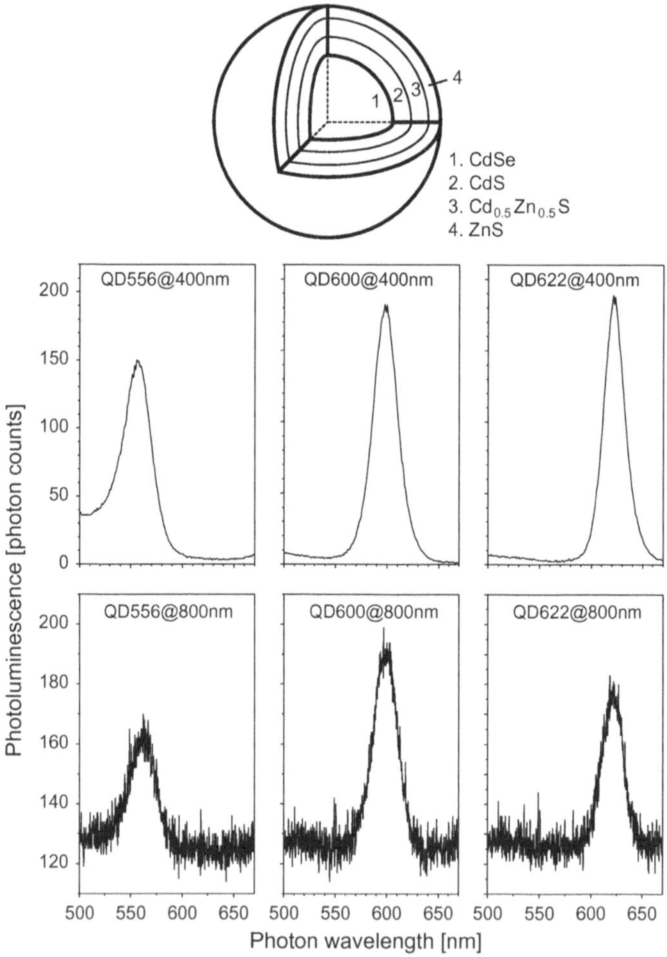

Fig. 4.27 Schematic material structure of core-multishell CdSe-CdS/$Cd_{0.5}Zn_{0.5}S$/ZnS quantum dots and one-photon (one-photon excitation) and two-photon (two-photon excitation) induced photoluminescence spectra excited by laser beams of wavelengths 400 and 800 nm, respectively

based, both during their chemical synthesis and applications, while the Stranski–Krastanov-grown InAs quantum dots in Fig. 4.22 are solid-state based (embedded in a solid-state matrix). Chemical synthesis is by now a common standard method to fabricate II-VI-based colloidal quantum dots, and the core/shell II-VI quantum dots obtained are generally rather robust. Table 4.2 lists the details about three types of core/multishell CdSe-based colloidal quantum dots, which are denoted as QD556, QD600 and QD622, respectively.

Photoluminescence spectroscopic measurements on these colloidal quantum dots are performed using a fluorescence spectrometer at excitation wavelength 400 nm (one-photon excitation), and two-photon-induced (two-photon excitation) fluorescence experiments are performed using a femtosecond laser with a pulse width of 130 fs, peak wavelength of 800 nm, FWHM of 7 nm, and average optical power output of about 630 mW at 800 nm at a pulse frequency of 76 MHz. The one-photon and two-photon induced photoluminescence spectra of the quantum dots in aqueous solutions are presented in Fig. 4.27. By removing a base level count of about 125, the two-photon induced photoluminescence spectra are strikingly identical to the one-photon spectra, both the shapes including the FWHMs and the relative photon counts.

We will return to the multiphoton excitation in Chap. 6 when we excite the photocurrent of a quantum well infrared photodetector (QWIP) device by a free electron laser (FEL) light source.

References

1. I. Vurgaftman, J.R. Meyer, L.R. Ram-Mohan, Band parameters for III-V compound semiconductors and their alloys. J. Appl. Phys. **89**, 5815–75 (2001)
2. C.F. Klingshirn, *Semiconductor Optics* (Springer, Berlin, 1995), p. 156
3. D.J. Olego, J.P. Faurie, S. Sivananthan, P.M. Raccah, Optoelectronic properties of Cd(1-x)Zn(x)Te films grown by molecular beam epitaxy on GaAs substrates. Appl. Phys. Lett. **47**, 1172–1174 (1985)
4. X.Q. Liu, W. Lu, Z.F. Li, Y.D. Chen, S.C. Shen, Y. Fu, M. Willander, H.H. Tan, S. Yuan, C. Jagadish, J. Zou, D.J.H. Cockayne, Spatially resolved luminescence investigation of AlGaAs/GaAs single quantum wires modified by selective implantation and annealing. Appl. Phys. Lett. **75**, 3339(3) (1999)
5. G. Turrel, J. Corset, *Raman Microscopy: Developments and Applications* (Academic Press, London, 1996)
6. P.W. Yu, D.C. Look, W. Ford, Photoluminescence in electrically reversible (semiconducting to semiinsulating) bulk GaAs. J. Appl. Phys. **62**(7) (1987)
7. X. Chen, Q. Zhuang, H. Alradhi, ZhM Jin, L. Zhu, X. Chen, J. Shao, Midinfrared photoluminescence up to 290 K reveals radiative mechanisms and substrate doping-type effects of InAs nanowires. Nano Lett. **17**, 1545–1551 (2017)
8. Y. Ji, G. Chen, N. Tang, Q. Wang, X.G. Wang, J. Shao, X.S. Chen, W. Lu, Proton-implantation-induced photoluminescence enhancement in self-assembled InAs/GaAs quantum dots. Appl. Phys. Lett. **82**, 2802(3) (2003)
9. W. Lu, Y.L. Ji, G.B. Chen, N.Y. Tang, X.S. Chen, S.C. Shen, Q.X. Zhao, M. Willander, Enhancement of room-temperature photoluminescence in InAs quantum dots. Appl. Phys. Lett. **83**, 4300(3) (2003)

10. Y. Fu, H. Ågren, L. Höglund, J.Y. Andersson, C. Asplund, M. Qiu, L. Thylén, Optical reflection from excitonic quantum-dot multilayer structure. Appl. Phys. Lett. **93**, 183117(3) (2008)
11. M. Goeppert Mayer, Elementary processes with two quantum jumps. Ann. Physik (Leipzig) **9**, 273–294 (1931)
12. P.A. Franken, A.E. Hill, C.W. Peters, G. Weinreich, Generation of optical harmonics. Phys. Rev. Lett. **7**, 118–119 (1961)
13. F. Helmchen, W. Denk, Deep tissue two-photon microscopy. Nat. Methods **2**, 932–940 (2005)

Chapter 5
Modulation Spectroscopy

Abstract Optical spectroscopy is studied when a periodic external perturbation, such as an electromagnetic field, temperature, and static pressure or uniaxial stress, is applied to the semiconductor during the process of experimental measurement. The modulation spectroscopy, i.e., the change of the optical spectrum produced by the external perturbation, rather than the absolute spectrum itself, reveals many fine structures in the electronic energy diagram of the semiconductor.

In Chap. 3 **Reflection and Transmission** and Chap. 4 **Photoluminescence**, we have learned that the optical spectrum of a material is closely related with the dielectric coefficient ϵ of the material, which in its turn is determined by various microscopic optical processes in the material. All these are expressed by (2.148) which is repeated below

$$\epsilon(\omega) = \epsilon_\infty \left(+ \sum_\ell \frac{\omega_{\mathrm{LT}\ell}}{\omega_\ell - \omega - i\gamma_\ell} \right) \tag{5.1}$$

where $\hbar\omega_\ell$ is the energy difference between two electron states that one electron transits from one state to the other, and $\omega_{\mathrm{LT}\ell}$ represents the optical transition strength. More specifically, for the optical transitions between conduction-band electron state $E_{\mathrm{c}\boldsymbol{k}}$ and valence-band hole state $E_{\mathrm{v}\boldsymbol{k}}$, the above expression can be written as

$$\frac{\epsilon(\omega)}{\epsilon_\infty} = 1 + \int \frac{\hbar\omega_{\mathrm{LT}\boldsymbol{k}}}{E_{\mathrm{c}\boldsymbol{k}} - E_{\mathrm{v}\boldsymbol{k}} - \hbar\omega - i\Gamma_{\boldsymbol{k}}} \frac{2\mathrm{d}\boldsymbol{k}}{(2\pi)^3} \tag{5.2}$$

where $\hbar\gamma_{\boldsymbol{k}}$, the relaxation energy, is now denoted as $\Gamma_{\boldsymbol{k}}$.

Consider the conduction- and valence-band minima at the Γ symmetry point in the $\boldsymbol{k}$ space

$$E_{\mathrm{c}\boldsymbol{k}} = E_{\mathrm{g}} + \frac{\hbar^2\boldsymbol{k}^2}{2m_{\mathrm{c}}^*}, \quad E_{\mathrm{v}\boldsymbol{k}} = -\frac{\hbar^2\boldsymbol{k}^2}{2m_{\mathrm{h}}^*}$$

for which

$$E_{\mathrm{c}\boldsymbol{k}} - E_{\mathrm{v}\boldsymbol{k}} = E_{\mathrm{g}} + \frac{\hbar^2\boldsymbol{k}^2}{2m_{\mathrm{c}}^*} + \frac{\hbar^2\boldsymbol{k}^2}{2m_{\mathrm{c}}^*} \equiv E_{\mathrm{g}} + \frac{\hbar^2\boldsymbol{k}^2}{2m_{\mathrm{r}}^*} \equiv E \tag{5.3}$$

W. Lu and Y. Fu, *Spectroscopy of Semiconductors*, Springer Series in Optical Sciences 215, https://doi.org/10.1007/978-3-319-94953-6_5

where m_r^* is the reduced effective mass. Here we put the potential energy of the valence bandedge to be zero so that the potential energy of the conduction bandedge is E_g the energy bandgap. Replace $\boldsymbol{k}$ with E using the above energy dispersion relationship in (5.2),

$$\frac{\epsilon(\omega)}{\epsilon_\infty} = 1 + \int \frac{\hbar\omega_{\mathrm{LT}E}}{E - \hbar\omega - i\Gamma_E} \frac{1}{4\pi^2} \left(\frac{2m_r^*}{\hbar^2}\right)^{3/2} \sqrt{E - E_g}\,\theta(E - E_g)\,\mathrm{d}E \\ \rightarrow \sqrt{\hbar\omega - E_g}\,\theta(\hbar\omega - E_g) \quad (5.4)$$

This is the dielectric coefficient for a bulk material discussed in Sect. 4.1. We also see in the above expression the three-dimensional density of associated electron states already expressed by (4.19) in Sect. 4.1.

In many electronics and photonics applications, the amplitudes of wave vector $\boldsymbol{k}$ of electrons involved in common semiconductor device functions are quite small, i.e., close to the conduction- and valence-band minima, the Γ points in (5.4), so that the dependence of $\omega_{\mathrm{LT}\boldsymbol{k}}$ on $\boldsymbol{k}$ is very weak (see Sect. 2.1). $\epsilon(\omega)$ in (5.4) as well as the optical spectrum, e.g., the photoluminescence in Fig. 4.3, reflect largely the density of electron states close to the Γ points, i.e., they are all proportional to $\sqrt{\hbar\omega - E_g}$. Such an optical spectrum is very smooth as a function of the energy band structure of the three-dimensional semiconductor, especially at room or higher temperature. However, for low-dimensional structures, the densities of electron states change much drastically as functions of the electron energy, see Fig. 4.10. The measurement or detection of a smooth function is more difficult than a strongly varying function. This is one of the many reasons low-dimensional structures have been extensively studied and developed for electronics and photonics applications.

Back to the three-dimensional bulk material, we notice that the differentiations of the dielectric coefficient, the density of electron states, as well as the photoluminescence spectrum with respect to the photon energy $\hbar\omega$, or with respect to E_g are

$$\frac{\mathrm{d}\sqrt{\hbar\omega - E_g}}{\mathrm{d}(\hbar\omega)}, \frac{\mathrm{d}\sqrt{\hbar\omega - E_g}}{\mathrm{d}E_g} \propto \frac{1}{\sqrt{\hbar\omega - E_g}} \quad (5.5)$$

i.e., they all diverge when $\hbar\omega = E_g$. This starts the so-called modulation spectroscopy with a much increased resolution of the spectral line otherwise hidden in the original smooth optical spectrum.

By (5.5), there are two ways to differentiate the optical spectrum, either we modify the photon energy $\hbar\omega$, we can also modify E_g. The modification of the photon energy $\hbar\omega$ is simple. We measure an optical spectrum then differentiate the optical spectrum, such as Fig. 4.3 in order to see more clearly the change at E_g. The second way is more fruitful that we add an extra electromagnetic field, or a strain field, or a temperature change on the semiconductor material of interest to modify E_g (we modify simultaneously also the effective mass of electrons involved in the optical transitions etc.). The differentiation of the optical spectra obtained before and after the modulation will unravel much rich information about electron states and optical transitions among these electron states at the microscopic level.

There is one critical aspect concerning modifying the energy band structure of a semiconductor material. The lattice crystal in a perfect bulk material is periodic in space. Such a periodicity is preserved when the material is uniformly heated (here the lattice constant is changed from one value to the other, while the lattice structure remains unchanged). The symmetry of the lattice will be broken down when an electromagnetic field is used to modify the energy band structure since the potential energy induced by the electromagnetic field is in the form of an electric dipole which is not periodic in space. Applying a mechanical strain field may modify the lattice structure differently depending on the spatial configuration of the strain field. We need to keep this in our minds when we study modification spectroscopies in the following sections.

5.1 Third-Derivative Modulation Spectroscopy

There are two main electric-field-based modulation spectroscopies, one is electro-modulation (electroreflectance) and the other is photo-modulation. Both utilize an electric field. In the electroreflectance an external electric field is applied, while it is the electric field of the modulating electromagnetic field in the photo-modulation spectroscopy. Theoretical analyses of the two modulation spectroscopies are exactly the same, while practically the advantage of optical-spectrum-based material characterization, i.e., the photo-modulation spectroscopy in the present context, is its noninvasiveness, as we have learned before, while electrodes need to be mounted for electroreflectance measurement.

We repeat the Schrödinger equation of electrons in an intrinsic bulk material

$$\left[-\frac{\hbar^2\nabla^2}{2m_0}+V(\boldsymbol{r})\right]\Psi_{\ell\boldsymbol{k}}(\boldsymbol{r})=E_{\ell\boldsymbol{k}}\Psi_{\ell\boldsymbol{k}}(\boldsymbol{r}) \tag{5.6}$$

where $V(\boldsymbol{r})$ is the periodic lattice potential energy, ℓ is the energy band index, $E_{\ell\boldsymbol{k}}$ is the energy band dispersion relationship, $\Psi_{\ell\boldsymbol{k}}(\boldsymbol{r})=e^{i\boldsymbol{k}\cdot\boldsymbol{r}}u_{\ell\boldsymbol{k}}(\boldsymbol{r})/\sqrt{N}$ is its corresponding eigen wave function. For the intrinsic material, the valence-band states are totally occupied while the conduction-band states are empty.

The imaginary part of the dielectric coefficient of the intrinsic material due to an external probing electric field

$$\boldsymbol{E}\,e^{i(\boldsymbol{s}\cdot\boldsymbol{r}-\omega t)}+\text{c.c.} \tag{5.7}$$

is given by (2.125) in the form of

$$\epsilon''(\omega)=\frac{1}{\omega^2}\int|\boldsymbol{e}_s\cdot\boldsymbol{p}_{\text{cv}\boldsymbol{k}}|^2\,\delta(E_{\text{c}\boldsymbol{k}}-E_{\text{v}\boldsymbol{k}}-\hbar\omega)\,\text{d}\boldsymbol{k} \tag{5.8}$$

apart from physical constants, when electrons transit from occupied valence-band states $\Psi_{vk}(\boldsymbol{r})$ to empty conduction-band states $\Psi_{ck}(\boldsymbol{r})$. $\boldsymbol{e}_s$ is the unit vector of the electric field of (5.7). And for the intrinsic material, we let the relaxation energy goes to zero so that the Lorentzian function in (2.125) becomes a delta function in the above equation to simplify mathematical expressions.

Equation (5.8) gives us the optical spectrum of a semiconductor material as a function of the photon energy $\hbar\omega$ of the probing electric field of (5.7). (We can obtain $\epsilon'(\omega)$ from $\epsilon''(\omega)$, right?)

Now we add a modulating electric field $E_0 z_0$ along the z axis, where z_0 is the unit vector of the z axis (alternatively we may define the direction of the modulating electric field as the z direction). Equation (5.6) becomes

$$\left[-\frac{\hbar^2\nabla^2}{2m_0} + V(\boldsymbol{r}) - eE_0 z\right]\Phi_{\ell\nu\boldsymbol{k}_{xy}}(\boldsymbol{r}) = \eta_{\ell\nu\boldsymbol{k}_{xy}}\Phi_{\ell\nu\boldsymbol{k}_{xy}}(\boldsymbol{r}) \tag{5.9}$$

where ℓ is the energy band index, $\boldsymbol{k}_{xy}$ is the wave vector in the plane perpendicular to the modulating electric field $E_0 z_0$, ν is the new quantum index in the z axis. Since the modulating electric field applies along the z axis, the periodicity in the xy plane is preserved, and $\boldsymbol{k}_{xy}$ in (5.9) is still a good quantum index in the xy plane. Thus, the eigen function $\Phi_{\ell\nu\boldsymbol{k}_{xy}}(\boldsymbol{r})$ of the above equation can be expressed as

$$\Phi_{\ell\nu\boldsymbol{k}_{xy}}(\boldsymbol{r}) = \sum_{k_z} C_{\ell\nu}(\boldsymbol{k})\Psi_{\ell\boldsymbol{k}}(\boldsymbol{r}) \tag{5.10}$$

where $\boldsymbol{k} = (\boldsymbol{k}_{xy}, k_z)$. Here we have neglected the optical transition between different energy bands that may be caused by $E_0 z_0$. In other words, we assume that $E_0 z_0$ is very weak and $-eE_0 z$ in (5.9) is only a perturbation (the value of z is limited in practice such as in nanostructures).

Note that the eigen solution of (5.6) is $\Psi_{\ell\boldsymbol{k}}(\boldsymbol{r}) = e^{i\boldsymbol{k}\cdot\boldsymbol{r}}u_{\ell\boldsymbol{k}}(\boldsymbol{r})/\sqrt{N}$, i.e., the Bloch theorem in Chap. 2. Insert (5.10) into (5.9) results in secular equations for $C_{\nu n}(\boldsymbol{k})$

$$\left[E_{\ell\boldsymbol{k}} - \eta_{\ell\nu\boldsymbol{k}_{xy}} - ieE_0\frac{\partial}{\partial k_z}\right]C_{\ell\nu}(\boldsymbol{k}) = 0 \tag{5.11}$$

having a general solution in the form of

$$C_{\ell\nu}(\boldsymbol{k}) = \frac{1}{\sqrt{N_z}}\exp\left[\frac{i}{eE_0}\int_0^{k_z}\left(\eta_{\ell\nu\boldsymbol{k}_{xy}} - E_{\ell\boldsymbol{k}_{xy},k'_z}\right)\mathrm{d}k'_z\right] \tag{5.12}$$

In (5.11), we replace z by its quantum mechanical counterpart $i\partial/\partial k_z$. As before, N_z is the number of unit cells in the z direction.

The eigen value $\eta_{\ell\nu\boldsymbol{k}_{xy}}$ of (5.9) can be expressed as

$$\eta_{\ell\nu\boldsymbol{k}_{xy}} = \frac{2\pi\nu eE_0}{K_z} + \tilde{E}_{\ell\boldsymbol{k}_{xy}} \tag{5.13}$$

where K_z is the period of the $\boldsymbol{k}$ space in the k_z direction, i.e.,

$$C_{\ell v,\boldsymbol{k}+K_z\boldsymbol{z}_0} = C_{\ell v\boldsymbol{k}}$$

so that

$$\tilde{E}_{\ell \boldsymbol{k}_{xy}} = \frac{1}{K_z}\int_0^{K_z} E_{\ell \boldsymbol{k}_{xy}k_z}\,\mathrm{d}k_z \tag{5.14}$$

(The periodicity of the energy dispersion relationship and the wave function in the $\boldsymbol{k}$ space are obvious because of the periodic lattice structure in the $\boldsymbol{r}$ space.)

Knowing the wave functions and their eigen values a lengthy mathematical analysis leads to the expression of the imaginary part of the dielectric coefficient under modulation field $E_0\boldsymbol{z}_0$

$$\begin{aligned}\epsilon''(\omega, E_0\boldsymbol{z}_0) = \frac{1}{\omega^2}\int\Bigg\{\int_{-\infty}\left(\boldsymbol{e}_s\cdot \boldsymbol{p}_{\mathrm{cv},\boldsymbol{k}+s\boldsymbol{z}_0}\right)\left(\boldsymbol{e}_s\cdot \boldsymbol{p}^*_{\mathrm{cv},\boldsymbol{k}-s\boldsymbol{z}_0}\right)\\ \times \exp\left[\frac{i}{eE_0}\int_{-s}^{s}\left(E_{\mathrm{c},\boldsymbol{k}+v\boldsymbol{z}_0} - E_{\mathrm{v},\boldsymbol{k}+v\boldsymbol{z}_0} - \hbar\omega\right)\mathrm{d}v\right]\mathrm{d}s\Bigg\}\,\mathrm{d}\boldsymbol{k}\end{aligned} \tag{5.15}$$

where s and v are wave numbers in the k_z direction.

Under normal device function conditions, the amplitude of $\boldsymbol{k}$ is small so that we can neglect high orders in k_z in the following Taylor expansion

$$E_{\ell,\boldsymbol{k}+k_z\boldsymbol{z}_0} = E_{\ell\boldsymbol{k}} + k_z\frac{\partial E_{\ell\boldsymbol{k}}}{\partial k_z} + \frac{k_z^2}{2!}\frac{\partial^2 E_{\ell\boldsymbol{k}}}{\partial k_z^2} + \cdots \tag{5.16}$$

and

$$\int_{-s}^{s}\left(E_{\mathrm{c},\boldsymbol{k}+v\boldsymbol{z}_0} - E_{\mathrm{v},\boldsymbol{k}+v\boldsymbol{z}_0} - \hbar\omega\right)\mathrm{d}v = 2s\left(E_{\mathrm{c}\boldsymbol{k}} - E_{\mathrm{v}\boldsymbol{k}} - \hbar\omega\right) + \frac{s^3}{3}\frac{\partial^2\left(E_{\mathrm{c}\boldsymbol{k}} - E_{\mathrm{v}\boldsymbol{k}}\right)}{\partial k_z^2} \tag{5.17}$$

Here we have used the condition of $\nabla_{\boldsymbol{k}}E_{\ell\boldsymbol{k}} = 0$ due to the periodicity in the real space, i.e., (2.9).

For spherical and parabolic conduction and valence bands,

$$E_{\mathrm{c}\boldsymbol{k}} = E_\mathrm{c} + \frac{\hbar^2\boldsymbol{k}^2}{2m_\mathrm{c}^*}, \quad E_{\mathrm{v}\boldsymbol{k}} = E_\mathrm{v} - \frac{\hbar^2\boldsymbol{k}^2}{2m_\mathrm{v}^*} \tag{5.18}$$

so that we have

$$\frac{\partial^2\left(E_{\mathrm{c}\boldsymbol{k}} - E_{\mathrm{v}\boldsymbol{k}}\right)}{\partial k_z^2} = \frac{\hbar^2}{m_\mathrm{e}^*} + \frac{\hbar^2}{m_\mathrm{h}^*} = \frac{\hbar^2}{m_\mathrm{r}^*} \tag{5.19}$$

m_r^* in the above expression is the well-known reduced effective mass we encountered before, e.g., in (2.129). The exponential term in (5.15) becomes now

$$\exp\left[\frac{i}{eE_0}2s\Big(E_{\mathrm{c}\boldsymbol{k}}-E_{\mathrm{v}\boldsymbol{k}}-\hbar\omega\Big)+\frac{i\hbar^2s^3}{3eE_0m_\mathrm{r}^*}\right]$$
$$\approx\exp\left[\frac{i}{eE_0}2s\Big(E_{\mathrm{c}\boldsymbol{k}}-E_{\mathrm{v}\boldsymbol{k}}-\hbar\omega\Big)\right]+\frac{i\hbar^2s^3}{3eE_0m_\mathrm{r}^*}\exp\left[\frac{i}{eE_0}2s\Big(E_{\mathrm{c}\boldsymbol{k}}-E_{\mathrm{v}\boldsymbol{k}}-\hbar\omega\Big)\right] \tag{5.20}$$

Apart from physical constants, it is easy to see that the second term in the above expression is equivalent to

$$\frac{\partial^3}{\partial\omega^3}\left\{\exp\left[\frac{i}{eE_0}2s\Big(E_{\mathrm{c}\boldsymbol{k}}-E_{\mathrm{v}\boldsymbol{k}}-\hbar\omega\Big)\right]\right\} \tag{5.21}$$

When $\boldsymbol{p}_{\mathrm{cv}}$ is weakly dependent on $\boldsymbol{k}$, which we have assumed before since the amplitude of $\boldsymbol{k}$ involved in the optical transitions of our interest is small, and let

$$B=\int\Big(\boldsymbol{e}_\mathrm{s}\cdot\boldsymbol{p}_{\mathrm{cv}\boldsymbol{k}}\Big)\Big(\boldsymbol{e}_s\cdot\boldsymbol{p}_{\mathrm{cv}\boldsymbol{k}}^*\Big)\left\{\int_{-\infty}^{\infty}\exp\left[\frac{2is}{eE_0}\Big(E_{\mathrm{c}\boldsymbol{k}}-E_{\mathrm{v}\boldsymbol{k}}-\hbar\omega\Big)\right]\mathrm{d}s\right\}\mathrm{d}\boldsymbol{k} \tag{5.22}$$

Equation (5.15) now becomes

$$\epsilon''(\omega,Ez_0)=\frac{1}{\omega^2}\left(B+\frac{\partial^3B}{\partial\omega^3}\right) \tag{5.23}$$

Since mathematically

$$\int_{-\infty}^{\infty}\exp\left[\frac{2is}{eE_0}\Big(E_{\mathrm{c}\boldsymbol{k}}-E_{\mathrm{v}\boldsymbol{k}}-\hbar\omega\Big)\right]\mathrm{d}s=eE_0\delta\Big(E_{\mathrm{c}\boldsymbol{k}}-E_{\mathrm{v}\boldsymbol{k}}-\hbar\omega\Big) \tag{5.24}$$

B in (5.22) is actually $\omega^2\epsilon''(\omega)$ and we therefore obtain

$$\epsilon''(\omega,E_0z_0)=\epsilon''(\omega)+eE_0\frac{1}{\omega^2}\frac{\partial^3\Big[\omega^2\epsilon''(\omega)\Big]}{\partial\omega^3} \tag{5.25}$$

by (5.8). The above expression shows that the effect of the photo-modulation on the dielectric coefficient is mathematically the third-order differentiation of the original dielectric coefficient [1, 2].

Figure 5.1 shows the real (ϵ', left column) and imaginary (ϵ'', right column) parts of the dielectric coefficient of Ge as well as their derivatives [3]. Figure 5.1a is the dielectric coefficient obtained from high-resolution ellipsometry measurements and the Kramers–Kronig relationship (recall Sect. 3.4 that we use the Kramers–Kronig relationship to calculate the phase factor from directly measured amplitude of the reflection coefficient and thereafter the dielectric coefficient, both its real and imaginary parts). (b)–(d) are the first-order, the second-order, and the

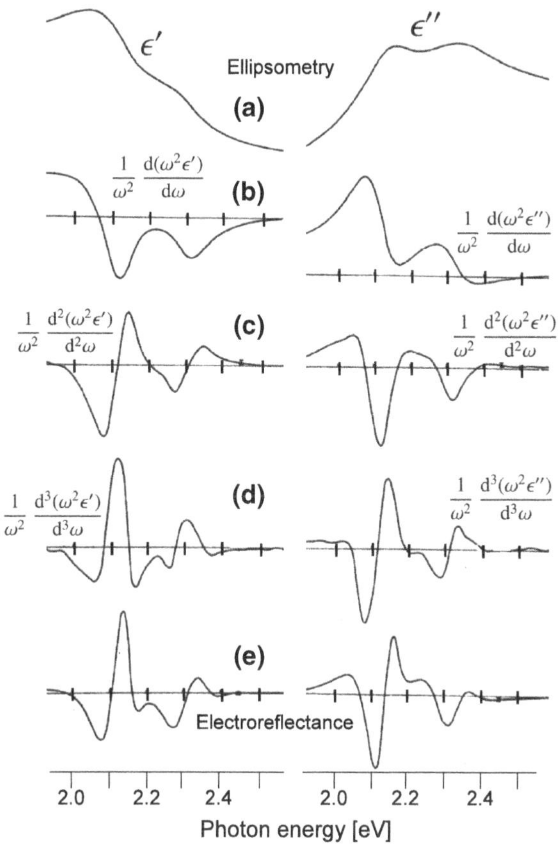

Fig. 5.1 Dielectric coefficient $\epsilon = \epsilon' + i\epsilon''$ of Ge [3]. Left column: real part ϵ', right column: imaginary part ϵ''. **a** are obtained from high-resolution ellipsometry measurements and the Kramers–Kronig relationship, **b**–**d** are the first-, second-, and third-order derivatives of ϵ' and ϵ''. **e** Experimental electroreflectance spectra. (Reprinted with permission from D. E. Aspnes, Phys. Rev. Lett. vol. 28, pp. 168–171, 1972. Copyright 1972 by the American Physical Society)

third-order derivatives, and (e) are the experimental electroreflectance spectra. It is easy to appreciate the agreement between (d) and (e), thereafter directly verify (5.25). Here we see the challenges in resolving fine structures in both ϵ' and ϵ'' in Fig. 5.1a, while the peaks and valleys in Fig. 5.1e are unambiguously clear.

5.2 Photoreflectance Spectroscopy

By Chaps. 2 and 3 we have learned the relationship between the dielectric coefficient $\epsilon = \epsilon' + i\epsilon''$ and the reflectance or transmittance. The macroscopic quantity related to microscopic electron transitions is the imaginary part ϵ'' of the dielectric coefficient, i.e., (5.8), also (2.125), which is repeated below

$$\epsilon''(\omega) = \sum_{qk} \frac{\pi \hbar^2 e^2}{m_0^2 \omega^2 \Omega} |\langle \Psi_q | \boldsymbol{e}_s \cdot \nabla | \Psi_k \rangle|^2 \frac{\Gamma_{qk}}{(E_q - E_k \pm \hbar\omega)^2 + \Gamma_{qk}^2} \Big[f(E_q) - f(E_k) \Big] \tag{5.26}$$

It is observed in Chap. 3 that the direct way to determine ϵ'' is the transmission measurement. However, the reflectance spectrum at normal incidence is the most practical measurement setup, which depends on both the real and imaginary parts of the dielectric coefficient. By (3.10),

$$R = \frac{(n-1)^2 + \kappa^2}{(n+1)^2 + \kappa^2} \tag{5.27}$$

where $\epsilon' = n^2 - \kappa^2$ and $\epsilon'' = 2n\kappa$, i.e., (2.77). For this, we need to know not only ϵ'' but also ϵ', the two of them are correlated by the Kramers–Kronig relationship, see Sect. 3.4.

For photo-modulated reflectance (photoreflectance) measurement, the modulation of ϵ'' is studied in the previous section. The modulation of ϵ'' can be transferred to ϵ' through the differential Kramers–Kronig relationship

$$\Delta\epsilon'(\omega) = \frac{2}{\pi} \mathrm{P} \int_0^\infty \frac{\omega' \Delta\epsilon''(\omega')}{\omega'^2 - \omega^2} \, d\omega' \tag{5.28}$$

which is easily derived from the Kramers–Kronig relationship of (3.28) in Sect. 3.4 by writing $\epsilon'' + \Delta\epsilon''$ under the integral. We thus know both $\Delta\epsilon'(\omega)$ and $\Delta\epsilon''(\omega)$ by which we in principle are able to obtain the mathematical description about the photoreflectance spectrum via (5.27) by differentiating the reflectance with respect to ϵ' and ϵ''

$$\frac{\Delta R}{R} = \beta' \Delta\epsilon' + \beta'' \Delta\epsilon'' \tag{5.29}$$

where

$$\beta' = \frac{\partial \ln R}{\partial \epsilon'}, \quad \beta'' = \frac{\partial \ln R}{\partial \epsilon''}$$

are Seraphin coefficients [4]. Note that in (5.29) the reflectance modification is expressed as a function of ϵ' and ϵ'', while modulation of (5.5) is expressed as a function of the photon energy.

It is easy to realize that β', β'', ϵ', and ϵ'' are very complicated functions of microscopic electron transitions. One way is to treat them as material parameters of the semiconductor, which however do not give us direct insights about microscopic electron transitions.

The following empirical expression

$$\frac{\Delta R}{R} = \sum_\ell \mathrm{Re} \left[\frac{f_\ell \, e^{i\phi_\ell}}{(E_\ell \pm \hbar\omega + i\hbar\gamma_\ell)^{m_\ell}} \right] \tag{5.30}$$

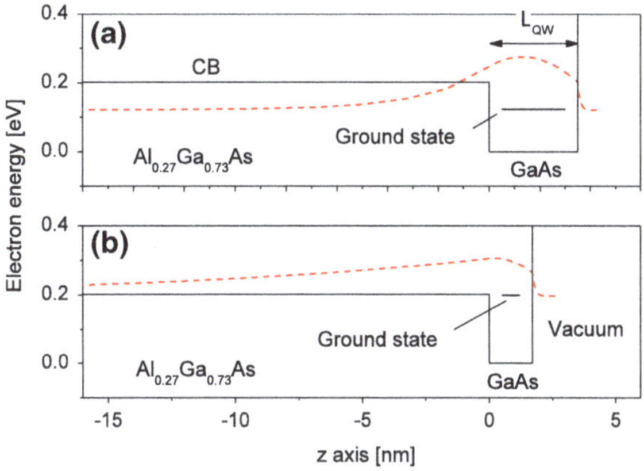

Fig. 5.2 Conduction band (CB) profiles (solid black lines) of two GaAs quantum wells grown on $Al_xGa_{1-x}As$ substrates. **a** GaAs quantum well width $L_{QW} = 3.5$ nm, **b** $L_{QW} = 1.7$ nm. Short-thick lines in the GaAs quantum well region mark the ground state energy levels in their respective wells, and red-dashed lines are the amplitudes of the ground-state envelope wave functions

is commonly accepted and used to describe the photoreflectance spectrum. Here $\hbar\omega$, f_ℓ, ϕ_ℓ, E_ℓ and $\hbar\gamma_\ell$ are the photon energy, effective oscillator strength, effective oscillator phase, electron transition energy, and broadening factor, respectively. m_ℓ is an empirical parameter describing the character of the electron transition. For inter-band transitions of extended states, $m_\ell = 3$, while for quantum confined states such as in a narrow quantum well, $m_\ell = 1$.

Reader may see the general rational of this empirical expression by comparing it with (5.26).

In the following we study the photoreflectance spectra of three GaAs/AlGaAs well/barrier based samples.

Figure 5.2 shows the conduction bandedge of a GaAs quantum well grown on an $Al_{0.27}Ga_{0.73}As$ substrate. We start from the Schrödinger equation, (4.35) in Sect. 4.2. For the one-dimensionally confined electron states in the GaAs quantum well, (4.35) becomes

$$\left[\frac{-\hbar^2}{2m^*}\frac{d^2}{dz^2} + V(z)\right]\psi_i(z) = E_i\psi_i(z) \tag{5.31}$$

Here we only discuss electrons in the conduction band so the band index is neglected. In the above equation, i is the quantum number along the z direction. $V(z)$ is the z-dependent conduction bandedge of the quantum well. Refer to Fig. 2.6, for AlAs and GaAs, the energy bandgaps at Γ symmetry points and VBOs are [5],

$$\begin{aligned} E_{g\Gamma,\mathrm{AlAs}} &= 3.099 - \frac{0.885\times10^{-3}T^2}{530+T}, \quad \mathrm{VBO_{AlAs}} = -1.33 \\ E_{g\Gamma,\mathrm{GaAs}} &= 1.519 - \frac{0.5405\times10^{-3}T^2}{204+T}, \quad \mathrm{VBO_{GaAs}} = -0.80 \end{aligned} \tag{5.32}$$

in the unit of eV. Here T is the temperature (unit = K). For $Al_xGa_{1-x}As$ ternary alloy,

$$\begin{aligned} E_{g\Gamma,\mathrm{Al}_x\mathrm{Ga}_{1-x}\mathrm{As}} &= xE_{g\Gamma,\mathrm{AlAs}} + (1-x)E_{g\Gamma,\mathrm{GaAs}} \\ \mathrm{VBO}_{\mathrm{Al}_x\mathrm{Ga}_{1-x}\mathrm{As}} &= x\mathrm{VBO}_{\mathrm{AlAs}} + (1-x)\mathrm{VBO}_{\mathrm{GaAs}} \end{aligned} \tag{5.33}$$

from which we can construct $V(z)$ in the $Al_xGa_{1-x}As$ substrate and the GaAs surface well which are numerically presented as solid black lines in Fig. 5.2.

One end of the GaAs quantum well, $z = 0$ in Fig. 5.2, is in contact with the $Al_xGa_{1-x}As$ substrate while the other end, $z = L_{QW}$, is exposed to air (these quantum wells are also termed as surface wells). Here L_{QW} is the GaAs quantum well thickness. Not to make things too complicated (surface relaxations and oxidation etc.), we assume that the GaAs quantum well is abruptly ended at $z = L_{QW}$. At the semiconductor interface, an electron needs a certain amount energy in order to move from the conduction bandedge to the vacuum. This energy is known as the electron affinity denoted as EA. For GaAs, the commonly accepted value of EA is 4.07 eV. For AlAs, it is 3.50 eV.

With all these data, we obtain $V(z)$ [eV] for our GaAs quantum well depicted in Fig. 5.2

$$V(z) = \begin{cases} 0.202 & z < 0 \\ 0.0 & 0 \le z < L_{QW} \\ 4.07 & z \ge L_{QW} \end{cases} \tag{5.34}$$

at room temperature ($T = 300$ K). And for an eigen value E_i for (5.31), the corresponding envelope wave function

$$\psi_i(z) = \begin{cases} A_1 e^{k_1 z} & z < 0 \\ A_2 e^{ik_2 z} + B_2 e^{-ik_2 z} & 0 \le z < L_{QW} \\ A_3 e^{-k_3 z} & z \ge L_{QW} \end{cases} \tag{5.35}$$

and its first-order derivative are continuous across the sample. Here

$$\frac{\hbar^2 k_1^2}{2m^*} = 0.202 - E_i, \quad \frac{\hbar^2 k_2^2}{2m^*} = E_i, \quad \frac{\hbar^2 k_3^2}{2m^*} = 4.07 - E_i \tag{5.36}$$

In Fig. 5.2, the short-thick horizontal lines mark the energy positions of the ground states E_i in the quantum wells, and red-dashed lines are amplitudes of the ground-state envelope wave functions. Theoretical estimation shows that the ground state will be aligned to the conduction bandedge of the $Al_{0.27}Ga_{0.73}As$ substrate when L_{QW} is reduced to 1.7 nm, see Fig. 5.2b.

Figure 5.3 shows the photoreflectance spectra of four GaAs quantum wells at room temperature. Four quantum wells of $L_{QW} = 1.5$, 2.5, 3.5 and 4.5 nm show quite similar spectral peaks. First of all, there is a third-order-derivative peak at 1.75 eV, independent of L_{QW}. Since the energy of this peak agrees with the energy of the optical transition between conduction- and valence bandedges of the $Al_{0.27}Ga_{0.73}As$ substrate, and it is in the form of a third order derivative for a bulk material that we studied in the previous section, the peak at 1.75 eV is unambiguously identified to be originated from the $Al_{0.27}Ga_{0.73}As$ substrate.

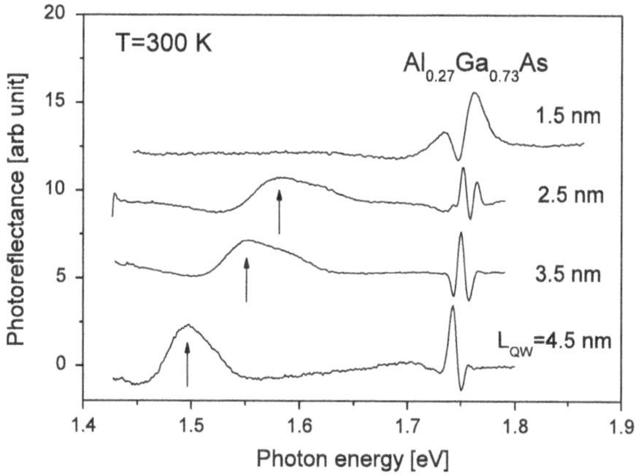

Fig. 5.3 Room-temperature photoreflectance spectra of four GaAs surface wells. Vertical arrows mark the optical transitions between conduction-band and valence-band states confined in the GaAs surface wells. Signals at about 1.75 eV comes from $Al_{0.27}Ga_{0.73}As$ substrates

There is one peak below the bandedge transition of the $Al_{0.27}Ga_{0.73}As$ substrate marked by vertical arrows in the energy range of 1.5–1.7 eV, its peak energy increases when we reduce the well width, similar to the energy level in Fig. 5.2. However, the peak shape certainly bears no resemblance to the third-order derivative form. We know that the corresponding energy levels are confined in the GaAs quantum well. The question therefore is: can (5.30) describe the photoreflectance spectrum of the one-dimensional quantum confinement?

Let us take a close look at the optical transitions between conduction- and valence-band states confined in the GaAs quantum well in the range of 1.5–1.7 eV. The black solid lines in both Fig. 5.4a, b are the same spectrum of $L_{\text{QW}} = 2.5$ nm from Fig. 5.3. When we assume a one-transition process, the numerically fitted spectrum using (5.30) is shown as the red dashed line in Fig. 5.4a, and the three harmonic-oscillator parameters are $\hbar\omega_1 = 1.563$ eV, $\hbar\gamma_1 = 31$ meV, $f_1 = 0.07$. Assume that the photoreflectance peak involves two transitions, the fitting spectrum agrees much better with the experimental data, and the harmonic-oscillator parameters are: (1.554 eV, 30 meV, 0.06) and (1.643 eV, 58 meV, 0.06), see Fig. 5.4b. A close theoretical calculation shows that the low-energy transition (1.554 eV) corresponds to the transition between the conduction-band ground state (E1) and the valence-band heavy hole ground state (H1), and the high-energy (1.643 eV) is between the conduction-band ground state and the valence-band light hole ground state (L1), see the energy diagram in Fig. 4.12, and the photoluminescence spectral features of Fig. 4.13.

The transitions between confined states in a quantum well, including the quantum well at the surface in Fig. 5.4, are intuitively easy to understand. Quantum confinement at a surface barrier is however also possible due to the wave functions' boundary conditions. Reader is encouraged to do theoretical study and numerical analysis by following the discussion of the GaAs quantum well. In the following we introduce the photoreflectance spectra of four different $Al_{0.24}Ga_{0.76}As$ barriers grown on semi-insulating (001)-oriented GaAs substrates, with different barrier thickness $L_{\text{barrier}} = 10$, 15, 25, and 35 nm, see Fig. 5.5a. We use the 632.8-nm line of a He-Ne

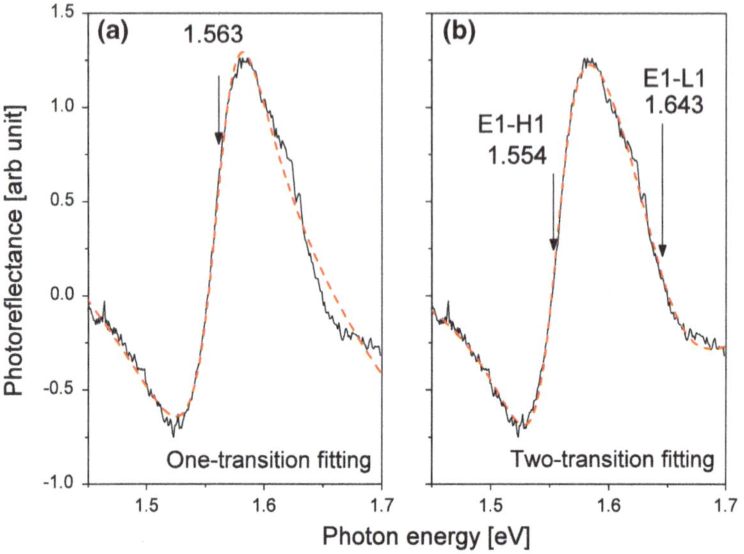

Fig. 5.4 Photoreflectance spectrum (black solid lines) of a GaAs quantum well (well width $L_{QW} = 2.5$ nm) grown on the $Al_{0.27}Ga_{0.73}As$ substrate and its numerical fitting spectra (red-dashed lines). **a** One transition fitting, **b** two-transition fitting

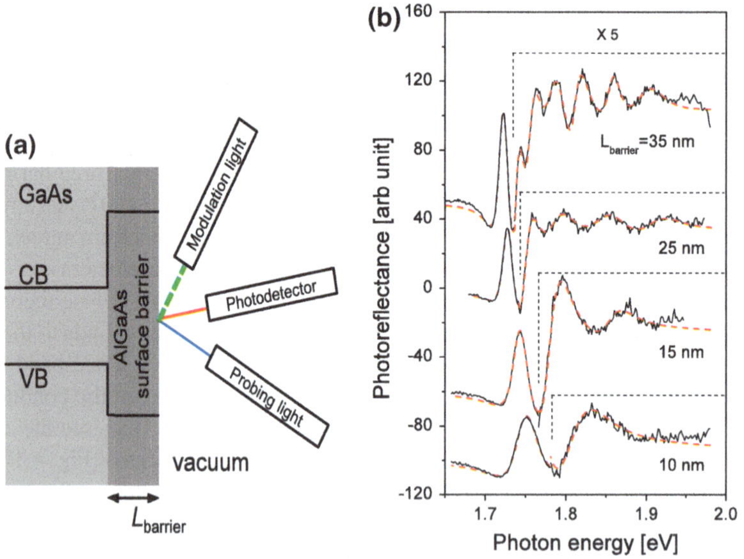

Fig. 5.5 **a** $Al_{0.24}Ga_{0.76}As$ surface barrier on GaAs substrate and photoreflectance measurement. CB = conduction band, and VB = valence band. **b** Photoreflectance spectra (black solid lines) of four $Al_{0.24}Ga_{0.76}As$ barriers. The parts of the spectra marked by vertical and horizontal dashed lines are magnified for better visualization. Red dashed lines are numerical fitting spectra

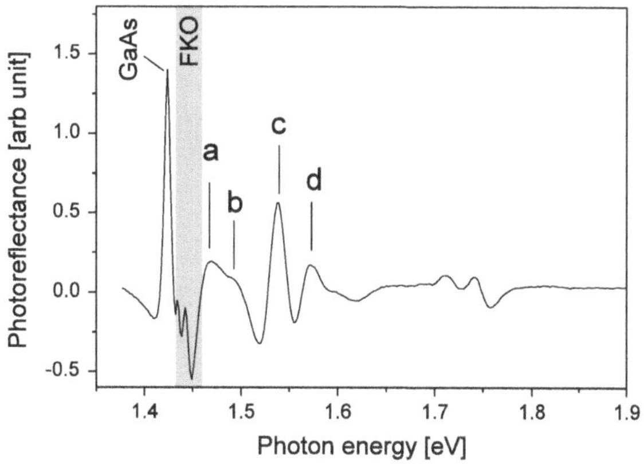

Fig. 5.6 Room-temperature photoreflectance spectrum of a multiple GaAs/AlGaAs well/barrier structure

laser as the photo-modulating light with a chopping frequency at 300 Hz. The probing beam from a tungsten-halogen lamp impinges on the sample with a light spot size of about 3 mm in diameter. And the photoreflectance spectra of the four barriers are shown in Fig. 5.5b. It is clearly observed here that there are quantum confinements in the surface barrier.

Figure 5.6 presents a room-temperature photoreflectance spectrum of a GaAs/AlGaAs multiple quantum well structure grown on a GaAs substrate. The peak at 1.42 eV is originated from the GaAs substrate. The oscillations just above the GaAs bandgap are most probably the so-called Franz–Keldysh oscillations (FKO) that the wave functions of the GaAs bandedge states, in both the conduction band and the valence band, penetrate into the $Al_{0.24}Ga_{0.76}As$ barrier regions induced by the electric field of the modulating light. Peaks marked by a, b, c and d are related to the optical inter-band transitions in the multiple quantum wells.

The application of photoreflectance spectroscopy has been largely restricted to short-wavelength infrared spectral region, e.g., it is ca $\in (1.4, 1.9)$ eV, i.e., $\in (700, 900)$ nm in Figs. 5.6 and 5.5. (The conversion between photon energy $\hbar\omega$ in eV and wavelength λ in μm is $\lambda = 1.2398/\hbar\omega$.) The spectral region was pushed to long wave infrared in 2006 about 9 μm [6], and very-long wave infrared of up to 20 μm in 2009 [7]. Figure 5.7a show a photoreflectance spectrum of arsenic-doped $Hg_{0.77}Cd_{0.23}Te$ at 77 K in a spectral range from 2 to 22 μm. By directly comparing with the photoluminescence spectrum in Fig. 5.7b, it is seen clearly that the photoreflectance spectrum is much information richer than the photoluminescence spectrum. Moreover, the spectral resolution is of critically importance for resolving narrow photoreflectance features, see Fig. 5.7c [8].

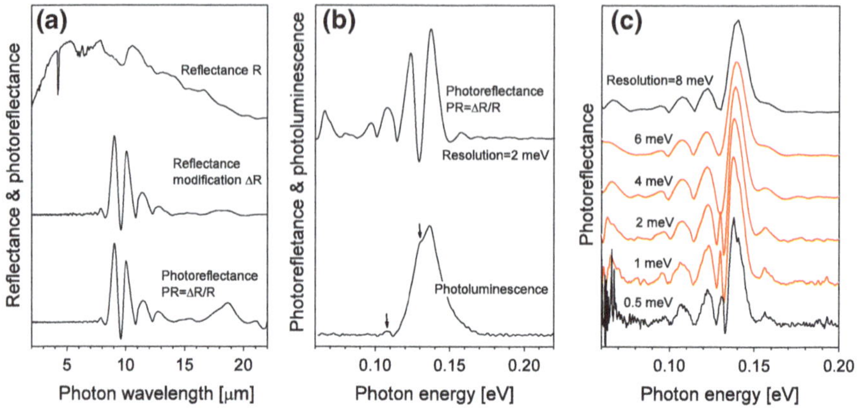

Fig. 5.7 **a** Reflectance R and reflectance modulation ΔR spectra recorded simultaneously for arsenic-doped $Hg_{0.77}Cd_{0.23}Te$ at 77 K, resulting in the photoreflectance $PR = \Delta R/R$. **b** The PR spectrum (same in **a** but now in a linear energy [eV] scale) and photoluminescence spectrum. Vertical arrows two resolvable weak photoluminescence features. **c** Photoreflectance spectra of the $Hg_{0.77}Cd_{0.23}Te$ at 95 K of different spectral resolutions

5.3 Thermo-Modulation Spectroscopy

Thermo-modulation spectrum is obtained when we add an oscillating thermal field, induced normally by an oscillating electric bias or a periodically chopped laser beam at a certain frequency, to the sample. In such an operation, the periodicity of the crystal lattice is not modified so that the dielectric coefficient of the sample is still expressed by (5.4). However, an increased temperature normally results in a volume expansion, i.e., a larger lattice constant, which in its turn, means weaker interactions among atoms in the lattice, thereafter a reduced energy bandgap in general, see Sect. 2.1. Moreover, we mentioned before that the electron-phonon interaction is the major energy relaxation process that largely determines the relaxation energy. The first consequence of increasing the sample temperature is the increased occupation probabilities of electron, photon and phonon states of high energies. Secondly, phonons are bosons so that the number of phonons per one phonon state is not limited, resulting in a higher density of phonons. All these point to a larger relaxation energy at a higher material temperature.

In general, thermo-modulation signal is therefore

$$\frac{d\epsilon(\omega, E_g, \Gamma)}{dT} = \frac{\partial\epsilon(\omega, E_g, \Gamma)}{\partial E_g}\frac{dE_g}{dT} + \frac{\partial\epsilon(\omega, E_g, \Gamma)}{\partial \Gamma}\frac{d\Gamma}{dT} \tag{5.37}$$

where for common zincblende semiconductors [9]

$$\frac{dE_g}{dT} \approx -4.5 \times 10^{-4}\ \text{eV/K}, \quad \frac{d\Gamma}{dT} \approx 1.5 \times 10^{-4}\ \text{eV/K} \tag{5.38}$$

Consider the dielectric coefficient in (5.4) which is repeated below

$$\frac{\epsilon(\omega)}{\epsilon_\infty} = 1 + \int \frac{\hbar\omega_{\mathrm{LT}}}{E - \hbar\omega - i\Gamma} \frac{1}{4\pi^2} \left(\frac{2m_{\mathrm{r}}^*}{\hbar^2}\right)^{3/2} \sqrt{E - E_{\mathrm{g}}}\, \theta(E - E_{\mathrm{g}})\, \mathrm{d}E \quad (5.39)$$

Neglect physical constants,

$$\epsilon(\omega, E_{\mathrm{g}}, \Gamma) \propto \int \frac{1}{E - \hbar\omega - i\Gamma} N_3^{\mathrm{a}}(E - E_{\mathrm{g}})\mathrm{d}E \quad (5.40)$$

where $N_3^{\mathrm{a}}(E - E_{\mathrm{g}})$ is the density of associated electron states in the three-dimensional bulk material, see (4.19). Write $\epsilon(\omega, E_{\mathrm{g}}, \Gamma) = \epsilon'(\omega, E_{\mathrm{g}}, \Gamma) + \epsilon''(\omega, E_{\mathrm{g}}, \Gamma)$,

$$\begin{aligned} \epsilon'(\omega, E_{\mathrm{g}}, \Gamma) &\propto \int \frac{E - \hbar\omega}{(E - \hbar\omega)^2 + \Gamma^2} N_3^{\mathrm{a}}(E - E_{\mathrm{g}})\mathrm{d}E \\ \epsilon''(\omega, E_{\mathrm{g}}, \Gamma) &\propto \int \frac{\Gamma}{(E - \hbar\omega)^2 + \Gamma^2} N_3^{\mathrm{a}}(E - E_{\mathrm{g}})\mathrm{d}E \end{aligned} \quad (5.41)$$

Look closely at

$$A = \frac{E - \hbar\omega}{(E - \hbar\omega)^2 + \Gamma^2}, \quad B = \frac{\Gamma}{(E - \hbar\omega)^2 + \Gamma^2} \quad (5.42)$$

A is significant when $\hbar\omega \neq E$, while B is significant when $\hbar\omega = E$. Moreover, Γ is normally quite small so that we may neglect Γ in the denominator in the expression of $\epsilon'(\omega, E_{\mathrm{g}}, \Gamma)$ and $(E - \hbar\omega)$ in the denominator in $\epsilon''(\omega, E_{\mathrm{g}}, \Gamma)$ so that

$$\epsilon'(\omega, E_{\mathrm{g}}, \Gamma) \propto \int \frac{N_3^{\mathrm{a}}(E - E_{\mathrm{g}})}{E - \hbar\omega} \mathrm{d}E, \quad \epsilon''(\omega, E_{\mathrm{g}}, \Gamma) \propto \frac{N_3^{\mathrm{a}}(\hbar\omega - E_{\mathrm{g}})}{\Gamma} \quad (5.43)$$

from which we see that ϵ' is proportional to $N_3^{\mathrm{a}}(E)/(E - E_{\mathrm{g}})$ while ϵ'' is proportional to N_3^{a}. In other words, ϵ' is largely determined by E_{g}, while ϵ'' is a function of both E_{g} and Γ.

Figure 5.8 shows schematically one spatial-gradient-based thermo-modulation spectroscopy. Here, the temperature of the sample is not changing in time, while one end of the sample is heated by a heating resistor, and the other end is fixed on a heat sink, by which a thermal gradient is formed across the sample. By moving the movable mirror to sweep the excitation laser spot along the thermal gradient, i.e., optical path 1 and 2, we measure the thermo-modulation spectrum. The principle advantage of such a spatial-gradient-based thermo-modulation spectroscopy is that the thermal modulation is formed in space so that the heat capacity of the sample does not play a role here (how fast to heat the sample up and how fast to cool down the sample), and the sweeping of the excitation laser spot across the sample can be

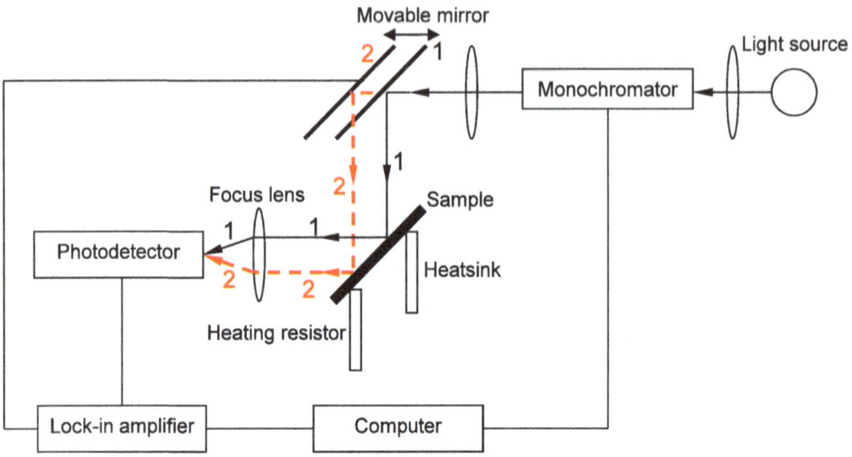

Fig. 5.8 Schematics of a spatial-gradient-based thermo-modulation spectroscopy setup

largely sample-independent. Normally the sweeping frequency of the excitation laser spot is about 100 Hz. In Fig. 5.8, the monochromatic excitation light beam incidents to the movable mirror then reaches and sweeps across the sample, and the difference between the beams from different spatial locations on the sample, e.g., beam 1 and 2 in Fig. 5.8, gives us the thermo-modulation spectroscopic spectrum.

Figure 5.9 presents the spatial-gradient-based thermo-modulation spectroscopic spectra of a semi-insulating (100) GaAs bulk material without and with a temperature gradient. Figure 5.9a shows that when the photon energy becomes larger than E_g, the energy bandgap of GaAs, the reflectance suddenly reduces to the switch on of the inter-band absorption. Applying a spatial temperature gradient, more information becomes revealed in Fig. 5.9b where transitions of E_g, $E_g + \Delta_{so}$ (optical transitions between the valence bandedge and spin-split-off band and the conduction bandedge at Γ point), E_1 and $E_1 + \Delta_1$ (high-energy bands at [111] directions) are clearly displayed. Detailed energy band structure of the GaAs bulk material is presented in Fig. 5.9c.

Figure 5.10 compares thermo-modulation reflectance spectra of a bulk $Hg_{0.8}Cd_{0.2}Te$ material and an epitaxial $Hg_{0.77}Cd_{0.23}Te$ film (thickness 0.5 mm) grown on a ⟨111⟩ CdZnTe substrate by liquid phase epitaxy method. We see the shifts of E_1 and $E_1 + \Delta_1$ due to the difference in the mole fraction x of Cd in the two materials. Most profoundly we also see the much decreased amplitude of $\Delta R/R$ at $E_1 + \Delta_1$ in the epitaxial film. The most possible reason is the strain between the epitaxial layer and the substrate (see next section).

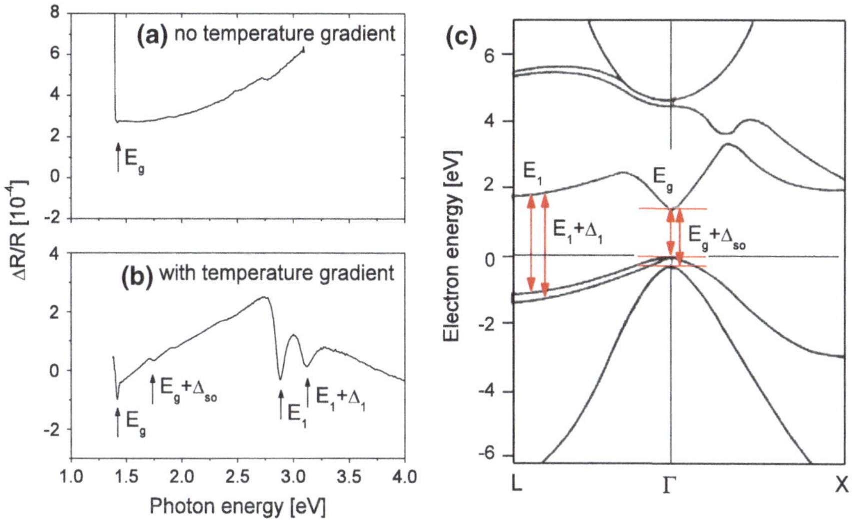

Fig. 5.9 Thermo-modulation reflectance spectra of a semi-insulating (100) GaAs bulk material without (**a**) and with (**b**) a temperature gradient. **c** Energy band structure of the GaAs bulk material

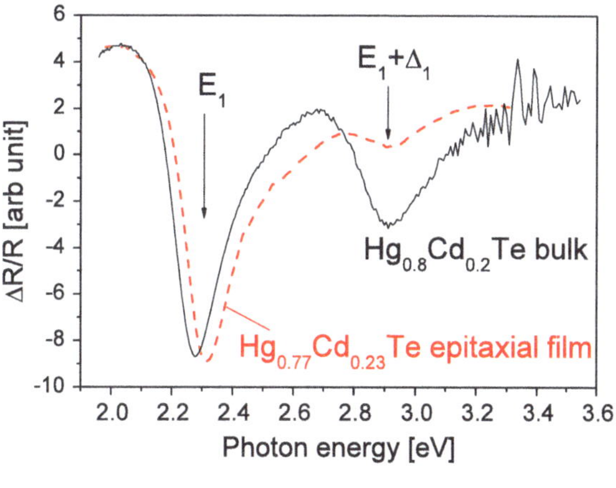

Fig. 5.10 Thermo-modulation reflectance spectrum of a bulk $Hg_{0.8}Cd_{0.2}Te$ material (solid black line) and an epitaxial $Hg_{0.77}Cd_{0.23}Te$ epitaxial film (dashed red line)

5.4 Piezoreflectance Spectroscopy

Piezoreflectance (PzR) spectroscopy is another modulation technique. It is performed by attaching the sample to a piezoelectric material. By applying an oscillating electric bias on the piezoelectric material a strain field is generated in the sample, which will induce modifications in the energy band structure of the sample, resulting in the piezo-modulation spectroscopy.

Let us study the piezo-modulation spectroscopy from the fundamental solid-state physics. Consider a unit cell located at (x, y, z). Under an external mechanical force defined by a three-dimensional Cauchy stress tensor $[\sigma_{ij}]$, $i, j = x, y, z$, the unit cell is displaced to $(x + u, y + v, z + w)$. The three-dimensional strains are

$$\varepsilon_{xx} = \frac{\partial u}{\partial x}, \quad \varepsilon_{yy} = \frac{\partial v}{\partial y}, \quad \varepsilon_{zz} = \frac{\partial w}{\partial z} \tag{5.44}$$

Note that we use the conventional notations ε for the strain in this section, which is to be distinguished from the dielectric coefficient ϵ. And the shear strains are

$$\varepsilon_{yz} = \frac{1}{2}\left(\frac{\partial v}{\partial z} + \frac{\partial w}{\partial y}\right), \quad \varepsilon_{xz} = \frac{1}{2}\left(\frac{\partial w}{\partial x} + \frac{\partial u}{\partial z}\right), \quad \varepsilon_{xy} = \frac{1}{2}\left(\frac{\partial u}{\partial y} + \frac{\partial v}{\partial x}\right) \tag{5.45}$$

Stress $[\sigma_{ij}]$ and strain $[\varepsilon_{ij}]$ are related by a fourth-order covariant tensor

$$\sigma_{ij} = \sum_{k\ell} \lambda_{ijk\ell}\varepsilon_{k\ell} \tag{5.46}$$

Pure mathematically there are 81 elements in $[\lambda_{ijk\ell}]$, also known as elastic coefficients. However, due to the symmetry of the stress and strain, there are only 21 independent elastic coefficients. Furthermore, for zincblende semiconductors, the number of independent elastic coefficient is further reduced to only 3. They are $\lambda_{xxxx} = C_{11}$, $\lambda_{xxyy} = C_{12}$, and $\lambda_{xyxy} = C_{44}$. For wurtzite lattice, there are 5 independent coefficients: $\lambda_{xxxx} = C_{11}$, $\lambda_{zzzz} = C_{33}$, $\lambda_{xxyy} = C_{12}$, $\lambda_{xxzz} = C_{13}$, and $\lambda_{xzxz} = C_{44}$.

Consider a zincblende semiconductor with a lattice constant of a. Assume that the numbers of unit cells along the x, y, and z direction be N so that the spatial extensions in the three directions are all Na, and the total volume is $(Na)^3$. Under an external force, the spatial extensions in the x, y, and z directions are modified by δ_x, δ_y, and δ_z, respectively. The displacements and strains are

$$\begin{aligned} u &= \frac{x}{Na}\delta_x, \quad v = \frac{y}{Na}\delta_y, \quad w = \frac{z}{Na}\delta_z \\ \varepsilon_{xx} &= \frac{\partial u}{\partial x} = \frac{\delta_x}{Na}, \quad \varepsilon_{yy} = \frac{\delta_y}{Na}, \quad \varepsilon_{zz} = \frac{\delta_z}{Na} \end{aligned} \tag{5.47}$$

And the Cauchy stress tensor

$$\begin{aligned} \sigma_{xx} &= C_{11}\varepsilon_{xx} + C_{12}\varepsilon_{yy} + C_{12}\varepsilon_{zz} \\ \sigma_{yy} &= C_{12}\varepsilon_{xx} + C_{11}\varepsilon_{yy} + C_{12}\varepsilon_{zz} \\ \sigma_{zz} &= C_{12}\varepsilon_{xx} + C_{12}\varepsilon_{yy} + C_{11}\varepsilon_{zz} \\ \sigma_{yz} &= C_{44}\varepsilon_{yz} \\ \sigma_{zx} &= C_{44}\varepsilon_{zx} \\ \sigma_{xy} &= C_{44}\varepsilon_{xy} \end{aligned} \tag{5.48}$$

Other components of ε and σ are zero.

The stress will induce modifications in the energy band structure of the semiconductor. The conduction band is normally approximated by a single band, and the modification due to an external mechanical force is linear. The valence band is rather complicated and is described by three bands, light-hole, heavy-hole, and spin-split-off band, as well as interactions among them. For many semiconductor materials that are commonly used for photonics, the energy bandgaps are around 1 eV so that we need to include the interactions between the conduction band and the valence band. The most commonly applied theoretical model to describe all these bands and their interactions is the so-called eight-band $\boldsymbol{k} \cdot \boldsymbol{p}$ theory. In this model, we choose eight basis functions

$$S\uparrow, X\uparrow, Y\uparrow, Z\uparrow, S\downarrow, X\downarrow, Y\downarrow, Z\downarrow \tag{5.49}$$

where S is the s-type orbital, X, Y and Z are p-type, $\uparrow$ and $\downarrow$ denote spin up and down. The $\boldsymbol{k} \cdot \boldsymbol{p}$ Hamiltonian is [10]

$$H_0 + H_{\text{so}} + D(\varepsilon) + H_1(\boldsymbol{k}_\varepsilon) + H_2(\boldsymbol{k}_\varepsilon) \tag{5.50}$$

where H_0 are eigen values of the eight basis functions, H_{so} is the spin-orbit interaction, D is the lattice-deformation potential energy (i.e., lattice deformation due to an external mechanical force), H_1 and H_2 are first- and second-order $\boldsymbol{k} \cdot \boldsymbol{p}$ interactions. The mathematical expressions of these Hamiltonians are:

$$H_0 = \begin{bmatrix} E_g & 0 & 0 & 0 & 0 & 0 & 0 & 0 \\ 0 & -\frac{\Delta_{\text{so}}}{3} & 0 & 0 & 0 & 0 & 0 & 0 \\ 0 & 0 & -\frac{\Delta_{\text{so}}}{3} & 0 & 0 & 0 & 0 & 0 \\ 0 & 0 & 0 & -\frac{\Delta_{\text{so}}}{3} & 0 & 0 & 0 & 0 \\ 0 & 0 & 0 & 0 & E_g & 0 & 0 & 0 \\ 0 & 0 & 0 & 0 & 0 & -\frac{\Delta_{\text{so}}}{3} & 0 & 0 \\ 0 & 0 & 0 & 0 & 0 & 0 & -\frac{\Delta_{\text{so}}}{3} & 0 \\ 0 & 0 & 0 & 0 & 0 & 0 & 0 & -\frac{\Delta_{\text{so}}}{3} \end{bmatrix} \tag{5.51}$$

$$H_{\text{so}} = \frac{\Delta_{\text{so}}}{3} \begin{bmatrix} 0 & 0 & 0 & 0 & 0 & 0 & 0 & 0 \\ 0 & 0 & -i & 0 & 0 & 0 & 0 & 1 \\ 0 & i & 0 & 0 & 0 & 0 & 0 & -i \\ 0 & 0 & 0 & 0 & 0 & -1 & i & 0 \\ 0 & 0 & 0 & 0 & 0 & 0 & 0 & 0 \\ 0 & 0 & 0 & -1 & 0 & 0 & i & 0 \\ 0 & 0 & 0 & -i & 0 & -i & 0 & 0 \\ 0 & 1 & i & 0 & 0 & 0 & 0 & 0 \end{bmatrix} \tag{5.52}$$

$$D(\varepsilon) = \begin{bmatrix} a_c(\varepsilon_{xx} + \varepsilon_{yy} + \varepsilon_{zz}) & 0 & 0 & 0 \\ 0 & D_3(\varepsilon) & 0 & 0 \\ 0 & 0 & a_c(\varepsilon_{xx} + \varepsilon_{yy} + \varepsilon_{zz}) & 0 \\ 0 & 0 & 0 & D_3(\varepsilon) \end{bmatrix} \quad (5.53)$$

$$D_3(\varepsilon) = \begin{bmatrix} l_\varepsilon \varepsilon_{xx} + m_\varepsilon(\varepsilon_{yy} + \varepsilon_{zz}) & n_\varepsilon \varepsilon_{xy} & n_\varepsilon \varepsilon_{xz} \\ n_\varepsilon \varepsilon_{yx} & l_\varepsilon \varepsilon_{yy} + m_\varepsilon(\varepsilon_{xx} + \varepsilon_{zz}) & n_\varepsilon \varepsilon_{yz} \\ n_\varepsilon \varepsilon_{zx} & n_\varepsilon \varepsilon_{zy} & l_\varepsilon \varepsilon_{zz} + m_\varepsilon(\varepsilon_{yy} + \varepsilon_{xx}) \end{bmatrix} \quad (5.54)$$

$$H_1(\boldsymbol{k}_\varepsilon) = \begin{bmatrix} H_4(\boldsymbol{k}_\varepsilon) & 0 \\ 0 & H_4(\boldsymbol{k}_\varepsilon) \end{bmatrix}$$
$$H_4(\boldsymbol{k}_\varepsilon) = \begin{bmatrix} 0 & ipk_{x\varepsilon} & ipk_{y\varepsilon} & ipk_{z\varepsilon} \\ -ipk_{x\varepsilon} & 0 & 0 & 0 \\ -ipk_{y\varepsilon} & 0 & 0 & 0 \\ -ipk_{z\varepsilon} & 0 & 0 & 0 \end{bmatrix} \quad (5.55)$$

$$H_2(\boldsymbol{k}_\varepsilon) = \begin{bmatrix} \frac{\hbar^2}{2m^*}(k_{x\varepsilon}^2 + k_{y\varepsilon}^2 + k_{z\varepsilon}^2) & 0 & 0 & 0 \\ 0 & S(\boldsymbol{k}_\varepsilon) & 0 & 0 \\ 0 & 0 & \frac{\hbar^2}{2m^*}(k_{x\varepsilon}^2 + k_{y\varepsilon}^2 + k_{z\varepsilon}^2) & 0 \\ 0 & 0 & 0 & S(\boldsymbol{k}_\varepsilon) \end{bmatrix} \quad (5.56)$$

$$S(\boldsymbol{k}_\varepsilon) = \begin{bmatrix} Lk_{x\varepsilon}^2 + M(k_{y\varepsilon}^2 + k_{z\varepsilon}^2) & Nk_{x\varepsilon}k_{y\varepsilon} & Nk_{x\varepsilon}k_{z\varepsilon} \\ Nk_{y\varepsilon}k_{x\varepsilon} & Lk_{y\varepsilon}^2 + M(k_{z\varepsilon}^2 + k_{x\varepsilon}^2) & Nk_{y\varepsilon}k_{z\varepsilon} \\ Nk_{z\varepsilon}k_{x\varepsilon} & Nk_{z\varepsilon}k_{y\varepsilon} & Lk_{z\varepsilon}^2 + M(k_{x\varepsilon}^2 + k_{y\varepsilon}^2) \end{bmatrix} \quad (5.57)$$

In the above expressions, E_g is the energy bandgap of strain-free material. Δ_{so} is the spin-orbit interaction energy, and

$$(S|p_x|X) = (S|p_y|Y) = (S|p_z|Z) = i(m_0/\hbar)p_{cv}$$

where p_{cv} is the optical matrix element between the conduction band and the valence band that appear many times in the previous chapters. Moreover, under an external mechanical force, the lattice is deformed from strain-free lattice position $\boldsymbol{r}$ to deformed lattice position $\boldsymbol{r}_\varepsilon$, i.e.,

$$\boldsymbol{r}_\varepsilon = (1 + \varepsilon)\boldsymbol{r}$$

And the corresponding lattice wave vector is similarly modified

$$\boldsymbol{k}_\varepsilon = (1 - \varepsilon)\boldsymbol{k}$$

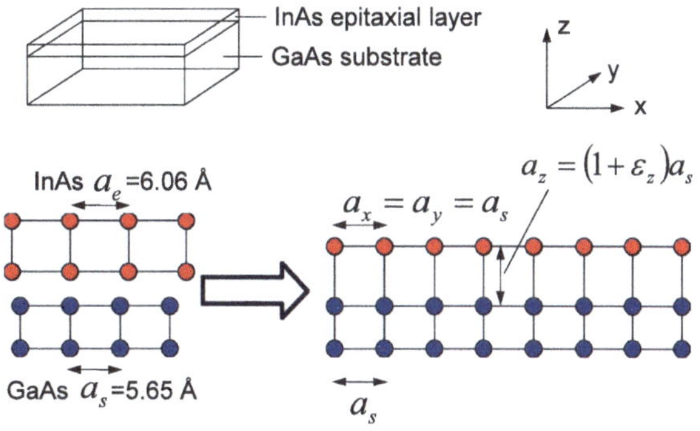

Fig. 5.11 Lattice structure and strain in an InAs epitaxial layer grown on GaAs substrate

In the piezo-modulation spectroscopy, what we do is to modify the deformation potential energy D, which is described by three coefficients a, b, and d given by ℓ_ε, m_ε, and n_ε as below [11]

$$a = \frac{\ell_\varepsilon + 2m_\varepsilon}{3}, \quad b = \frac{\ell_\varepsilon - m_\varepsilon}{3}, \quad d = \frac{n_\varepsilon}{\sqrt{3}} \tag{5.58}$$

All these parameters can be found in literature, see e.g., [5].

Consider an InAs epitaxial layer grown on a GaAs substrate, schematically shown in Fig. 5.11. At room temperature, the lattice constant of the GaAs is $a_s = 5.65$ Å. It is $a_e = 5.65$ Å for InAs. Here we added subscript "s" for substrate and "e" for epitaxial. The stress tensor is minimal when the lattice constants, a_x, a_y and a_z along the x, y, and z direction, of the InAs epitaxial layer are

$$a_x = a_y = a_s, \quad a_z = \left(1 + \varepsilon_{zz}\right)a_s \tag{5.59}$$

Under such a situation, the stress in the InAs epitaxial layer is

$$\varepsilon_{xx} = \varepsilon_{yy} = \frac{a_e - a_s}{a_s}, \quad \varepsilon_{zz} = -\frac{2C_{12}}{C_{11}}\varepsilon_{xx} \tag{5.60}$$

Other components of the stress tensor are zero. In the above equations, C_{12} and C_{11} are elastic coefficients of the InAs material.

It is thus observed that in two-dimensionally extended (in the xy plane) epitaxial layers such as an $In_xGa_{1-x}As$ quantum well grown on a GaAs substrate, the epitaxial layer is biaxially stressed in the xy plane and uniaxially stressed along the sample growth direction (z direction). One effect is the modification in the energy bandgap, and the second effect is the reduction in the degeneracy of the heavy-hole and light-hole bands in the valence band at the Γ symmetry point in the $\boldsymbol{k}$ space. Let "HH" denote heavy hole, "LH" as light hole, and "so" as spin-orbit split band, we construct a new set of basis functions for the valence band from the original basis functions of (5.49)

$$
\begin{aligned}
|\mathrm{HH}_1\rangle &\Rightarrow \left|\frac{3}{2},+\frac{3}{2}\right\rangle = \frac{1}{\sqrt{2}}|(X+iY)\uparrow\rangle \\
|\mathrm{HH}_2\rangle &\Rightarrow \left|\frac{3}{2},-\frac{3}{2}\right\rangle = \frac{1}{\sqrt{2}}|(X-iY)\downarrow\rangle \\
|\mathrm{LH}_1\rangle &\Rightarrow \left|\frac{3}{2},+\frac{1}{2}\right\rangle = \frac{1}{\sqrt{6}}|(X+iY)\downarrow\rangle - \sqrt{\frac{2}{3}}|Z\uparrow\rangle \\
|\mathrm{LH}_2\rangle &\Rightarrow \left|\frac{3}{2},-\frac{1}{2}\right\rangle = \frac{-1}{\sqrt{6}}|(X-iY)\uparrow\rangle - \sqrt{\frac{2}{3}}|Z\downarrow\rangle \\
|\mathrm{so}_1\rangle &\Rightarrow \left|\frac{1}{2},+\frac{1}{2}\right\rangle = \frac{1}{\sqrt{3}}|(X+iY)\downarrow\rangle + \frac{1}{\sqrt{3}}|Z\uparrow\rangle \\
|\mathrm{so}_2\rangle &\Rightarrow \left|\frac{1}{2},-\frac{1}{2}\right\rangle = \frac{-1}{\sqrt{3}}|(X-iY)\uparrow\rangle + \frac{1}{\sqrt{3}}|Z\downarrow\rangle
\end{aligned}
\tag{5.61}
$$

where the digits in the above new basis functions are the angular momentum J and its z component J_z, i.e., $|J, J_z\rangle$.

By using the above basis functions to formulate its Hamiltonian, the deformation potential $D_3(\varepsilon)$ in (5.54) of the eight-band $\boldsymbol{k}\cdot\boldsymbol{p}$ theory becomes now

$$
D_3(\varepsilon) = \begin{array}{c} \quad |3/2,3/2\rangle \qquad |3/2,1/2\rangle \qquad |1/2,1/2\rangle \\ \begin{pmatrix} -\delta E_H + \delta E_S & 0 & 0 \\ 0 & -\delta E_H - \delta E_S & -\sqrt{2}\delta E_S \\ 0 & -\sqrt{2}\delta E_S & -\delta E_H - \Delta_{\mathrm{so}} \end{pmatrix} \end{array}
\tag{5.62}
$$

where δE_H and δE_S are energy shifts due to the biaxial stress in the xy plane and uniaxial stress along the z direction, respectively, and by (5.53) and (5.54) [12]

$$
\delta E_H = -2a\left(1-\frac{C_{12}}{C_{11}}\right)\varepsilon_{xx}, \quad \delta E_S = -b\left(1+\frac{2C_{12}}{C_{11}}\right)\varepsilon_{xx}
\tag{5.63}
$$

from which we obtain the energy shifts of the light-hole, heavy-hole, and conduction-band electron state

$$
\begin{aligned}
\Delta E_{\mathrm{LH}} &= E_{\mathrm{v}} + \delta E_H - \frac{2\Delta_{\mathrm{so}} + 3\delta E_S}{12} + \frac{1}{2}\sqrt{\Delta_{\mathrm{so}}^2 + \Delta_{\mathrm{so}}\delta E_S + \frac{9}{4}\delta E_S^2} \\
\Delta E_{\mathrm{HH}} &= E_{\mathrm{v}} + \delta E_H + \frac{\Delta_{\mathrm{so}}}{3} - \frac{\delta E_S}{2} \\
\Delta E_{\mathrm{e}} &= E_{\mathrm{c}} + \Delta E_S + \frac{\Delta_{\mathrm{so}}}{3} - \delta E_H
\end{aligned}
\tag{5.64}
$$

E_{c} and E_{v} are conduction- and valence bandedges respectively, as before. Here the change of the basis functions from (5.49) to (5.61) is to obtain explicit relationships between the strain field and the energy shift.

Figure 5.12a shows a piezoreflectance spectrum of a GaAs single quantum well embedded between two $\mathrm{Al_{0.29}Ga_{0.71}As}$ barriers, inset shows the fine spectral

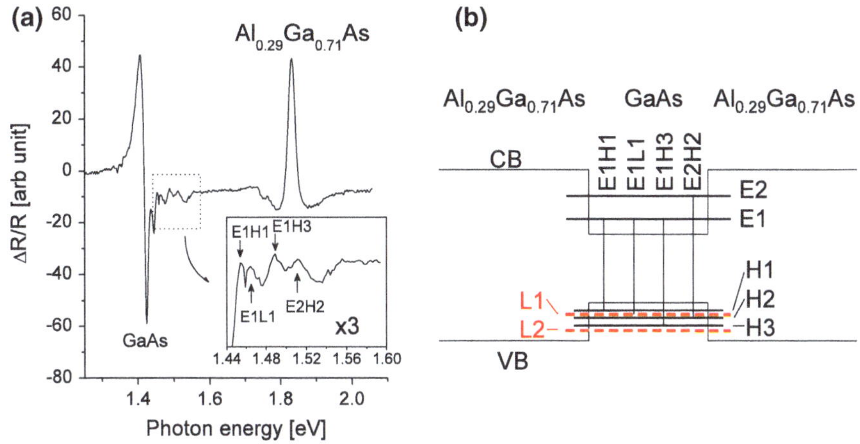

Fig. 5.12 **a** Room-temperature piezoreflectance of a GaAs single quantum well embedded between $Al_{0.29}Ga_{0.71}As$ barriers. **b** Schematic energy diagram and optical transitions

structures. The sample is grown on a semi-insulating GaAs substrate consisting of a 50-nm GaAs buffer layer, a 50-nm $Al_{0.29}Ga_{0.71}As$ barrier, a 25-nm GaAs single quantum well, and finally a 40-nm $Al_{0.29}Ga_{0.71}As$ cap layer. Here we observe not only inter-band transitions from the valence-band light hole and heavy hole state to the conduction-band electron state with $\Delta n = 0$, i.e., E1L1, E1H1 and E2H2, we also see the weak E1H3 transition ($\Delta n \neq 0$), see Fig. 5.12b, where n is the quantum number of the confined state along the z direction.

We have studied the inter-band photoluminescence spectrum of the quantum well in Sect. 4.2. Refer to (4.66), the optical transition between an electron state $\psi_j(z)$ with a quantum number j in the conduction band and a hole state $\psi_i(z)$ with a quantum number i in the valence band is proportional to $\left|\langle\psi_j(z)|\psi_i(z)\rangle\right|$. Δn in the above paragraph refers to $|j - i|$. For simplicity let us assume that the potential of the energy barrier surrounding the quantum well is infinitely high so that the wave functions of the electron states are totally confined within the quantum well. In this case, we study the following equation for the conduction-band electron

$$\left[\frac{-\hbar^2}{2m^*}\frac{\mathrm{d}^2}{\mathrm{d}z^2} + V(z)\right]\psi_j(z) = E_j\psi_j(z) \tag{5.65}$$

with a potential energy

$$V(z) = \begin{cases} \infty & z < 0 \\ 0 & 0 \leq z < L_{\mathrm{QW}} \\ \infty & z \geq L_{\mathrm{QW}} \end{cases} \tag{5.66}$$

where L_{QW} is the quantum well width. And the solutions are

$$E_j = \frac{j^2\hbar^2\pi}{2m^*L_{\mathrm{QW}}^2}$$
$$\psi_j(z) = \sqrt{\frac{2}{L_{\mathrm{QW}}}}\sin\left(\frac{j\pi z}{L_{\mathrm{QW}}}\right) \tag{5.67}$$

Similarly, the envelope wave function for the valence band state with a quantum number i is

$$\psi_i(z) = \sqrt{\frac{2}{L_{\mathrm{QW}}}}\sin\left(\frac{i\pi z}{L_{\mathrm{QW}}}\right) \tag{5.68}$$

Note that the envelope wave functions in the above two equations are only defined for $z \in (0, L_{\mathrm{QW}})$. Outside the quantum well, the envelope functions are zero.

The optical transition is easily calculated

$$\langle\psi_j(z)|\psi_i(z)\rangle = \frac{2}{L_{\mathrm{QW}}}\int_0^{L_{\mathrm{QW}}}\sin\left(\frac{j\pi z}{L_{\mathrm{QW}}}\right)\sin\left(\frac{i\pi z}{L_{\mathrm{QW}}}\right)\mathrm{d}z \tag{5.69}$$

which is nonzero ($= 1$) only when $j = i$, i.e., $\Delta n = |j - i| = 0$. This is known to be a quantum selection rule about optical inter-band transitions in a quantum well. This is mostly reflected in Fig. 5.12.

However, there is also the weak E1H3 transition ($\Delta n \neq 0$) in Fig. 5.12. Below are a few major reasons to understand the weak 13H transition. First of all, the envelope wave functions in (5.67) and (5.68) are obtained under the assumption of (5.66) which are less accurate for high-energy states because the envelope wave functions of high-energy states will penetrate more into energy barrier regions. Second, the valence band is more complicated to be modelled properly by a similar equation as (5.65) which is the major foundation of the eight-band $\boldsymbol{k} \cdot \boldsymbol{p}$ theory. Moreover, the strain applied during the piezoreflectance measurement can modify the interaction between the conduction band and the valence band.

References

1. Y. Fu, K.A. Chao, Validity of the third-derivative modulation spectroscopy. Phys. Rev. B **40**, 10712–10716 (1989)
2. D.E. Aspnes, P. Handler, D.F. Blossey, Interband dielectric properties of solids in an electric field. Phys. Rev. **166**, 921–933 (1968)
3. D.E. Aspnes, Direct verification of the third-derivative nature of electroreflectance spectra. Phys. Rev. Lett. **28**, 168–71 (1972)
4. B.O. Seraphin, N. Bottka, Band-structure analysis from electro-reflectance studies. Phys. Rev. **145**, 628–36 (1966)
5. I. Vurgaftman, J.R. Meyer, L.R. Ram-Mohan, Band parameters for III-V compound semiconductors and their alloys. J. Appl. Phys. **89**, 5815–5875 (2001)

6. J. Shao, F. Yue, X. Lü, W. Lu, W. Huang, Z. Li, S. Guo, J. Chu, Photomodulated infrared spectroscopy by a step-scan Fourier transform infrared spectrometer. Appl. Phys. Lett. **89**, 182121(3) (2006)
7. J. Shao, L. Chen, X. Lü, W. Lu, L. He, S. Guo, J. Chu, Realization of photoreflectance spectroscopy in very-long wave infrared of up to 20 μm. Appl. Phys. Lett. **95**, 041908(3) (2009)
8. L. Ma, J. Shao, X. Lü, S. Guo, W. Lu, Spectral resolution effects on the lineshape of photoreflectance. Chin. Phys. Lett. **28**, 047801(4) (2011)
9. S. Gopalan, P. Lautenschlager, M. Cardona, Temperature dependence of the shifts and broadenings of the critical points in GaAs. Phys. Rev. B **35**, 5577–5584 (1987)
10. P. Enders, A. Bärwolff, M. Woerner, D. Suisky, $\boldsymbol{k} \cdot \boldsymbol{p}$ theory of energy bands, wave functions, and optical selection rules in strained tetrahedral semiconductors. Phys. Rev. B **51**, 16695–16704 (1995)
11. T.B. Bahder, Eight-band $k \cdot p$ model of strained zinc-blende crystals. Phys. Rev. B **41**, 11992–12001 (1990)
12. M. Chandrasekhar, F.H. Pollak, Effects of uniaxial stress on the electroreflectance spectrum of Ge and GaAs. Phys. Rev. B **15**, 2127–2144 (1977)

Chapter 6
Photocurrent Spectroscopy

Abstract Applying an external electric bias on a semiconductor can produce an electric current. Impinging a light on a semiconductor already having an external electric bias can generate an extra electric current called photocurrent whose amplitude depends on the optical power and the wavelength of the light. In this chapter we study the photocurrent spectra of quantum well infrared photodetector and solar cells using quantum dots. We also study the photocurrent induced by multiphoton excitation.

In Chap. 4 we have studied the photoluminescence spectrum where we sent a probing light beam of relatively high-energy photons to the sample then collected photons of different energies emitted from the sample. The microscopic mechanism of the photoluminescence of an intrinsic semiconductor is that one electron, initially occupying one valence-band state, absorbs one incident photon (with an energy $\hbar\omega'$ larger than the energy bandgap E_g of the semiconductor) to transit to one electron state in the empty conduction band, leaving a hole in the valence band; The electron in the conduction band and the hole in the valence band relax to the conduction and valence bandedges, respectively; The electron in the conduction bandedge transits to the empty state (hole) in the valence bandedge by emitting one photon with an energy $\hbar\omega = E_g$ (Fig. 4.1).

The valence band of an intrinsic semiconductor in darkness is completely occupied and the conduction band completely empty, implying that the intrinsic semiconductor is not conductive, see Chap. 2. When we apply an external electric bias across this intrinsic semiconductor in darkness, we do not expect any significant electric current. Now we impinge a light of $\hbar\omega'$ ($> E_g$) on the semiconductor, electrons in the conduction band and holes in the valence band are generated after absorbing photons of energy $\hbar\omega'$. The electrons will be driven towards the positive electrode and the holes towards the negative electrode of the external electric bias during the same time they relax and recombine. When the electric bias is high enough so that electrons and holes can reach electrodes before they recombine to emit photons, an electric current forms. This electric current is called a photocurrent. And the photocurrent spectrum is the relation between the photocurrent and the wavelength of the incident light beam that generates the photocurrent.

W. Lu and Y. Fu, *Spectroscopy of Semiconductors*, Springer Series in Optical Sciences 215, https://doi.org/10.1007/978-3-319-94953-6_6

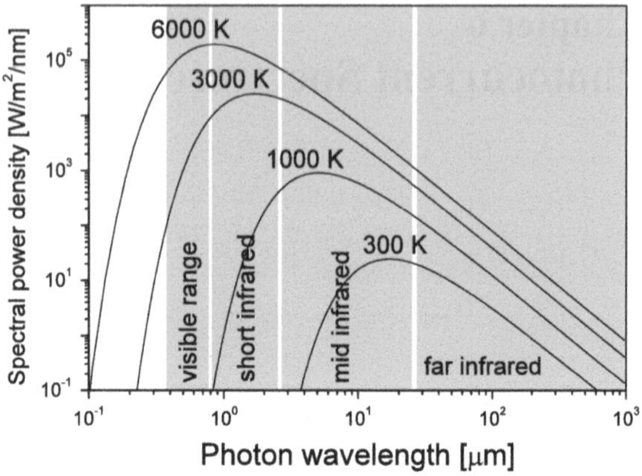

Fig. 6.1 Spectral power density of electromagnetic radiation emitted by a black body in thermal equilibrium at temperature T

6.1 Basics of Quantum Well Infrared Photodetector

As mentioned before, the infrared spectrum is usually divided into three regions: the short- (wavelength ca 0.8–2.5 μm, also known as near infrared region), the middle- (2.5–25 μm) and the far-infrared (25–1000 μm) regions. The importance of the infrared spectrum is reflected by Planck's law for the spectral power density of electromagnetic radiation emitted by a black body in thermal equilibrium at temperature T

$$w(\omega, T) = \frac{2\hbar\omega^3}{\pi c^3} \frac{1}{e^{\hbar\omega/k_B T} - 1} \tag{6.1}$$

which is plotted in Fig. 6.1.

An infrared photodetector absorbs infrared radiations in the infrared range then outputs an electric signal, commonly an electric current under an external electric bias. (Under zero bias, an electric current output of the photodetector is not possible, a voltage builds up instead, which is known as the photovoltaic voltage.) When the density of incident photons is high enough, the output electric signal is directly a photocurrent. When the photon density is too low, the electric signal is to be amplified, and the eventual output is normally quantified in terms of a photon count.

For a very long time, II–VI ternary $Hg_yCd_{1-y}Te$-on-$Cd_{1-x}Zn_xTe$-based narrow-bandgap bulk semiconductors, see Sect. 4.1, have been dominating in the field of infrared photodetections simply because of their narrow energy bandgaps which match with the energies of photons in the three infrared regions. Following the development of the nanotechnology, we now have also semiconductor nanostructures with all possible energy band structures. In 1993 Levine reported the development of $GaAs/Al_xGa_{1-x}As$-based quantum well infrared photodetector (QWIP) [1].

Figure 6.2a is a schematics of the common QWIP material structure. By alternatively growing GaAs and $Al_{0.30}Ga_{0.70}$ layers, GaAs quantum wells sandwiched between thick $Al_{0.30}Ga_{0.70}As$ barriers are formed with discrete energy states in both the conduction band and the valence band. Thick barriers are normally used for

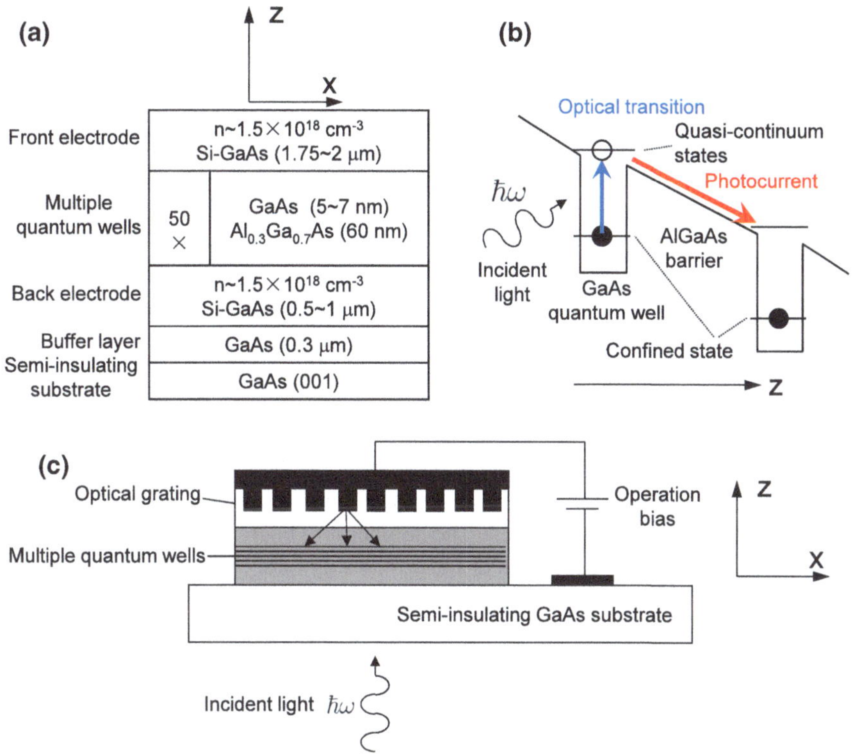

Fig. 6.2 **a** Schematic material structure of the QWIP. **b** Schematic electron transfer in an operating QWIP, and **c** QWIP device operation structure

QWIP applications so that discrete energy states confined in different GaAs quantum wells are not coupled with each other. By tuning the barrier height, i.e., x the mole fraction of Al in $Al_xGa_{1-x}As$ barriers, and/or the width of the GaAs quantum well can we adjust the energy positions of the discrete energy states confined in the GaAs quantum well.

In Fig. 6.2a, with $Al_{0.30}Ga_{0.70}As$ potential barriers of thickness 60 nm, there is only one discrete energy state in the conduction band of the 5-nm wide GaAs quantum well, with an energy distance of ca 125 meV to the conduction bandedge of the $Al_{0.30}Ga_{0.70}As$ barrier. When the discrete energy state is occupied by electrons (note that this energy state is discrete only in the z direction), the electrons can absorb photons then transit to high-energy levels which extend to $Al_{0.30}Ga_{0.70}As$ barrier regions, known as the quasi-continuum states, when the photon energies are larger than 125 meV (corresponding to an optical wavelength of 8 μm, i.e., a middle-infrared light). Here we see that the optical transition occurs between two states in the same conduction band, it is thus called optical intra-band transition. Because the high-energy levels in $Al_{0.30}Ga_{0.70}As$ barrier regions are quite extended and the $Al_{0.30}Ga_{0.70}As$ barrier regions are commonly doped, an external electric field will induce the photo-excited electrons to form the photocurrent, see Fig. 6.2b. Figure 6.2c shows the device structure of a common multiple quantum well infrared photodetector. As to be studied shortly (Sect. 6.1), the electron cannot absorb a photon propagating along the z direction to transit between two energy sublevels in a quantum well structure grown along the same z direction (the so-called quantum selection rule), there is an optical grating to diffract the incident light coming from the substrate side of the QWIP device.

Normally we wish to have only one state confined in the GaAs quantum well region so that the QWIP responses efficiently to only one photon energy (in the above example, the response wavelength is 8 μm). We may modify the Al mole fraction x in the in $Al_xGa_{1-x}As$ barriers, and/or the width of the GaAs quantum well to adjust the energy position of the discrete energy state confined in the GaAs quantum well in order to obtain different response wavelengths. Assume perfect interfaces between the GaAs quantum well and its surrounding $Al_xGa_{1-x}As$ barriers so that we may approximate the potential energy profile of the GaAs quantum well as square. There is a simple but elegant formula about the number of confined energy states in such a square quantum well. The number n of energy states confined in the square quantum well (i.e., energies of those states that are below the barrier height V_0) is given by [2]

$$n = 1 + \mathrm{Int}\left(\sqrt{\frac{2m^* m_0 V_0 L_{QW}^2}{\pi^2 \hbar^2}}\right) = 1 + \mathrm{Int}\left(0.163\sqrt{m^* L_{QW}^2 V_0}\right) \tag{6.2}$$

where mathematical operation Int(A) results in an integer whose magnitude is the largest that does not exceed the magnitude of A. m^* is the electron effective mass. On the right side of the second equals sign, the effective mass is in terms of m_0 ($m^* = 0.067$ for electrons in GaAs), L_{QW} is the quantum well width in units of Å, and the unit of V_0 is eV. The above equation shows that there is minimally one confined state in a square well. For quantum well based infrared photodetector, we wish to have only one confined state in the GaAs quantum well so what we need to do is to adjust L_{QW} and/or V_0 properly so that $0.163\sqrt{m^* L_{QW}^2 V_0}$ is less than one.

Multiple quantum wells are commonly grown so that the photons that are not absorbed by the electrons in the first quantum well have more chances to be absorbed in the second and the third quantum wells. Technically, however, only a limited number GaAs quantum wells can be grown epitaxially with fairly well controlled uniformity (for one-color response) of the quantum well structures (the widths of GaAs wells and $Al_xGa_{1-x}As$ barriers as well as x in $Al_xGa_{1-x}As$).

Figure 6.2b shows the ideal photocurrent formation in a QWIP device. In addition, there are a few more electric current components that exist even when the QWIP is in darkness, which are then denoted as dark currents. They are thermionic emission, thermally assisted tunneling, and direct tunneling. Refer to Fig. 6.3a, the thermionic emission is that at finite temperature, electrons can transit from low-energy states to high-energy states by absorbing thermal energies (i.e., phonons), while tunneling is totally quantum mechanical. There can be first a thermal excitation then a tunneling,

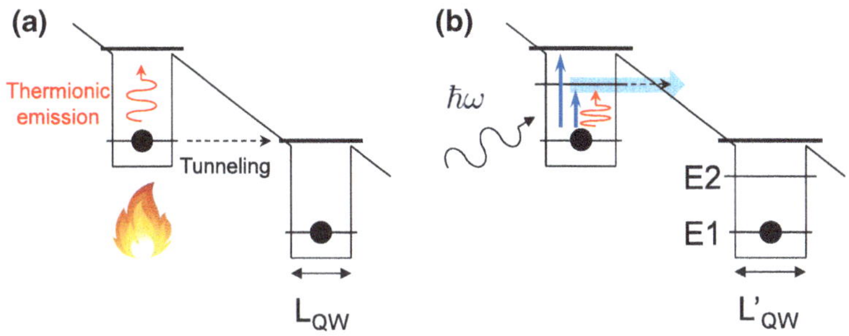

Fig. 6.3 **a** Dark current components: thermionic emission and tunneling. **b** Extra transition paths of electrons from the GaAs quantum well to extended quasi-continuum states in $Al_xGa_{1-x}As$ barriers when the GaAs quantum well is a bit thick so that there are multiple confined states in the GaAs quantum well

commonly known as thermal-assisted tunneling. Thermionic emission is temperature dependent which can be suppressed by lowering the device temperature, and the quantum mechanical tunneling can be suppressed by increasing the barrier thickness.

Knowing the dark currents, we realize the other reason for having only one confined state in the quantum well. Refer to Fig. 6.3b, the combination of the optical absorption, the thermal excitation, from the ground state to the excited state, and the tunneling from the excited state to extended quasi-continuum states in $Al_xGa_{1-x}As$ barriers may contribute significantly to the total electric current, making the analysis of the photocurrent more complicated.

One important aspect about the QWIP device structure is that the GaAs quantum wells are not doped, while the $Al_xGa_{1-x}As$ barriers are n-type doped. Electrons from the n-doped $Al_xGa_{1-x}As$ barriers migrate to the confined energy states in the GaAs quantum wells, ready to be photoexcited. The electric field induced by the external bias applies largely across the $Al_xGa_{1-x}As$ barriers due to the fact that there are very few mobile electrons (thus high resistivity and high electric field) in $Al_xGa_{1-x}As$ barrier regions.

The QWIP is much advantageous in many ways as compared with HgCdTe/CdZnTe-based infrared photodetector. Most noticeably is the uniformity in the xy plane (see discussions in Sect. 4.1) and excellence in the III–V epitaxial techniques, which are extremely important for large-scale focal plane array (FPA) structure [3, 4]. There are already commercialized 512 × 512, 640 × 480, 640 × 512, even 1024 × 1024 GaAs/$Al_xGa_{1-x}As$-based QWIP FPA thermal imaging cameras. Disadvantages of the QWIP are also obvious. Most prominent is the result of the quantum selection rule about the optical intra-band transitions, i.e., the necessary optical grating. The next issue is the relatively low density of the two-dimensional electron states in the quantum well structure as compared with ones in three-dimensional bulk material, see Fig. 4.10, resulting in a low concentration of the carriers that are ready to interact with the incident photons. We already mention one way to overcome the last issue, i.e., multiple quantum wells, which is however limited by the epitaxial growth technique.

Another way is to lower the device operation temperature. This is why QWIP-based thermal imaging camera comes with a built-in cooler.

Anyway, GaAs/Al_xGa_{1-x}As-based thermal imaging device has been steadily improved in parallel with the development of epitaxial techniques such as molecular beam epitaxy (MBE) and metal organic chemical vapor deposition (MOCVD). It is emerging as a complimentary partner to the traditional HgCdTe/CdZnTe-based infrared photodetector in middle- and far-infrared optical ranges [5].

The QWIP device is designed to absorb incident photons then output an electric current. The photocurrent is thus the most fundamental parameter to characterize the QWIP device. Figure 4.14 already includes the peak wavelength of the photocurrent, which is obtained under the operation of the QWIP device with an external bias. We now study carefully the absorption and photocurrent in the next section.

6.2 Photon Absorption and Photocurrent

We focus on an n-type QWIP structure shown in Fig. 6.4. The multiple GaAs quantum wells are designed in such a way that there is only one confined (ground) state in the GaAs quantum well region and the next excited states are correlated with continuum states in the conduction bands of the Al_xGa_{1-x}As barriers. GaAs quantum wells are non-intentionally doped while Al_xGa_{1-x}As barriers are n-type doped. The electrons from the n-type dopants in the Al_xGa_{1-x}As barriers migrate to the confined states in GaAs quantum wells, resulting in a high resistivity in Al_xGa_{1-x}As barriers so that the external bias applies largely in the Al_xGa_{1-x}As barrier regions. Upon the incidence of photons, the electrons occupying the ground state in the GaAs quantum well transit to continuum states above the Al_xGa_{1-x}As conduction bandedge which will be driven by the external bias to form the photocurrent.

We start from the Schrödinger equation (4.33), in Sect. 4.2. For the one-dimensionally confined electron states in the quantum well, (4.33) becomes

$$\left[\frac{-\hbar^2}{2m^*}\frac{\mathrm{d}^2}{\mathrm{d}z^2} + V(z)\right]\psi_i(z) = E_i\psi_i(z) \tag{6.3}$$

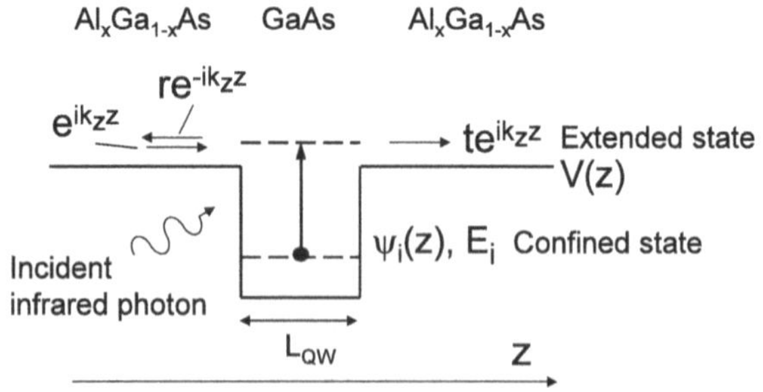

Fig. 6.4 Electron states, optical transition and transports in an n-type GaAs/Al_xGa_{1-x}As QWIP

Here we only discuss electrons in the conduction band so the conduction band index is neglected. In the above equation, i is the quantum number along the z direction, and the envelope function is normalized in the quantum well (width L_{QW})

$$\int_{L_{\mathrm{QW}}} |\psi_i(z)|^2 \mathrm{d}z = 1 \tag{6.4}$$

As discussed before, the final electron state must be an continuum state ψ_{k_z} above the conduction bandedge of the $\mathrm{Al}_x\mathrm{Ga}_{1-x}\mathrm{As}$ barriers in order to form the photocurrent. Consider the electron transport in the $\mathrm{Al}_x\mathrm{Ga}_{1-x}\mathrm{As}$ region. Refer to Fig. 6.4, the boundary conditions for the final state is: it comes as a plane wave $e^{ik_z z}$ from the left side of the GaAs quantum well, which will be reflected back to the $\mathrm{Al}_x\mathrm{Ga}_{1-x}\mathrm{As}$ barrier due to the GaAs quantum well in the form of $re^{-ik_z z}$ so that the final electron state is

$$\psi_{k_z}(z) = e^{ik_z z} + re^{-ik_z z} \tag{6.5}$$

The electron will reach the right side of the GaAs quantum well in the form of

$$\psi_{k_z}(z) = te^{ik_z z} \tag{6.6}$$

which is the transmitted wave. To simplify mathematical operations we have assumed here a very small external bias so the corresponding electric field is neglected in (6.3).

By (4.63), the total wave functions of the initial and the final electron states are

$$\begin{aligned} \Psi_{ik_x k_y} &= \sqrt{\frac{\Omega_{\mathrm{cell}}}{S_{\mathrm{QW}}}}\, \psi_i(z)\, e^{i(k_x x + k_y y)} u(\boldsymbol{r}) \\ \Psi_{\boldsymbol{k}} &= \sqrt{\frac{\Omega_{\mathrm{cell}}}{S_{\mathrm{QW}}}}\, \psi_{k_z}(z)\, e^{i(k_x x + k_y y)} u(\boldsymbol{r}) \end{aligned} \tag{6.7}$$

where $u(\boldsymbol{r})$ is the periodic Bloch function of the conduction band, and S_{QW} is the surface area of the QWIP device in the xy plane. $\psi_i(z)$ and $\psi_{k_z}(z)$ are normalized along the z axis. Here we utilize directly the vertical transition requirement in the $\boldsymbol{k}$ space about the optical excitation and radiative recombination, i.e., the wave numbers k_x and k_y in the xy plane of the initial and final electron states must be the same.

Similar to the derivations of (4.64), we obtain the matrix element of the optical transition from the initial to the final electron states

$$\begin{aligned} \langle \Psi_{\boldsymbol{k}} | \boldsymbol{e}_s \cdot \nabla | \Psi_{ik_x k_y} \rangle &= \frac{\Omega_{\mathrm{cell}}}{S_{\mathrm{QW}}} \boldsymbol{e}_s \cdot \int_{\Omega} \psi_{k_z}^*(z) \psi_i(z) u^*(\boldsymbol{r}) \nabla u(\boldsymbol{r}) \mathrm{d}\boldsymbol{r} \\ &- \frac{i(e_x k_x + e_y k_y)\Omega_{\mathrm{cell}}}{S_{\mathrm{QW}}} \int_{\Omega} \psi_{k_z}^*(z) \psi_i(z) u^*(\boldsymbol{r}) u(\boldsymbol{r}) \mathrm{d}\boldsymbol{r} \\ &+ \frac{e_z \Omega_{\mathrm{cell}}}{S_{\mathrm{QW}}} \int_{\Omega} \psi_{k_z}^*(z) \frac{\partial \psi_i(z)}{\partial z} u^*(\boldsymbol{r}) u(\boldsymbol{r}) \mathrm{d}\boldsymbol{r} \end{aligned} \tag{6.8}$$

where $\boldsymbol{e}_s = (e_x, e_y, e_z)$ is the direction of the electric field of the incident light beam. The second term on the right side of the equals sign in the above equation is zero because of the wave function's orthogonality. Note that different from (4.64), here the initial and final electron states contain the same periodic Bloch function of the conduction band. The first and the third terms can be further evaluated by replacing $\boldsymbol{r}$ of the slow-varying envelope functions as the sum of lattice position $\boldsymbol{R}$ of the unit cell and $\boldsymbol{r}'$ within the unit cell (we did so before). Eventually, the first term is zero since it contains the spatial integration of the product of u and ∇u (the parities of u and ∇u are opposite so that the spatial integration is zero). The third term contains

$$\int_{\text{cell}} u^*(\boldsymbol{r})u(\boldsymbol{r})\mathrm{d}\boldsymbol{r}$$

which is 1 (normalization of the Bloch function). Finally,

$$\left\langle \Psi_{\boldsymbol{k}} \left| \boldsymbol{e}_s \cdot \nabla \right| \Psi_{ik_xk_y} \right\rangle = e_z \int \psi_{k_z}^*(z) \frac{\partial \psi_i(z)}{\partial z}\, \mathrm{d}z \equiv e_z\, W_{k_z i} \tag{6.9}$$

which is known as the optical intra-band transition. In Chap. 4 we study the inter-band photoluminescence of a quantum well, where the optical transition occurs between a valence-band hole state and a conduction-band electron state.

As shown in Figs. 6.2 and 6.4, the electrons in a QWIP are confined along the z direction initially in the quantum well. Both the QWIP, and therefore the electrons are extended in the xy plane. By Sect. 2.3 we know that the electric field of a light beam is perpendicular to the propagation direction of the light beam. If the light beam comes into the device along the z direction, the electric field of the light beam has no component along the z direction, i.e., $e_z = 0$. The transition matrix element of (6.9) is thus zero. This is the so-called quantum selection rule mentioned in the previous section.

Therefore, n-type QWIP needs an optical coupling component such as a diffraction grating we studied in Sect. 1.2 to generate a significantly nonzero e_z component thereafter an optical response, see Fig. 6.2c. This is why the incident lights to the quantum wells in Figs. 6.2b and 6.4 are all schematically tilted. This quantum selection rule however does not apply to the inter-band photoluminescence shown in Fig. 4.12. Figure 6.5 shows the responsivity (the ratio between the photocurrent and the optical power of the incident radiation) enhanced significantly by the metal patch structure on a single-quantum-well infrared photodetector [5].

Refer to Fig. 6.4, let us assume that the potential energy of the conduction bandedge of the $Al_xGa_{1-x}As$ barriers is zero, and the energy distance of the confined state to the $Al_xGa_{1-x}As$ conduction bandedge is E_i. Since the electron occupying E_i is largely confined within the GaAs quantum well for which the electron effective mass is $m_1^* = 0.067m_0$, i.e., the electron effective mass in GaAs. The effective mass of the electron in the continuum states above the $Al_xGa_{1-x}As$ conduction

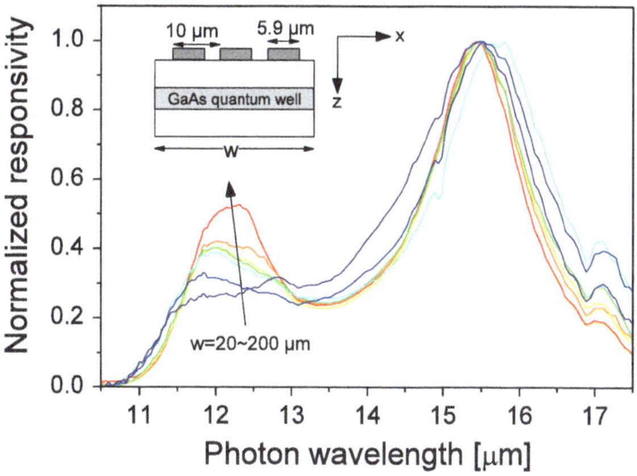

Fig. 6.5 Responsivity is enhanced significantly by the metal patch structure on the single-quantum-well infrared photodetector (207-nm thick GaAs quantum well embedded between $Al_{0.15}Ga_{0.85}As$ barriers). Metal patches are extended along the y axis and the light incidents along the z axis

bandedge assumes the effective mass of the $Al_xGa_{1-x}As$ conduction-band m_2^* (its value depends on the Al mole fraction). The energies of the initial and final electron states are therefore

$$E_{ik_xk_y} = \frac{\hbar^2(k_x^2 + k_y^2)}{2m_1^*} - E_i \quad , \quad E_{\boldsymbol{k}} = \frac{\hbar^2 \boldsymbol{k}^2}{2m_2^*} \tag{6.10}$$

With a nonzero e_z, inserting (6.9) into (2.123) in Chap. 2 will result in the absorption coefficient

$$\begin{aligned}
\alpha(\omega) &= \sum_{\boldsymbol{k},ik_xk_y} \frac{e^2\hbar^2 \left|\langle \Psi_{\boldsymbol{k}}(\boldsymbol{r})|\boldsymbol{e}_s \cdot \nabla |\Psi_{ik_xk_y}(\boldsymbol{r})\rangle\right|^2 \Gamma\left[f(E_{ik_xk_y}) - f(E_{\boldsymbol{k}})\right]}{2m_0^2 c\epsilon\omega\Omega\left[\left(E_{\boldsymbol{k}} - E_{ik_xk_y} - \hbar\omega\right)^2 + \Gamma^2\right]} \\
&= \frac{e_z e^2\hbar^2\Gamma}{2m_0^2 c\epsilon\omega\Omega} \sum_{k_z i} \int\int \frac{|W_{k_z i}|^2\left[f(E_{ik_xk_y}) - f(E_{\boldsymbol{k}})\right]}{\left(E_{\boldsymbol{k}} - E_{ik_xk_y} - \hbar\omega\right)^2 + \Gamma^2} \frac{2\mathrm{d}k_x\mathrm{d}k_y}{(2\pi)^2/S_{\mathrm{QW}}} \\
&= \frac{e_z e^2\hbar^2\Gamma S_{\mathrm{QW}}}{2m_0^2 c\epsilon\omega\Omega} \sum_{k_z i} \int\int \frac{|W_{k_z i}|^2\left[f(E_{ik_xk_y}) - f(E_{\boldsymbol{k}})\right]}{\left(E_{\boldsymbol{k}} - E_{ik_xk_y} - \hbar\omega\right)^2 + \Gamma^2} \frac{2\mathrm{d}k_x\mathrm{d}k_y}{(2\pi)^2}
\end{aligned} \tag{6.11}$$

Note that (2.123) represents in general the optical gain, i.e., one electron transits from an occupied high-energy state to an empty low-energy state to emit a photon, while the above equation represents the photon absorption, with an opposite sign with respect to the optical gain, meaning that one electron transits from an occupied low-energy state $E_{ik_xk_y}$ to an empty high-energy state $E_{\boldsymbol{k}}$ by absorbing one photon.

As mentioned before, the quantum well structure is designed in such a way that there is only one confined state so we drop off the summation over i in the above equation. For the wave number k_z in the continuum $Al_xGa_{1-x}As$ conduction band, its density of electron states is $\mathrm{d}k_z/(2\pi/L_z)$ so that

$$\alpha(\omega) = \frac{e_z e^2 \hbar^2 \Gamma}{2m_0^2 c \epsilon \omega} \int \frac{|W_{k_z i}|^2 \left[f(E_{ik_x k_y}) - f(E_{\boldsymbol{k}}) \right]}{\left(E_{\boldsymbol{k}} - E_{ik_x k_y} - \hbar\omega\right)^2 + \Gamma^2} \frac{2\mathrm{d}\boldsymbol{k}}{(2\pi)^3} \tag{6.12}$$

Note $\Omega = S_{\mathrm{QW}} L_z$. The above expression represents the optical absorption for one GaAs quantum well surrounded by two $\mathrm{Al}_x\mathrm{Ga}_{1-x}\mathrm{As}$ barriers. One way to increase the optical absorption of the QWIP device is to increase the number of quantum wells, N_{QW}, so that

$$\alpha(\omega) = N_{\mathrm{QW}} \frac{e_z e^2 \hbar^2 \Gamma}{2m_0^2 c \epsilon \omega} \int \frac{|W_{k_z i}|^2 \left[f(E_{ik_x k_y}) - f(E_{\boldsymbol{k}}) \right]}{\left(E_{\boldsymbol{k}} - E_{ik_x k_y} - \hbar\omega\right)^2 + \Gamma^2} \frac{2\mathrm{d}\boldsymbol{k}}{(2\pi)^3} \tag{6.13}$$

This is what we depict in Fig. 6.2 that $N_{\mathrm{QW}} = 50$. In principle we may increase N_{QW} as much as we want, while in practice the growth of multiple identical quantum wells is not trivial.

Figure 6.6a shows the theoretical optical absorption spectrum of one GaAs/$\mathrm{Al}_x\mathrm{Ga}_{1-x}\mathrm{As}$ QWIP as a function of Γ (0.1, 1.0, 1.6, and 6.0 meV). Equation (6.13) represents the total probability that electrons initially occupying E_i will transit to the continuum states above the $\mathrm{Al}_x\mathrm{Ga}_{1-x}\mathrm{As}$ conduction bandedge by absorbing incident photons. It also represents the number of electrons that occupy the initially empty $\mathrm{Al}_x\mathrm{Ga}_{1-x}\mathrm{As}$ conduction band after absorbing photons, i.e., the photogenerated electrons, or simply photocarriers. The photocarriers will form the photocurrent when an external bias is applied, see Fig. 6.2b, c. Equation (6.13) therefore directly represents the photocurrent of the QWIP device. This can be observed by comparing

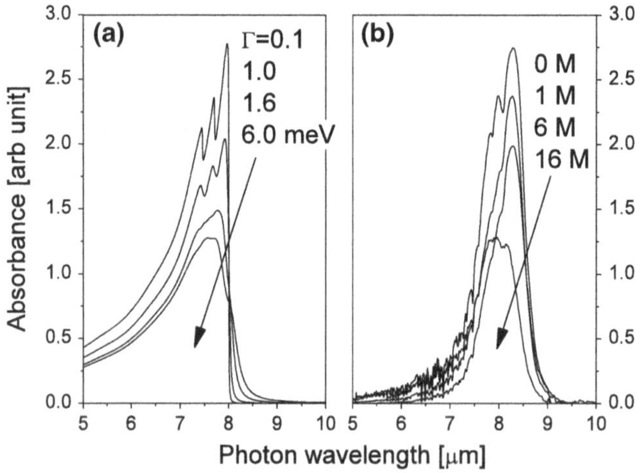

Fig. 6.6 **a** Theoretical absorption spectrum of a GaAs/$\mathrm{Al}_x\mathrm{Ga}_{1-x}\mathrm{As}$ QWIP as a function of relaxation energy $\Gamma = 0.1$, 1.0, 1.6, and 6.0 meV. **b** Photocurrent spectrum of the GaAs/$\mathrm{Al}_x\mathrm{Ga}_{1-x}\mathrm{As}$ QWIP device after γ irradiation of different doses (0, 1, 6, and 16 Mrad)

the theoretical absorption spectrum of Fig. 6.6a with the experimental photocurrent spectrum shown in Fig. 6.6b.

Figure 6.6a shows that the QWIP responses to a light beam with a wavelength below 8 μm. 8 μm is therefore denoted as the response wavelength λ_0 of the QWIP, which corresponds to the energy between E_i confined in the GaAs quantum well and the conduction bandedges of the $Al_xGa_{1-x}As$ barriers. It is a bit longer than the real response wavelength shown in Fig. 6.6b. There are two major reasons. The real GaAs quantum well width and the Al mole fraction are all a bit smaller than their designed values, see Table 3.1. Both of them decrease the energy separation between E_i and the conduction bandedges of the $Al_xGa_{1-x}As$ barriers, resulting in a longer λ_0. And the approximation of square quantum wells also introduces errors in theoretical predictions that in practice, there are also inter-diffusions of atoms across the interfaces between the barrier and the well.

Figure 6.6b presents also the photocurrent spectra of the QWIP device after γ irradiation. One important application field of infrared photodetection is air-born and space-born remote sensing for weather forecasts and environmental surveillance. γ irradiation is strong in space so it may induce negative effects on the device performance. The dose of the γ irradiation is measured in terms of Mrad. The photocurrent spectra of different γ irradiation dose presented in Fig. 6.6b are measured when the QWIP is at 80 K and biased by an external negative bias of 5 V. γ irradiation blue-shifts the response wavelength from 8.28, 8.0 to 7.8 μm.

A major effect of the γ irradiation is the increase of defects in the crystal lattices which reduces the lifetime of the photo-generated carriers, which is reversely proportional to the relaxation energy Γ of these photo-generated carriers. Theoretical absorption spectra of $\Gamma = 0.1, 1.0, 1.6$ and 6.0 meV are presented in Fig. 6.6a, which agree qualitatively with the experimental observation.

Before we close this section, let us take a look at our pre-assumption that we mostly focus on electron states with small amplitudes of wave vectors from a new perspective other than the cases that only electron states of small energies, thus small amplitudes of wave vectors, are occupied. Take a single GaAs QW embedded in $Al_xGa_{1-x}As$. The potential energy is written as

$$V(z) = \begin{cases} 0 & \text{for } z < 0 \\ -\Delta & \text{for } 0 \le z < L_{QW} \\ 0 & \text{for } z \ge L_{QW} \end{cases} \tag{6.14}$$

when we set the conduction bandedge of the AlGaAs barrier as zero. Due to the finite barrier height, the wave function of the confined state penetrates into the barrier region. As a first-order approximation, let us neglect the wave function penetration in the barrier and approximate the ground confined state as

$$\psi_1(z) = \begin{cases} 0 & \text{for } z < 0 \\ \sqrt{\frac{2}{L_{QW}}} \sin\left(\frac{\pi z}{L_{QW}}\right) & \text{for } 0 \le z < L_{QW} \\ 0 & \text{for } z \ge L_{QW} \end{cases} \tag{6.15}$$

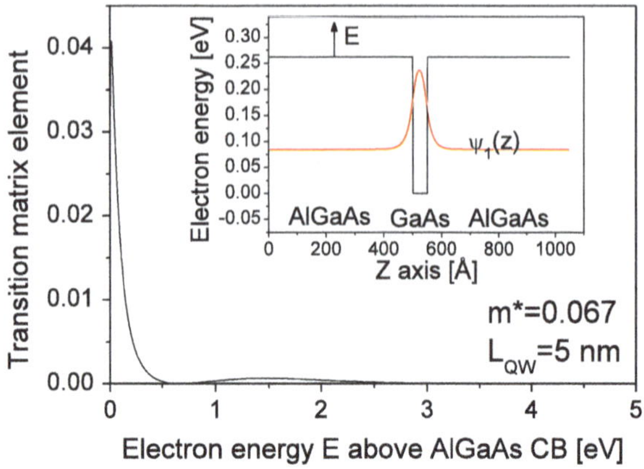

Fig. 6.7 Electrons at high-energy states are not optically coupled with the ground-state electron (red line in the inset)

This is a very good numerical approximation to the real case, which can be easily observed by comparing with the numerical solution when the finite barrier height is used, see the red line in the inset in Fig. 6.7.

By using the boundary conditions of (6.5) and (6.6) to calculate the wave function of the extended state $\psi_E(z)$, it is easy to obtain the optical transition matrix element between $\psi_1(z)$ and $\psi_E(z)$, i.e.,

$$\left| \int \psi_1^*(z) \frac{\mathrm{d}\psi_E(z)}{\mathrm{d}z} \, \mathrm{d}z \right|^2$$

as a function of E, where $E > 0$ is the kinetic energy of the electron in the extended state above the conduction bandedge of the AlGaAs barrier. The numerical result is presented in Fig. 6.7. It is observed here that the optical transition matrix decreases very quickly with E. The reason is numerical and very simple. The kinetic energy of state E in the quantum well region is E plus the barrier height (0.26 eV in Fig. 6.7), so is very high, resulting in a fast oscillation, while ψ_1 is a smooth function. The overlap between a fast oscillating function and a smooth function is small so that the electrons at high-energy states do not optically coupled with the ground-state electron. The implication of this result is that the approximation of the high-energy states in the conduction band of AlGaAs by a single effective mass of the electron at the conduction bandedge is of course very rough but still good enough for our studies.

6.3 Photocurrent of Solar Cell Using Quantum Dot

Common solar cell is based on the photovoltaic effect, and semiconductor *p*-*n* junction based solar cell is the most popular structure, which is schematically shown in Fig. 6.8. At equilibrium, free electrons provided by donors in the *n* region and free holes provided by acceptors in the *p* region will recombine with each other at the

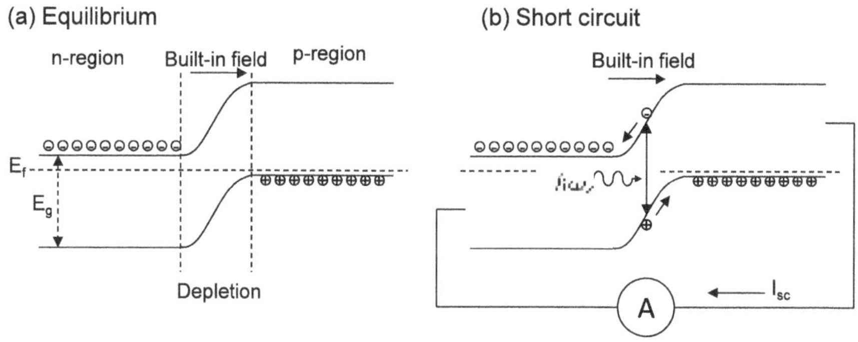

Fig. 6.8 **a** Energy band structure of a *p*-*n* junction at equilibrium in darkness. **b** Short circuit of a *p*-*n* junction under illumination. I_{sc} denotes the short-circuit current

Table 6.1 Basic GaAs *p*-*n* solar cell structure parameters

Layers	Doping (cm^{-3})	Thickness (nm)
p GaAs	2×10^{17}	500
i GaAs	Undoped	140
n^- GaAs	2×10^{16}	1860
n^+ GaAs	2×10^{18}	500

junction interface, resulting in a depletion region, where donor ions and acceptor ions form the built-in electric field. Fermi level E_f is flat everywhere and there is no net electric current and the voltage output is zero. Under light illumination, excess electrons and holes are generated after absorbing photons. In the depletion region, the excess photogenerated electrons and holes will be driven by the built-in field to drift into opposite regions. In the short-circuit case when the Fermi level is still flat across the structure, the photocurrent output reaches the maximal value I_{sc}.

Let us study a common GaAs *p*-*n* junction solar cell whose depletion region is embedded with InAs quantum dots, denoted as sample A, as well as a reference GaAs solar cell without embedding any nanostructures, sample B [6]. Table 6.1 lists the basic GaAs *p*-*n* solar cell structure parameters. There is also a top 100 nm *p*-doped $Al_{0.85}Ga_{0.15}As$ window layer and a back *n*-doped $Al_{0.2}Ga_{0.8}As$ surface field layer to prevent minority carriers from diffusing into the surfaces.

We approximate each quantum dot layer by an effective InAs quantum well layer with a thickness of 2.3 ML (0.7 nm, i.e., InAs quantum dot growth condition) along the sample growth direction (the z direction), while the density of electron and hole states in such a quantum well is approached by a single δ function to account for the discrete nature of the three-dimensional confined states. Energy band structures of the two solar cells at equilibrium are presented in Fig. 6.9 which are calculated theoretically.

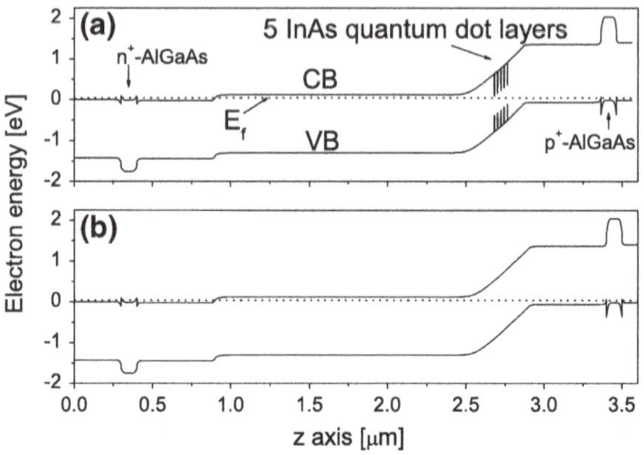

Fig. 6.9 One-dimensional energy band structures of the two *p-n* GaAs solar cells. **a** Sample A: 5 layers of InAs quantum dots separated by 20 nm thick GaAs layers embedded in the depletion region; **b** Sample B: reference solar cell

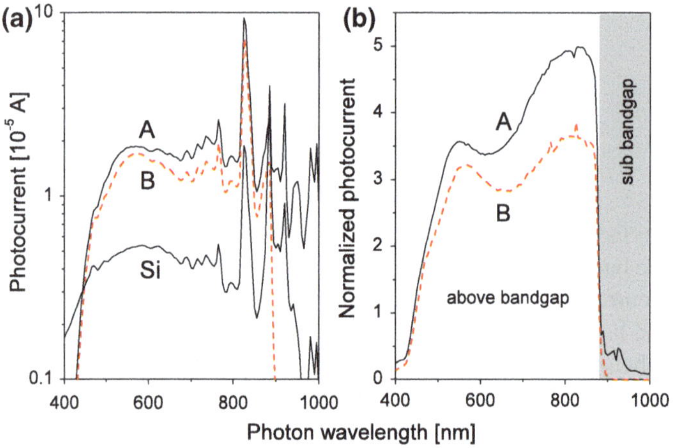

Fig. 6.10 **a** As-measured photocurrent spectra at 300 K under a xenon lamp. **b** Photocurrent spectra normalized with respect to the Si photodiode

As-measured photocurrent spectra shown in Fig. 6.10a are measured by a Keithley multimeter under a xenon lamp with a monochromator, where a Si photodiode is used as the reference. We have learned the importance of the reference measurement already in Sect. 1.5. While the as-measured photocurrent spectra are composed of many fine structures, the normalized ones are pristine. Here two main spectral differences are observed between the two samples. The first one is the sub-GaAs-bandgap photocurrent ($\lambda > 900$ nm), and the second is the photocurrent enhancement in sample A above the GaAs bandgap ($\lambda < 900$ nm).

The sub-GaAs-bandgap photocurrent ($\lambda > 900$ nm) is easily understood that it is due to the photon absorption of the embedded InAs quantum dots. The most

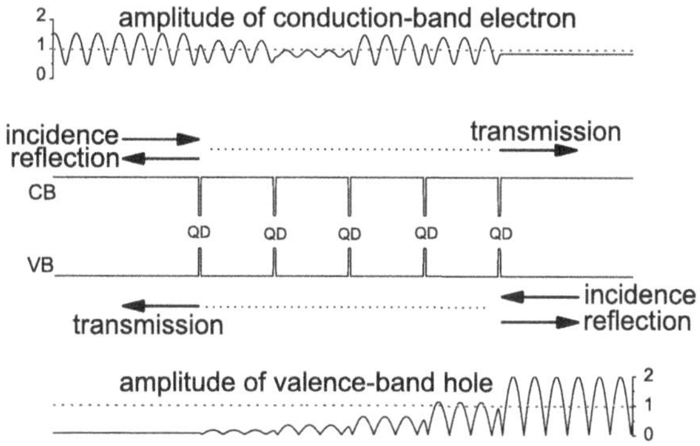

Fig. 6.11 Wave functions of conduction-band (CB) electron and valence-band (VB) hole with a transition energy $\hbar\omega = 1.593$ eV (above the GaAs energy bandgap). Also presented are the schematic CB and VB of GaAs embedded with InAs quantum dots modeled as one-dimensional quantum wells

interesting thing about Fig. 6.10b is the enhancement of the photocurrent in the above-GaAs-bandgap wavelength range of $\lambda < 900$ nm in quantum-dot-embedded sample A. The most possible mechanics of the photocurrent enhancement above the GaAs energy bandgap in sample A is the increased photon absorption due to quantum-dot induced reflections of the wave functions of an incident conduction-band electron from one side of the depletion region and an incident valence-band hole from the other side.

Refer to Fig. 6.11, for simplicity, we model the quantum-dot potential variations in the conduction and valence bands as one-dimensional quantum wells. The incident conduction-band electron, with an energy above the GaAs conduction bandedge, comes to the quantum-dot layers in a plane wave form

$$e^{ik_1 z}$$

it is partially reflected

$$r_1 e^{-ik_1 z}$$

and partially transmitted

$$t_1 e^{ik_1 z}$$

Similarly the wave function components for the valence-band hole are

$$e^{-ik_2 z}\,, \quad r_2 e^{ik_2 z}\,, \quad t_2 e^{-ik_2 z}$$

Note that there are other transmission cases such as both the conduction-band electron and the valence-band hole incident to the quantum-dot regions from the same side but they do not contribute to the photocurrent.

We first study sample B. There are no reflections without quantum dots so that

$$r_1 = r_2 = 0\,, \quad t_1 = t_2 = 1.0$$

The amplitudes of the two wave functions in sample B are shown as red dotted lines in Fig. 6.11. And the optical interband transition matrix element between these two wave functions is proportional to

$$W_{\text{sample B}} = \langle e^{ik_1 z} | e^{-ik_2 z} \rangle = \int e^{-i(k_1+k_2)z}\, \mathrm{d}z \tag{6.16}$$

The wave functions are modified when quantum dots are embedded. As shown in Fig. 6.11, the conduction-band electron transmits relatively well due to the small electron effective mass so that it can be well approximated as a perfect transmission and its wave function on the right side of the structure is

$$e^{ik_1 z}$$

The much heavier valence-band hole is greatly affected. For the wave function presented in Fig. 6.11, we can neglect the transmitted valence-band hole so that its wave function on the right side of the structure becomes

$$e^{-ik_2 z} + e^{ik_2 z}$$

The optical transition matrix element of sample A becomes now

$$W_{\text{sample A}} = \langle e^{ik_1 z} | e^{-ik_2 z} + e^{ik_2 z} \rangle = W_{\text{sample B}} + \int e^{-i(k_1-k_2)z} \mathrm{d}z \tag{6.17}$$

$W_{\text{sample B}}$ oscillates while the second term in $W_{\text{sample A}}$ can be very large when $k_1 = k_2$. The condition of $k_1 = k_2$ can easily be fulfilled which explains the photocurrent enhancement of sample A over the whole optical range above the GaAs energy bandgap. Note that the $W_{\text{sample A}}$ enhancement in the above analysis is very large as compared with experimental data due to the approximation of the three-dimensionally confined quantum-dot potentials by one-dimensionally confined quantum-well potentials, leading to a strongly exaggerated wave function reflection.

As briefly mentioned in the introduction of Chap. 1, colloidal quantum dots are very bright under optical excitation. This implies that (1) excitons in these quantum dots are easy to be photogenerated, (2) the easily photogenerated electrons and holes readily relax to their respective bandedge states, (3) then radiatively recombine to emit photons, see Fig. 4.1. A novel idea is to extract the electron and hole photogenerated in the quantum dot to an external electric circuit before they recombine, forming a photocurrent. When the solar light is used to photogenerate the exciton in the quantum dot, we have a so-called quantum dot sensitized solar cell [7], which is schematically represented in Fig. 6.12.

As will be shown in Fig. 7.1, the conduction bandedge of CdSe is lower than that of CdS, while the valence bandedge of CdSe is higher than that of CdS. So for a quantum dot consisted of a CdSe core and a CdS shell, the photogenerated electron and hole will all be confined in CdSe core region. In other words, CdSe is a quantum well for

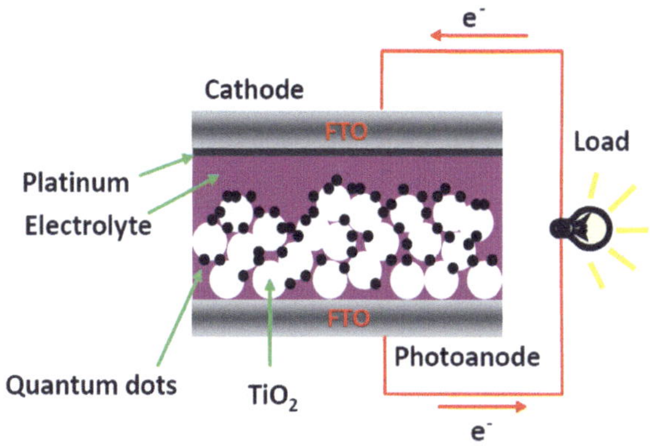

Fig. 6.12 Schematics of quantum dot sensitized solar cell. FTO = fluorine-doped tin oxide film

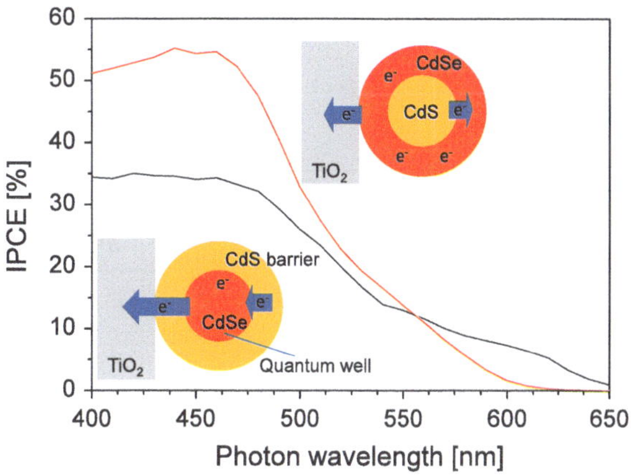

Fig. 6.13 Incident photon to current efficiency (IPCE) spectra of CdSe core-CdS shell (black line) and CdS core-CdSe shell (red line) structured QD sensitized solar cells

both the conduction-band electron and valence-band hole, while CdS is the energy barrier, forming so-called type-I heterostructure. When we apply CdSe-CdS core-shell quantum dots for photoluminescence purpose, the core-shell structure is perfect since both the photogenerated electron and hole are all confined in the CdSe core, resulting in an efficient radiative recombination. However, for solar cell applications, we want to extract photoexcited electron and hole for electric current, thus a CdS core-CdSe shell reversed type-I quantum dots will work better, since this reversed structure will facilitate the injection of photoexcited electrons from the quantum dot to the TiO_2 substrate, resulting in significantly improved electron injection efficiency. Figure 6.13 shows that the energy conversion efficiency of reversed type-I quantum dot sensitized solar cell is better than the common type-I structure. Here the energy conversion efficiency is presented in terms of incident photon to current efficiency (IPCE).

6.4 Multiphoton Induced Photocurrent

Recall the multiphoton-induced (multiphoton excitation) photoluminescence spectra in Sect. 4.6 where under the photo-excitation of an optical field

$$\sum_s h^s e^{-i\omega_s t}$$

the steady-state occupation $C_m(t)$ of electron state m up to the second order is in the form of

$$C_m(\infty) = C_m^{(0)} + C_m^{(1)}(\infty) + C_m^{(2)}(\infty) \tag{6.18}$$

where the electron occupies initially an electron state E_k with an occupation probability $C_k^{(0)}$,

$$\begin{aligned} C_m^{(0)} &= 0 \\ C_m^{(1)}(\infty) &= \sum_s \frac{h_{mk}^s C_k^{(0)}}{E_m - E_k - \hbar\omega_s + i\Gamma_{mk}} \\ C_m^{(2)}(\infty) &= -\sum_{s,i} \frac{h_{mi}^s h_{ik}^s C_k^{(0)}}{\left(E_i - E_k - \hbar\omega_s + i\Gamma_{ik}\right)\left(E_m - E_k - 2\hbar\omega_s + i\Gamma_{mk}\right)} \end{aligned} \tag{6.19}$$

i.e., (4.97) and (4.101).

Since the transition probabilities are proportional to the square of C_m,

$$\left|C_m^{(1)}(\infty)\right|^2 \propto \left|h^s\right|^2 \propto S\,, \quad \left|C_m^{*(2)}(\infty)\right|^2 \propto \left|h^s\right|^4 \propto S^2 \tag{6.20}$$

where S is the optical power of the excitation light, see (4.87) which links S and h^s. In other words, under weak excitation, perturbation theory predicts that the one-photon excitation will result in a linear relationship between the photocarrier density and excitation power, and the photocarrier density due to the two-photon excitation is proportional to the square of the excitation power.

Let us see the experimental results of the multi-photon generated photocurrent. The QWIP photo-pixel to be tested has a surface area of $D = 50 \times 50\ \mu\text{m}^2$. It has an optical coupling surface of 45°, see Fig. 6.14, so that the electric field of the incident electromagnetic field has a non-zero z-component. And the detailed structure of the QWIP is presented in Fig. 6.2a. This QWIP responses to infrared radiations in the range of 6–9 μm at 80 K. And the energy difference between the confined state in GaAs to the conduction bandedge of the $Al_xGa_{1-x}As$ barriers is 155 meV (corresponding to 8.0 μm response wavelength).

We use a free electron laser (FEL) light source that contains a train of macro pulse at a frequency of 3 Hz as the excitation light beam. Each macro pulse contains 10^4 micro pulses with a time duration of 4 ps. A series resistor of $R_L = 100\ \text{K}\Omega$ is connected to the QWIP pixel, and a driving bias of V_{ex} applies. Voltage at QWIP is

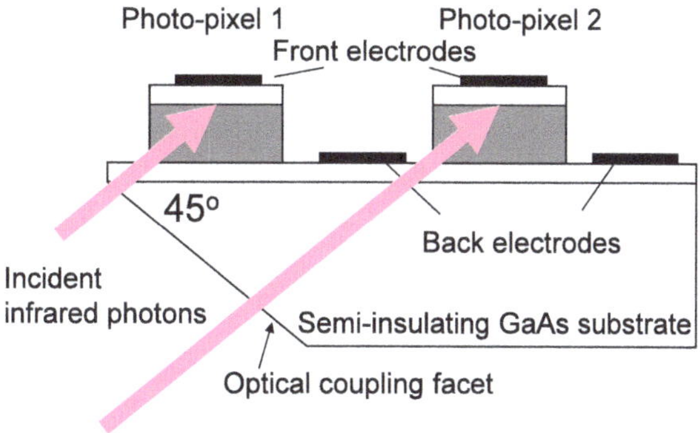

Fig. 6.14 Test structure of the QWIP device

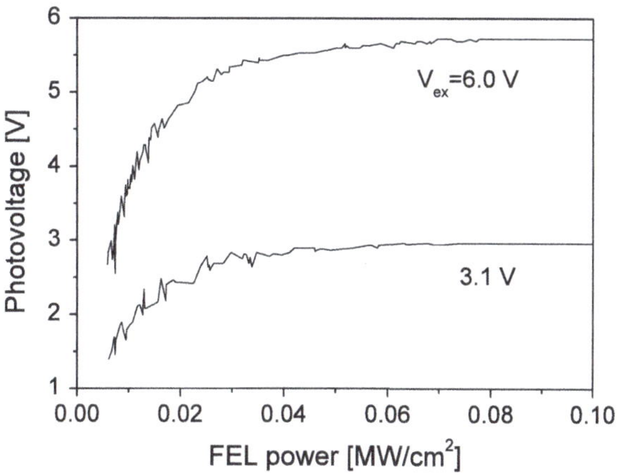

Fig. 6.15 The relationship of the photovoltage V_{ac} and the FEL excitation power. $T = 80$ K

measured and recorded. We first measure V_{dc} and therefore R_0 of QWIP in darkness. When $V_{ex} = 6.0$ V, $V_{dc} = 5.81$ V, $V_{ex} = 3.1$ V, $V_{dc} = 3.02$ V. We then introduce the 3-Hz FEL macro pulse train to measure the bias on QWIP. The difference of it from V_{dc} is the photovoltage V_{ac}, which is shown in Fig. 6.15 as a function of the FEL power when the FEL wavelength is 8.0 μm.

Let R_{FEL} be the resistance of the QWIP under the illumination of the FEL, R_0 is the resistance of the QWIP in darkness, it is easy to see that

$$R_{FEL} = \frac{w}{De\mu_n\left(n_{ph} + n_0\right)}, \quad R_0 = \frac{w}{De\mu_n n_0} \tag{6.21}$$

where $w = 2.79$ μm is the thickness of the multiple quantum wells (50 periods of alternating GaAs layers, each ca 5 nm, and $Al_xGa_{1-x}As$ layers, each ca 50 nm). $D = 50 \times 50$ μm^2 is the area of the optical pixel (mentioned before). n_0 is the density of carriers above the conduction bandedge of $Al_xGa_{1-x}As$ barriers, while n_{ph} is the density of photocarriers excited by the FEL light beam. μ_n is the carrier mobility, and e is the charge unit. We obtain directly the photocarrier density

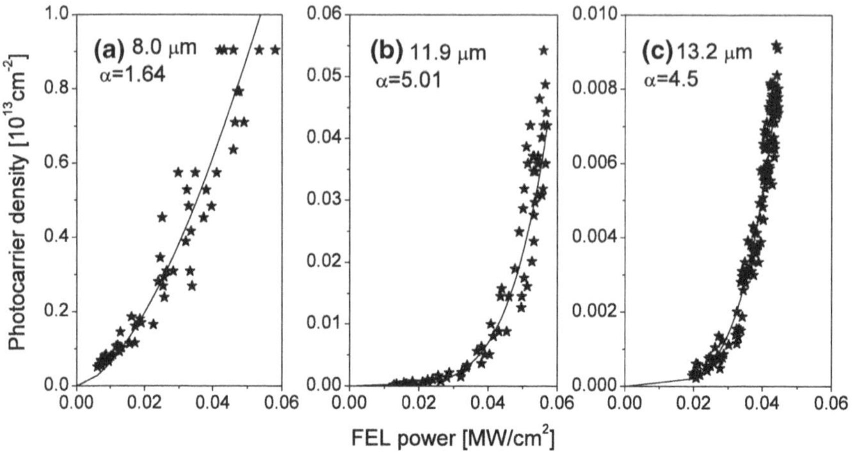

Fig. 6.16 The relationship between photocarrier density and FEL excitation power at 80 K. The wavelengths of the FEL excitation laser sources are 8.0, 11.9, and 13.2 μm, respectively. Solid lines are fitted curves S^{α}, where S is the FEL excitation power

$$n_{\text{ph}} = \frac{wV_{\text{ex}}V_{\text{ac}}}{De\mu_n R_L V_{\text{dc}}\left(V_{\text{dc}} - V_{\text{ac}}\right)} \tag{6.22}$$

For n-type $Al_{0.3}Ga_{0.7}As$, μ_n is normally expressed as

$$\mu_n = \mu_0\left[1 + \left(\frac{\mu_0 F_z}{v_s}\right)^2\right]^{-1/2} \tag{6.23}$$

where μ_0 is the carrier mobility at low field, ca 2000 cm^2/V s, v_s is the saturation velocity, ca $0.1 \sim 5 \times 10^6$ cm/s, F_z is the electric field in the $Al_xGa_{1-x}As$ barriers due to the external bias.

Take $v_s = 5 \times 10^6$ cm/s, Fig. 6.16 presents the photocarrier density obtained from (6.22) and Fig. 6.15. It is shown clearly that the relationship between the photocarrier density and the FEL excitation power is not linear. Using a relationship

$$n_{\text{ph}} \propto S^{\alpha} \tag{6.24}$$

to fit the data of Fig. 6.16, we obtain $\alpha = 1.64$ for the 8.0 μm photons, $\alpha = 5.01$ for the 11.0 μm photons, and $\alpha = 4.5$ for the 13.2 μm photons.

It can be observed in light of relationship of (6.20) and the QWIP response wavelength of 8 μm that there are clear differences between the one-photon ($\alpha = 1.64$ at photon wavelength 8.0 μm) and the multiphoton ($\alpha = 5.01$ at 11.9 and 4.5 at 13.2 μm) excitation processes. But the experimental data of multiphoton excitations do not agree with the two-photon excitation relationship in (6.20). One major difficulty is that the FEL excitation is composed of a train of macro pulse at a frequency of 3 Hz as the excitation light beam while each macro pulse contains 10^4 micro pulses with a time duration of 4 ps, so that the time-dependent solution of (4.85) is necessary. We will study the time-resolved photoluminescence in Sect. 7.2.

References

1. B.F. Levine, Quantum-well infrared photodetectors. J. Appl. Phys. **74**, R1(81) (1993)
2. G. Bastard, *Wave Mechanics Applied to Semiconductor Heterostructures* (Halsted Press, 1988), p. 4
3. S.C. Shen, Comparison and competition between MCT and QW structure material for use in IR detector. Microelectron. J. **25**, 713–739 (1994)
4. W. Lu, Y. Fu, Quantum well infrared detectors, in *Encyclopedia of Nanoscience and Nanotechnology*, vol. 9, ed. by H.S. Nalwa (American Scientific Publishers), p. 179
5. Y. Jing, Z. Li, Q. Li1, X. Chen, P. Chen, H. Wang, M. Li, N. Li, W. Lu, Pixel-level plasmonic microcavity infrared photodetector. Scientific Reports **6**, 25849 (2016). https://doi.org/10.1038/srep25849
6. X.-J. Shang, J.-F. He, M.-F. Li, F. Zhan, H.-Q. Ni, Z.-C. Niu, H. Pettersson, Y. Fu, Quantum-dot-induced optical transition enhancement in InAs quantum-dot-embedded $p-i-n$ GaAs solar cells. Appl. Phys. Lett. **99**, 113514 (2011)
7. Z. Ning, H. Tian, H. Qin, Q. Zhang, H. Ågren, L. Sun, Y. Fu, Wave function engineering of CdSe-CdS core-shell quantum dots for enhanced electron transfer to a TiO_2 substrate. J. Phys. Chem. C **114**, 15184–9 (2010)

Chapter 7 Optical Properties of Fluorescent Colloidal Quantum Dots

Abstract In the last chapter of the book we introduce the latest applications and developments of semiconductor spectroscopy in the field of bio-nano-photonics where semiconductor colloidal quantum dots are researched and developed as biomarkers using standard absorbance, fluorescence, time-resolved fluorescence (in ns time domain), fluorescence blinking (ms time domain) spectra, as well as bioimaging.

A tremendous amount of research and technical development in nanotechnology has been invested and published in the biological, medical and environmental fields. This chapter focuses on briefly introducing fluorescent colloidal quantum dots (QDs). Some of them are listed in Table 7.1.

For bioimaging applications, semiconductor colloidal quantum dot (QD) is commonly protected by a shell or multiple shells so that the electron and hole states confined in the QD core are not affected by the external environments. There are also biological reasons for the shells such as protecting heavy semi-metal atoms contained in the QD core from leaking. CdS and ZnS are commonly used as shell materials for CdSe-based QDs, as the energy bandgap of bulk CdSe is smaller than those of CdS and ZnS. CdSe-CdS/ZnS heterostructure is further known to be type-I, meaning that CdSe is the quantum well material while CdS/ZnS the quantum barrier for both electrons in the conduction band and holes in the valence band. CdSe therefore is principally used as the QD core material while CdS/ZnS the shell material. Figure 7.1 shows a schematic drawing of a core-shell CdSe-CdS QD and its energy band structure. Because of the small QD size, the quantum confinement induces energy state discreteness, i.e., the discrete energy states E_{c0}, E_{c1} ... in the conduction band and E_{v0}, E_{v1} ... in the valence band. The transition energy between ground levels E_{c0} and E_{v0} is thus a function of the QD size.

Many of these semiconductor QDs (CdS, CdSe, ZnSe, ZnO) are prepared by sol-gel method at 250–300 °C, thus denoted as colloidal [1, 2]. Here, metallic oxide is selected as the source metal due to its low toxicity. Moreover, 1-Octadecene (ODE) is selected as the non-coordinating solvent because of its relatively low melting point (below 20 °C), relatively high boiling point (about 320 °C), low cost, low toxicity, low reactivity to precursors, and excellent solvation power for many compounds at

W. Lu and Y. Fu, *Spectroscopy of Semiconductors*, Springer Series in Optical Sciences 215, https://doi.org/10.1007/978-3-319-94953-6_7

Table 7.1 Representative fluorescent quantum dots

Materials (core/shell)	Emission wavelength (nm)	FWHM (nm)	Quantum efficiency (%)
ZnSe/ZnS	400–440	<20	Up to 90
InP/ZnS	470–800	50–70	>50
CdS/ZnS	440–470	<25	>80
CdSe/ZnS	520–650	<30	60–90
PbS	800–2000		

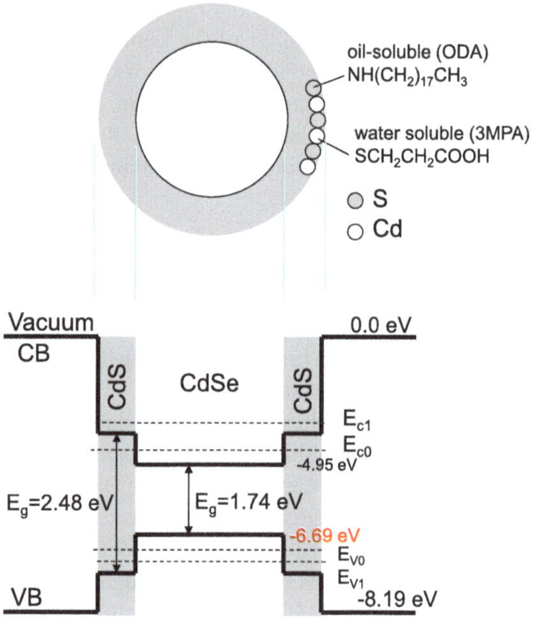

Fig. 7.1 Schematic diagram of a core-shell CdSe-CdS QD with surface ligand (either ODA or 3MPA) and its energy band structure

elevated temperatures. As-grown core-shell CdSe-CdS QDs are commonly coated with surface ligands octadecylamine (ODA, linear chemical formula $CH_3(CH_2)_{17}NH_2$) thus are dissolved in chloroform or toluene. To make them dispersible in water, the surface ligands are exchanged to, for example, 3-mercaptopropionic acids (3MPA, $HSCH_2CH_2COOH$), see Fig. 7.1.

7.1 Absorbance and Fluorescence

The size of the colloidal QD is commonly controlled by the reaction time, which is demonstrated by a series of absorbance and fluorescence spectra in Fig. 7.2 of CdSe reaction solutions sampled at different reaction times. The basic setups of absorbance

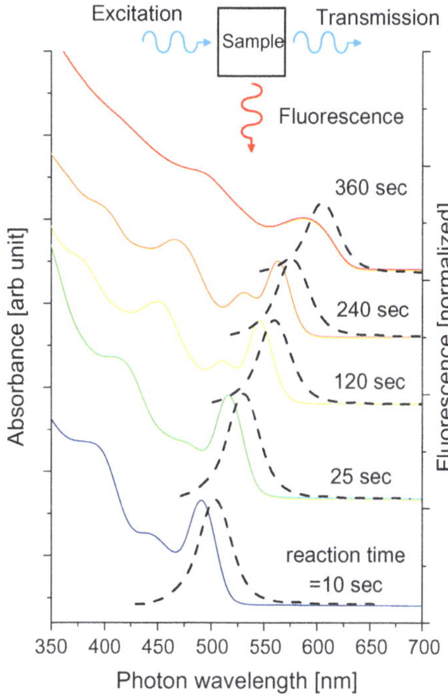

Fig. 7.2 Absorption and fluorescence spectra of a series of CdSe reaction solution samples collected at different reaction times. Inset shows the basic setup of absorbance (=difference between excitation and transmission spectra) and fluorescence measurements

and fluorescence measurements are shown in the inset where the absorbance is obtained as the difference between the excitation and transmission spectra while the fluorescence is measured normally at the perpendicular direction of the excitation. Here the correlation between the size of CdSe QDs in the reaction solution and the growth time is clearly reflected in the wavelengths of the absorbance and fluorescence peaks that the short growth time means a small QD size (a large quantum confinement and thus a short wavelength).

Interestingly, the word "photoluminescence" is used in previous chapters describing the photon emission from a matter after the absorption of photons. "Fluorescence" means basically the same thing. Photoluminescence is commonly used in semiconductor physics, while fluorescence is in biophysics and chemistry. In this chapter we adopt the word fluorescence as it is mostly used in the research and technical development of colloidal QDs.

Figure 7.3 shows the absorbance and fluorescence spectra of ODA-coated QDs of single component CdSe (average diameters ca 2.6 and 3.4 nm, respectively), core-shell CdSe-CdS QDs (average diameters of CdSe/CdS-1 and CdSe/CdS-2 are 4.4 nm and 5.7 nm, respectively) [2].

Things can be different depending on the QD applications. For most bioimaging applications, the electron in the conduction band and the hole in the valence band are to be confined in the QD core for best fluorescence. For solar energy applications, however, either the electron or the hole is to be extracted from the QD to the external

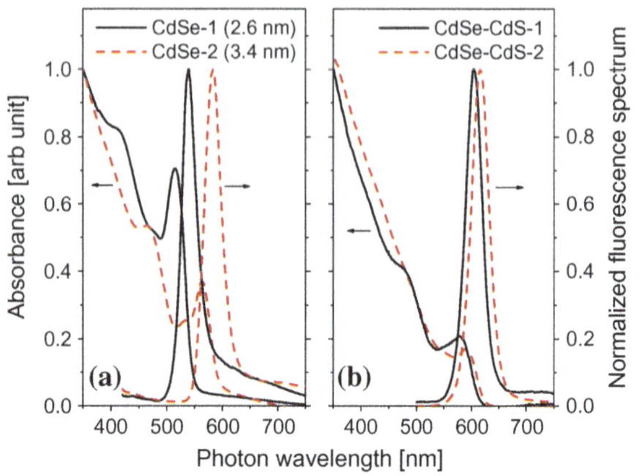

Fig. 7.3 Absorption and fluorescence spectra of **a** CdSe and **b** CdSe-CdS core-shell QDs in toluene. Two QDs are presented in each QD type: solid lines represent CdSe-1 (**a**) and CdSe-CdS-1 (**b**), dash lines represent CdSe-2 (**a**) and CdSe-CdS-2 (**b**)

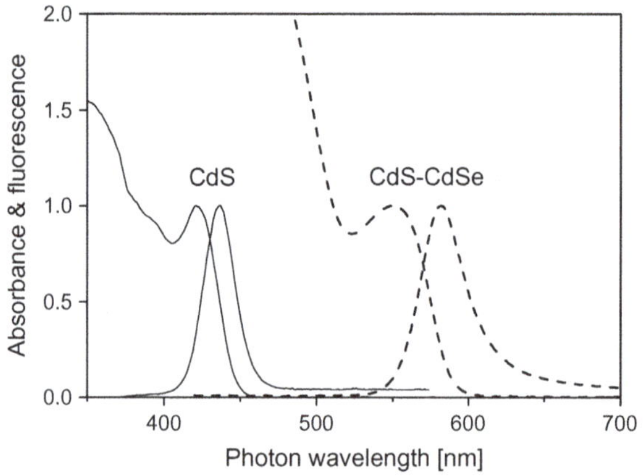

Fig. 7.4 Normalized absorbance (with respect to the first absorption peak) and fluorescence spectra of CdS and CdS-CdSe core-shell QDs in toluene normalized. A noticeable difference in wavelengths of the first absorption peak and the fluorescence peak is observed, known to be the Stokes shift

electric circuit as soon as it is photogenerated before it recombines with its counterpart (Sect. 6.3). For this, CdS is used as the QD core while CdSe as the QD shell. The absorbance and fluorescence spectra of these CdS-CdSe core-shell QDs in toluene are shown in Fig. 7.4. Here the diameter of the CdS core is about 4.0 nm, and 6.8 nm for whole CdS-CdSe QDs.

A noticeable difference in wavelengths of the first absorption peak and the fluorescence peak is always observed, known to be the Stokes shift, in Figs. 7.2 and 7.3, most clearly in Fig. 7.4 when comparing the absorbance spectrum normalized with respect to the first absorption peak with the normalized fluorescence. The understanding of the Stokes shift is not complete yet, see, e.g., [3].

In Figs. 7.2 and 7.3, as well as Fig. 7.4, the full width at half maximum (FWHM) of the fluorescence peaks of these QDs is all about 25 nm, it is a bit smaller for CdS QDs in Fig. 7.4. In aqueous solution the value of the FWHM is usually much larger. Such a broad peak is commonly attributed to the distribution of QD sizes in the QD solution. Referring to the schematic QD structure in Fig. 7.1, when E_{c0}

and E_{v0} denote the ground-state energies measured from the conduction-band and valence-band edges, $E = E_g + E_{c0} + E_{v0}$ is the ground-state transition energy of the QD, where E_g is the energy bandgap of the QD core material. For an infinite potential QD with a radius of r, $k^2 = 2m^* E_{c0}/\hbar^2$, $kr = \pi$ so that

$$r^2 = \frac{\pi^2\hbar^2}{2m^* E_{c0}} \quad , \quad 2r\delta r = \frac{\pi^2\hbar^2}{2m^*}\frac{\delta E_{c0}}{E_{c0}^2} \tag{7.1}$$

where m^* is the electron effective mass, δE_{c0} is the variation in E_{c0} induced by δr. Since E_{v0} depends weakly on the QD radius because of the large hole effective mass (about $0.5m_0$ for commonly used semiconductors, where m_0 is the electron rest mass), $\delta E \approx \delta E_{c0}$ so that [4]

$$\frac{\delta r}{r} = \frac{\delta E_{c0}}{2E_{c0}} = \frac{\delta E_{c0}}{2(E - E_g - E_{v0})} \approx \frac{\delta E}{2(E - E_g)} \tag{7.2}$$

Consider CdSe-based QDs emitting at 600 nm, $E_g = 1.74$ eV (Fig. 7.1) and $E = 1.24/0.6 = 2.07$ eV (600 nm). For $\delta E = 86$ meV (25 nm),

$$\frac{\delta r}{r} \approx \frac{\delta E}{2(E - E_g)} = \frac{0.086}{2(2.067 - 1.74)} = \frac{0.086}{2 \times 0.32667} = 0.13 \tag{7.3}$$

This agrees well with high-resolution transmission electron microscopy (HRTEM) imagings of QDs, e.g., see Fig. 7.5a–c where the statistic distribution of the QD diameters is presented in Fig. 7.5d. Thus, the statistic distribution of the QD size can be one reason for the finite FWHM of the QDs in solution. However, there may be other reasons.

As mentioned before, the absorbance and fluorescence spectra thus far presented are measured on many QDs in solution, i.e., a group of QDs which are well dispersed in solvent at microscopic level. At microscopic level, they may be totally dispersed as single QDs in solvent. However, they may also form clusters through various chemical interactions via their surface ligands.

One good way to measure the fluorescence spectra from single QDs is to deposit QDs on a microscope slide then study their optical properties using a fluorescence microscope. The sample preparation is schematically shown in Fig. 7.6a–c: First, use a capillary pipette to place a small drop of QD solution on the center of a microscope slide, carefully place a clean coverslip over the drop, seal the coverslip with nail polish. Allow the nail polish to dry before microscopy study. We first use an AxioObserver.D1 microscope (Carl Zeiss) equipped with a mercury lamp (HBO 100, Carl Zeiss), a filter set (Exciter: FF02-435/40-25, Dichroic: FF510-Di02-25 × 36, Emitter: FF01-500/LP-25, Semrock), an EMCCD camera (Andor), and a 100 × 1.4 NA oil immersion objective (Carl Zeiss) to observe and record a time series of QD fluorescence images. Figure 7.6d displays one typical image frame of two single CdSe-ZnS QDs with an emission peak wavelength at 655 nm (the exposure time for each frame is 20 ms and the readout time of the CCD is 30 ms).

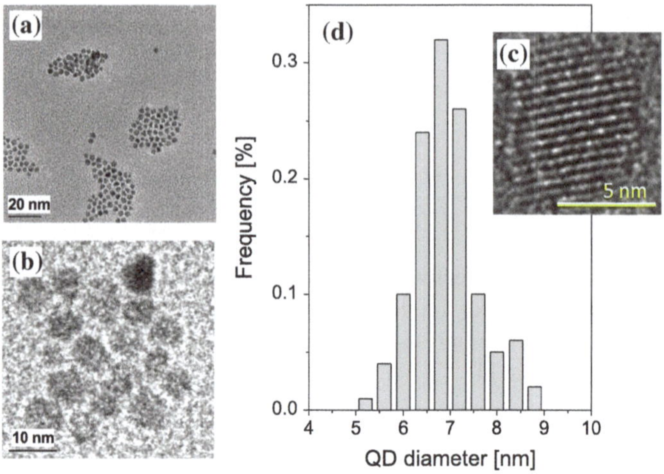

Fig. 7.5 High-resolution transmission electron microscopy (HRTEM) micrographs of one batch of ODA-coated CdSe-CdS QDs at resolution of **a** 20 nm, **b** 10 nm and **c** 5 nm where the atomic structure of a single CdSe-CdS QD is clearly depicted. **d** Statistic distribution of the QD diameters

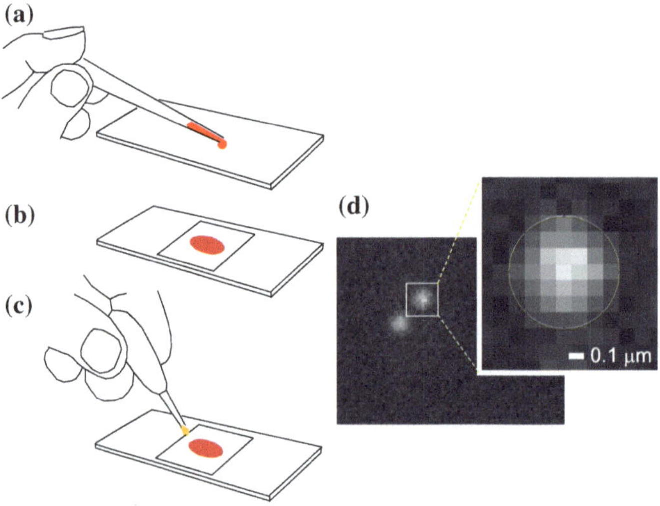

Fig. 7.6 **a** Use a capillary pipette to place a small drop of QD solution on the center of a microscope slide; **b** Carefully place a clean coverslip over the drop; **c** Seal with nail polish; Allow the nail polish to dry before microscopic study. **d** One representative frame of two single CdSe-ZnS QDs with an emission peak wavelength 655 nm

As mentioned before, the size of the QDs is about 7 nm in diameter, while the light spot in Fig. 7.6d is about 1.0 μm, which is no less than the wave nature of the light (recall that the emission peak wavelength of the CdSe-ZnS QDs is 655 nm). It is therefore not possible to assess whether the light source is a single QD or a group of QDs by only looking at one single image frame. However, it is empirically known that the fluorescence of a single QD irregularly switches between bright and dark under continuous irradiation at millisecond to second time domain depending on the QD materials (including surface ligands) and structures. This phenomenon is normally known as fluorescence blinking or intermittency (we will discuss it in Sect. 7.3).

Using this blinking effect, we can measure the fluorescence spectrum of single QDs. Let us now study water-soluble (3MPA-coated) CdSe-CdS/$Cd_{0.5}Zn_{0.5}S$/ZnS core-multishell QDs that have a fluorescence peak at 596 nm at room temperature. We first deposit QD solutions of various QD concentrations on microscope slides to find the right sample displaying QD blinkings using the AxioObserver.D1 microscope. Here the microscope uses a bandpass filter to obtain an excitation beam centered at 435 nm with a bandwidth 40 nm and a reflector bandpass filter that records all light with wavelengths above 500 nm. We then take the sample to a structured illumination microscope (Zeiss Elyra PS) with a 63 × 1.4NA oil immersion objective (Carl Zeiss), excite the sample using a 488 nm excitation laser, and record over a bandwidth of 562–637 nm (to cover the QD fluorescence peak) with a spectral resolution of 2.9 nm. More specifically, the microscope takes one image of a single QD in a wavelength window of 562–564.9 nm, takes another image in 564.9–577.8 nm and so on. We then calculate the brightness of the QD image per each wavelength window, which becomes the fluorescence spectrum of the single QD.

Figure 7.7a shows one image of quite diluted QDs (the QD concentration in solution is only 10 pM) so that the optical spectra of single QDs are obtained, while Fig. 7.7b shows the image of highly concentrated QDs.

The fluorescence spectra of 15 randomly chosen single QDs from the diluted sample (a) are presented in Fig. 7.7c as black solid lines, together with the optical spectrum of the QD ensemble from sample (b), indicating that the optical spectra of single QDs and the QD ensemble are very similar. Moreover, we further examine the spectral diffusion of the microscope by measuring the spectrum of the excitation laser light reflected from the microscope slide. (Two fundamental factors are involved here. The first is that the excitation laser wavelength was stated to be 488 nm, which is always to be understood to be the wavelength of a spectral peak with a finite FWHM, which is denoted as $FWHM_{laser}$. The second is that the spectral resolution of an optical instrument is finite so that the measured FWHM, denoted as $FWHM_{measure}$ is larger than $FWHM_{laser}$. The difference between the two FWHMs is the major cause of the spectral diffusion.) The FWHM of the reflected spectral peak was less than 5 nm shown in Fig. 7.7d, far narrower than the QD fluorescence peak. It is therefore safe to conclude that, in addition to the distributed QD size factor discussed before, the large FWHM, ca 33 nm in Fig. 7.7c, see also Table 7.1, of the single QD fluorescence peak can be intrinsic. This is principally due to the large ratio between

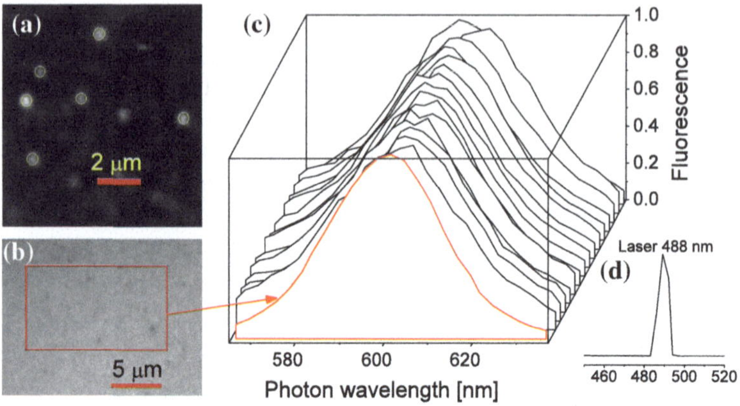

Fig. 7.7 **a** One image frame showing the spatial locations of single QDs in the diluted sample. **b** One image frame of the highly concentrated QDs. **c** Fluorescence spectra of single QDs (black lines) and the compact QD ensemble (red line). **d** Optical spectrum of the reflected excitation laser of 488 nm

the number of surface atoms to that of volume atoms and the abrupt change from large semiconductor atoms (Cd, Zn, Se, S etc.) to small organic ligand atoms (mostly H, C, O) in colloidal nano-size systems.

7.2 Time-Resolved Fluorescence

Photoluminescence or fluorescence spectroscopy is a steady-state technique that detects the photon emission from a sample under a continuous-wave laser excitation. We now introduce the time-resolved fluorescence spectroscopy that is used to study the dynamics of the QD fluorescence. Refer to Fig. 7.8a–d, the QD is initially in its ground state that all energy states in the valence band are occupied (we do not consider magnetic properties so that each energy state can be maximally occupied by two electrons, one spin up and the other spin down) while the conduction band is totally empty. In the electron picture (a–d), the photoexcitation of the QD occurs such that an electron originally occupying a valence-band state absorbs a photon to transit to an initially empty conduction-band state, leaving the valence-band state empty (i.e., a hole). The electron will relax to the ground state E_{c0} in the conduction band, while the hole will relax to the ground level E_{v0} in the valence band. The electron at E_{c0} transits to the empty E_{v0} (i.e., the hole at the valence-band ground state) to emit a photon (QD fluorescence), for which the conduction band states are all empty and the valence band states are occupied.

One critical factor in a QD is the quantum confinement for both the electron and hole. In this case, the electron in a conduction-band state and the hole in a valence-band state interact with each other via Coulombic interaction to form an electron-hole

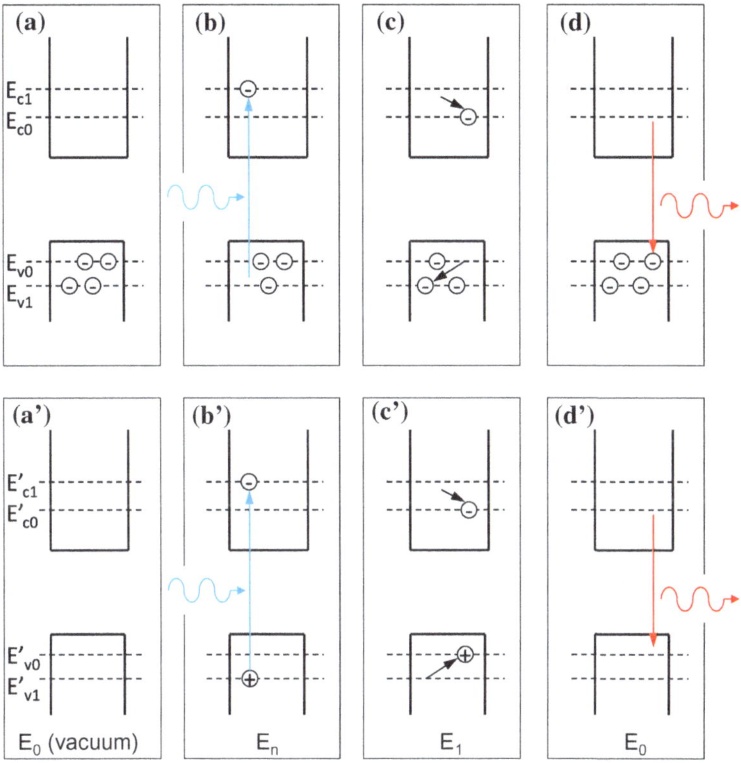

Fig. 7.8 **a** A QD at its ground state. **b** One electron in a valence-band state absorbs an excitation photon then transits to a conduction-band state. **c** Both the electron in the conduction band and the hole in the valence band relax to their respective ground states. **d** Radiative recombination of the electron and hole to emit a photon, i.e., fluorescence. **a'–d'** are the same as **a–d** but in the electron-hole picture with modified electron states and hole states

pair, i.e., the exciton. Note a few concepts here. The electron and the hole in the QD are always interacting with each other via Coulombic interaction which is very strong in the QD because of the small size of the QD (commonly less than 10 nm). Before photo-excitation, the QD is at its ground state, also known as vacuum state that the valence-band states in the QD are completely filled and the conduction-band states are completely empty (no electron-hole pair, i.e., no exciton, thus "vacuum"). The ground exciton is formed between an electron occupying E_{c0} and a hole occupying E_{v0}. Thus, the photoexcitation, energy relaxation and radiative recombination of the electron and hole in the QD is more precisely described in the electron-hole picture shown in Fig. 7.8a'–d'. Here we have denoted the electron states by E'_{c0} and E'_{c1}, and the hole states by E'_{v0} and E'_{v1} after including the Coulombic interaction between the electron and hole.

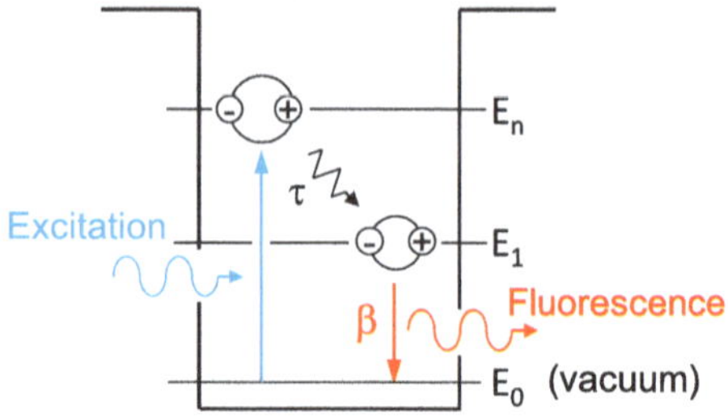

Fig. 7.9 Photoexcitation from QD ground (vacuum) state E_0 to an excited exciton state E_n, energy relaxation from E_n to exciton ground state E_1 at a rate of $1/\tau$, and radiative recombination from E_1 to E_0 at a rate of $1/\beta$

Theoretically the description of the energy states of the electron-hole pair is the Hamiltonian of (2.126) that we have studied in Sect. 2.4 which is repeated below

$$\left[-\frac{\hbar^2\nabla_e^2}{2m_e^*}-\frac{\hbar^2\nabla_h^2}{2m_h^*}-\frac{e^2}{4\pi\epsilon_\infty|\boldsymbol{r}_e-\boldsymbol{r}_h|}+V_e(\boldsymbol{r}_e)+V_h(\boldsymbol{r}_h)\right]\psi_n(\boldsymbol{r}_e,\boldsymbol{r}_h)=E_n\psi_n(\boldsymbol{r}_e,\boldsymbol{r}_h) \tag{7.4}$$

where subscript "e" denotes the electron and "h" denotes the hole, $V_e(\boldsymbol{r}_e)$ and $V_h(\boldsymbol{r}_h)$ are quantum confinement potential energies for the electron and the hole, respectively. $\psi_n(\boldsymbol{r}_e, \boldsymbol{r}_h)$ is the envelope function of the exciton composed of electron $\boldsymbol{r}_e$ and hole $\boldsymbol{r}_h$. In the exciton picture, the initial vacuum state of the QD is denoted by E_0; An incident photon generates an exciton at excited state E_n, which will relax to the ground exciton state E_1 then radiatively recombine to emit a photon thereafter return to vacuum state E_0. The transition processes in the exciton picture are summarized in Fig. 7.9 with transition rates indicated.

The time-resolved fluorescence spectrum is to study the energy relaxation time τ and radiative recombination lifetime β. Refer to Fig. 7.10, a picosecond laser pulse is led into a QD ensemble, and the QD fluorescence is detected as a function of time t, denoted as $F(t)$, i.e., the aforementioned time-resolved fluorescence spectrum. The temporal resolution can be obtained by many techniques. One of them is the time-to-amplitude-converter (TAC), which is briefly introduced in Fig. 7.10.

In order to quantitatively correlate the principal decay processes to the time-resolved fluorescence spectrum, we study the transition rates of the energy relaxation and radiative recombination of an exciton in the QD by the following principal decay processes:

1. Three principal exciton states: excited exciton state ψ_n (energy E_n), with its occupation n_n; ground exciton state ψ_1 (E_1) with occupation n_1; vacuum state ψ_0 (E_0) with occupation n_0.
2. Optical excitation to generate an exciton from vacuum to ψ_n.
3. ψ_n-exciton relaxes at a rate $1/\tau$ to ψ_1 nonradiatively.
4. Ground-state exciton transits radiatively to ψ_0 at a rate $1/\beta$.

Following the time-resolved fluorescence experimental procedure, the QD is initially at its vacuum state. One pulsed excitation excites the QD at $t = 0$ so that

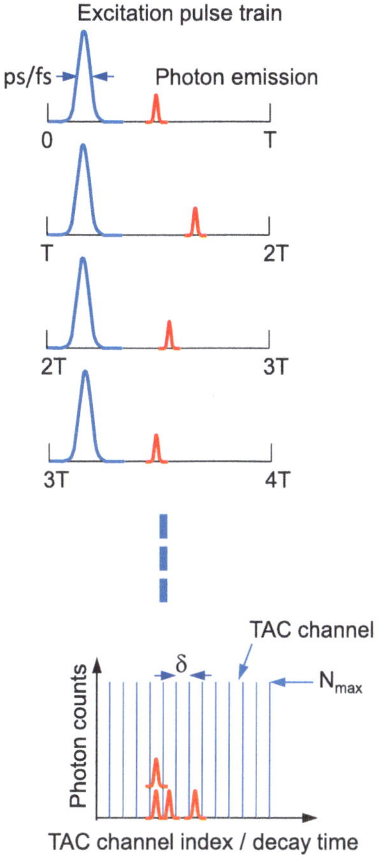

Fig. 7.10 A train of picosecond (or femtosecond) laser pulses (at a period of T, where $1/T$ is the frequency of the pulse train) is led into a QD ensemble, and the QD fluorescence is detected as a function of time t by time-to-amplitude-converter (TAC) channels (there are other techniques to obtain temporal resolutions). When one photon count in one TAC channel reaches $N_{\max}$, the measurement is then stopped. Each measurement setup has a certain number of TAC channels whose integration times δ can be varied depending on the decay time. $N_{\max}$ is in principle to be set as large as possible for the best signal-to-noise ratio while a large $N_{\max}$ means a long measurement time (to be considered such as the stabilities of the excitation laser as well as the fluorescence including photobleaching)

$n_n = 1$ and $n_1 = n_0 = 0$. The decay processes, schematically shown in Fig. 7.9, are described mathematically by the following rate equations:

$$
\begin{aligned}
\frac{\mathrm{d}n_n}{\mathrm{d}t} &= -\frac{n_n(1-n_1)}{\tau} \\
\frac{\mathrm{d}n_1}{\mathrm{d}t} &= \frac{n_n(1-n_1)}{\tau} - \frac{n_1(1-n_0)}{\beta} \\
\frac{\mathrm{d}n_0}{\mathrm{d}t} &= \frac{n_1(1-n_0)}{\beta}
\end{aligned}
\tag{7.5}
$$

The key factor here is the Pauli exclusion principle (fermion anticommutation relations of the electron creation and annihilation operators) that the occupation of one exciton state reduces the efficiencies of transitions to this exciton state since each exciton state can be occupied by only one exciton. As a consequence, the transition efficiency of an exciton from an initial exciton state to a final exciton state is

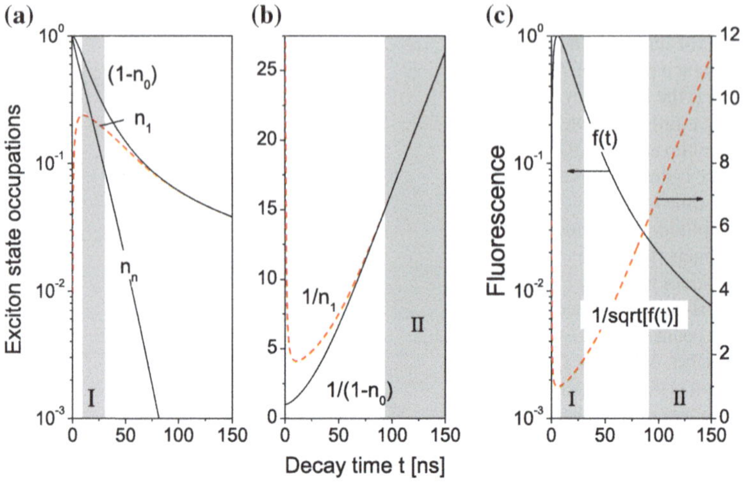

Fig. 7.11 **a** Temporal developments of n_n, n_1, $(1-n_0)$. **b** $1/n_1$ and $1/(1-n_0)$. **c** Theoretical time-resolved fluorescence spectrum $f(t)$ (left vertical axis) and $1/\sqrt{f(t)}$ (right vertical axis). $\tau = 10.0$ ns, $\beta = 5.0$ ns, $\delta = 0.1$ ns

determined by both the occupation of the initial state and the un-occupation of the final state, see (2.116).

We perform a numerical simulation of (7.5) using $\tau = 10.0$ ns and $\beta = 5.0$ ns. Temporal developments of n_n, n_1, n_0, and the theoretical time-resolved fluorescence spectrum

$$f(t)|_{t=\ell\delta} = \int_{(\ell-1)\delta}^{\ell\delta} \frac{n_1(1-n_0)}{\beta}\, dt \tag{7.6}$$

are presented in Fig. 7.11. Here we try to mimic the measurement procedure that δ is the integration time of the TAC channels. We observe two distinct decay characters in $f(t)$, a rather fast decay directly after the optical excitation (time region I) and a slow decay when t is long (time region II).

In time region I, see Fig. 7.11a and c, i.e., a very short time directly after the optical excitation, ca $t \in (5, 22)$ ns, both n_1 and n_0 are very small compared with n_n. $f_1(t)$ is well characterized by a single decay term

$$f_1(t) \propto e^{-t/\tau'} \tag{7.7}$$

Subscript "1" in $f_1(t)$ indicates the time-resolved fluorescence spectrum of short t. And τ' depends on both τ and β.

For $t > 100$ ns, i.e., time region II, see Fig. 7.11b and c, n_n is negligibly small so (7.5) reduce to

$$\frac{dn_1}{dt} = -\frac{n_1(1-n_0)}{\beta}, \quad \frac{dn_0}{dt} = \frac{n_1(1-n_0)}{\beta} \tag{7.8}$$

Let $1 - n_0 = n'_0$. Since $n_1 = n'_0$ (i.e., $n_0 + n_1 = 1$ for negligibly small n_n), the solution of the above equations is

$$\frac{1}{n_1} = \frac{1}{n'_0} = \frac{t+a}{\beta} \tag{7.9}$$

where a is a constant. The above solution is confirmed by the numerical results of $1/n_1$ and $1/(1-n_0)$ presented in Fig. 7.11b. The long-time fluorescence decay in time region II is thus

$$f_2(t)|_{t=\ell\delta} = \int_{(\ell-1)\delta}^{\ell\delta} \frac{n_1(1-n_0)}{\beta}\,dt \approx \left.\frac{\beta\delta}{(t+a)^2}\right|_{t=\ell\delta} \tag{7.10}$$

since the variations of n_1 and $(1-n_0)$ in t is very small in the time duration $\in \Big((\ell-1)\delta, \ell\delta\Big)$. Equivalently,

$$1/\sqrt{f_2(t)} = \frac{t+a}{\sqrt{\beta\delta}} \tag{7.11}$$

In other words, $1/\sqrt{f_2(t)}$ is linear in t. This is confirmed numerically in Fig. 7.11c by the linear relationship between $1/\sqrt{f(t)}$ and t in time region II presented using the right vertical axis with a slope of $1/\sqrt{\beta\delta}$, where δ is the integration time (determined by the measurement setup) and $1/\beta$ is the radiative recombination rate of the exciton ground state.

Let us analyze the experimental data of the time-resolved fluorescence spectrum of water-soluble CdSe-CdS/$Cd_{0.5}Zn_{0.5}S$/ZnS core-multishell QDs (fluorescence peak at 596 nm at room temperature measured by using a time-correlated single-photon counting machine (FluoroMax-3, Horiba Jobin Yvon). A spectral line centered at 495 nm with a 2 nm bandpass from a pulsed light-emitting diode (peak wavelength 495 nm and 30 nm FWHM) is led to the cuvette containing the QD aqueous solution in the form of a train of pulses (pulse duration is ca 1.4 ns) at 1 MHz. The detector is set at 596 nm (QD fluorescence peak wavelength) with a bandpass of 2 nm. There are 2048 TAC channels with a variable integration time δ. A typical time-resolved fluorescence spectrum versus TAC channel index ℓ of the QDs is shown in Fig. 7.12a, where $N_{\max} = 10^4$ and $\delta = 0.1148971$ ns. The relationship of $F(t) - t$ after converting TAC channel index ℓ to decay time t $(= \ell\delta)$ is presented in Fig. 7.12a'. Here we can easily identify time region I where $F(t)$ decays by a single exponential model.

For further spectral analysis, we normalize $F(\ell)$

$$f_n(\ell) = \frac{F(\ell)}{N_{\max}} \tag{7.12}$$

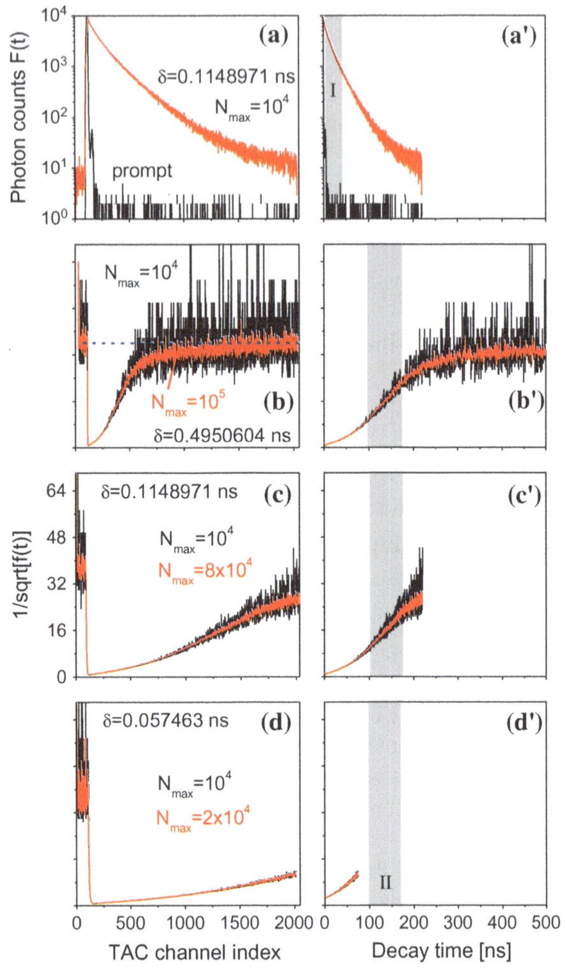

Fig. 7.12 **a** Time-resolved fluorescence spectrum $F(\ell)$ versus TAC channel index ℓ; **a'** $F(t) - t$ where $t = \ell\delta$ is the decay time. **b/b'**, **c/c'**, **d/d'** $\sqrt{1/f(\ell\delta)}$ versus decay time $t = \ell\delta$ of different δ

$1/\sqrt{f_n(\ell)}$ of various $N_{\max}$ and δ are presented in Fig. 7.12b/b'–d/d'. The dashed horizontal line in Fig. 7.12b shows that the noise levels before and long after the excitation pulse are aligned, ensuring that the QD fluorescence is excited by a single excitation pulse.

In order to efficiently utilize the TAC channels, we modify δ and the results are presented in Fig. 7.12b–d. δ in Fig. 7.12b is too long, while it is too short in (d). Measurement setup parameters of Fig. 7.12c/c' are the most proper that unravel time region II that is closely correlated to the exciton radiative recombination process, β in (7.11).

To verify (7.10), (7.11), we tune the detection wavelength of the time-resolved fluorescence measurement setup to scan QD fluorescence peak with a bandpass of only 1 nm which is much smaller than the FWHM of the QDs. The as-measured time-resolved fluorescence spectra $F(\ell)$ at various detection wavelengths are presented in Fig. 7.13a. Here $\delta = 0.4950604$ ns.

We normalize $F(\ell)$ to obtain $f_n(\ell)$ by (7.12), which are presented in Fig. 7.13b showing a strong detection-wavelength dependence of the fluorescence decay, especially at the long decay time, which is much better visualized in $1/\sqrt{f_n(\ell)}$ in Fig. 7.12c. More profoundly, the profile of the long-time $1/\sqrt{f_n(\ell)}$ versus the detection wavelength is identical to the fluorescence spectrum of the QDs shown as the inset in Fig. 7.13a, i.e., (7.15). Applying (7.10), (7.11) to fit the time-resolved fluorescence spectra of Fig. 7.13 results in the data presented in Fig. 7.14 together with the QD fluorescence spectrum.

We have discussed many times the width of a Lorentzian peak. In Sect. 2.4 we know that this width is mainly due to the simultaneous transition of the electron from the excited state back to the ground state when the electron is being photo-excited from the ground state to the excited state, see (2.138), (2.139). By the scattering theory and the generalized Fermi's golden rule, the temporal development $\hat{T}$ of exciton state ψ_1 is described by

$$\langle\psi_1|\hat{T}(t)|\psi_1\rangle \approx e^{-wt/2} \tag{7.13}$$

where $1/w = \beta$ is the decay time of ψ_1, which is given as

$$w(\hbar\omega) = \frac{2\pi}{\hbar}\left|\langle\psi_0|H'|\psi_1\rangle\right|^2 \frac{\Gamma}{\Gamma^2 + (E_1 - E_0 - \hbar\omega)^2} \tag{7.14}$$

E_1 and E_0 are energies of ψ_1 and ψ_0, respectively. $\hbar\omega$ is the photon energy. H' is the light-matter interaction between ψ_1 and ψ_0, see (2.107). Γ is the relaxation energy. In other words,

$$\frac{1}{\beta(\hbar\omega)} \propto \frac{1}{\Gamma^2 + (E_1 - E_0 - \hbar\omega)^2} \quad \rightsquigarrow \quad \beta(\hbar\omega) \propto \Gamma^2 + (E_1 - E_0 - \hbar\omega)^2 \tag{7.15}$$

which is confirmed exactly by the β curve in Fig. 7.14. Meanwhile, by Chap. 2 we know that the fluorescence spectrum is proportional to $w(\hbar\omega)$. Therefore, the fluorescence lifetime spectrum in Fig. 7.14 is well described by the microscopic exciton transition model of Fig. 7.9.

The physics of the energy relaxation and its lifetime τ are much more involved. It involves not only multiple energy-relaxation processes but also the radiative recombination process (i.e., β), which is beyond the scope of the book. Reader may refer to [3] for a brief introduction.

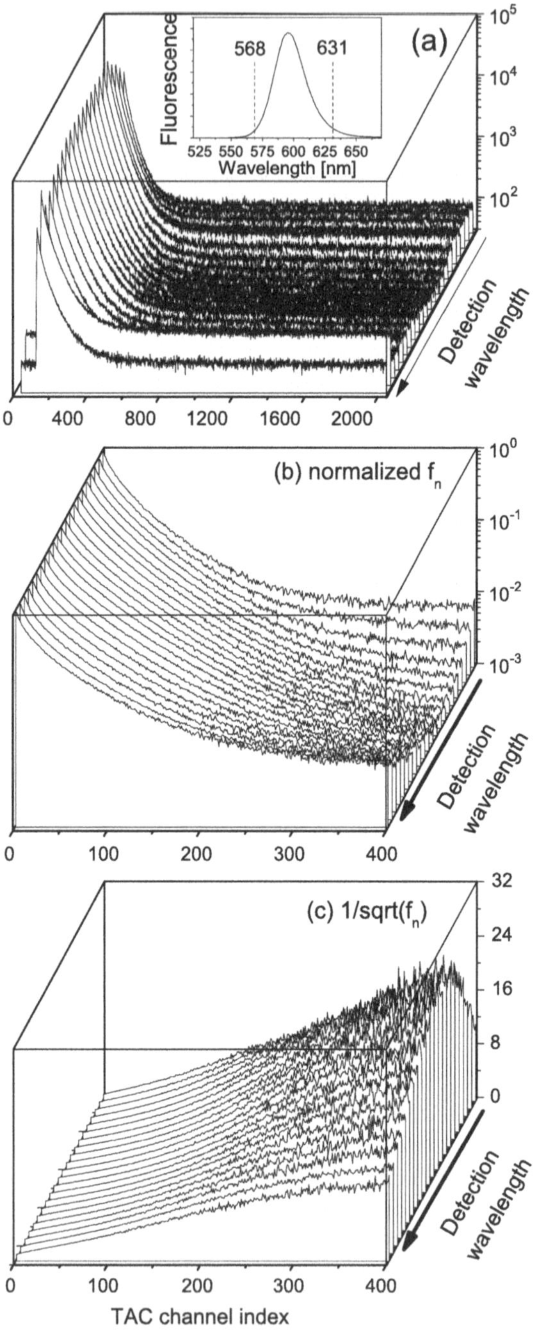

Fig. 7.13 **a** As-measured time-resolved fluorescence spectra $F(\ell)$ of QDs measured at different detection wavelengths $\in (568, 631)$ nm covering the fluorescence peak while the detection bandpass was fixed to be 1 nm. Inset shows the fluorescence spectrum of the QD solution. **b** Normalized $f_n(\ell)$. **c** $1/\sqrt{f_n(\ell)}$. $\delta = 0.4950604$ ns

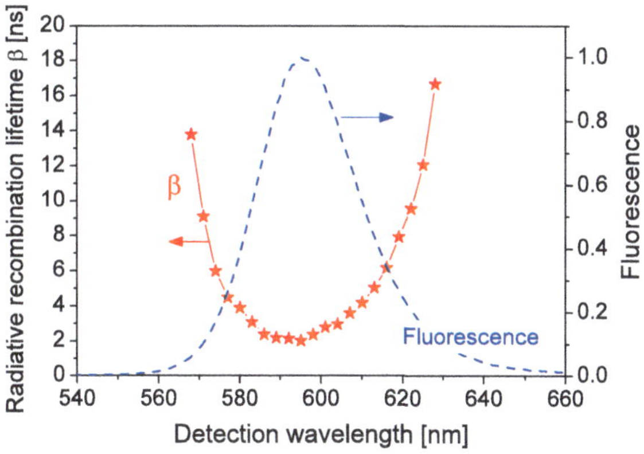

Fig. 7.14 Radiative recombination lifetime spectrum versus detection wavelength ∈ (568, 631) nm. Blue dashed line (right vertical axis) shows the fluorescence spectrum of the QD solution

7.3 Fluorescence Blinking

We briefly mentioned the fluorescence blinking in Sect. 7.1 that QDs irregularly switch between bright and dark under continuous irradiation. Fluorescence blinking is not a specific property for QDs; it is also observed in fluorescence proteins [5], and single organic molecules [6]. Nonblinking or blinking-suppressed QDs have been reported by coating the QDs with thick shells [7, 8], or by forming gradient compositions [9]. It is strongly believed that the fluorescence blinking is related to the surface states of the QD.

A typical time series of one ODA-coated CdSe-$(CdS)_2/(ZnS)_{1.5}$ QD fluorescence is presented in Fig. 7.15a, where a threshold level between the on and off states can be easily identified, see the occurrence profile (the number of times that the QD emits a certain number of photons per frame) of Fig. 7.15c, from which we calculate the on and off time durations, i.e., how long the QD stay at one state before switching to the other state. Here the notation about the QD means that the QD core is CdSe and there are two shell layers, one is composed of two monolayers of CdS and the other is 1.5 monolayer of ZnS. In a close picture Fig. 7.15b of the time series (a), we see the first on-state time duration ∈ (1, 27) frames, i.e., $\text{on}_1 = 27$ frames, followed by the first off-state $\text{off}_1 = 1 \in (28, 28)$ frames, $\text{on}_2 = 12 \in (29, 40)$ frames, $\text{off}_2 = 13 \in (41, 53)$ frames, ... We count the number of on-states which have a time duration of 1 frame, 2 frames, etc. resulting in the diagram about the numbers of on/off events shown in Fig. 7.15d. Here we observe that within 50 000 frames there are totally 73 times the QD stays at the on state for only 1 frame before it switches to the off state, there are 34 times that the QD stays at the on state for 2 frames, and so on. Dividing these numbers by the total number of switches between on- and off-states in the whole fluorescence trajectory gives us the so-called on/off probability density distributions shown in Fig. 7.15e. A larger probability density means that the QD dwells more frequently at this state. Denoting P_{off} and P_{on} as the off and on probability density distributions, respectively, Fig. 7.15e shows clearly an almost linear relationship between $\log[P_{\text{off}}(t)]$ and $\log(t)$ (see the dashed straight line), while the relationship

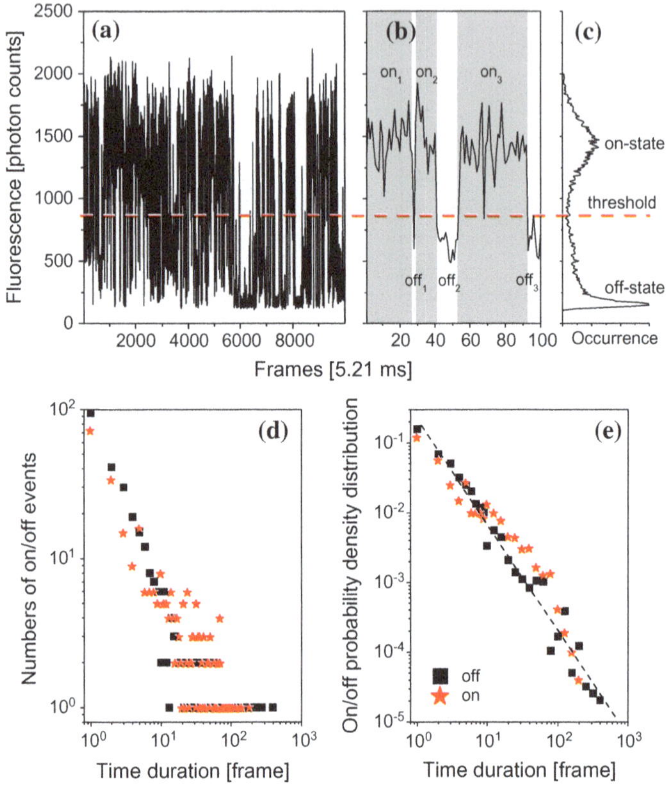

Fig. 7.15 **a** One fluorescence time series of a single ODA-coated CdSe-$(CdS)_2/(ZnS)_{1.5}$ QD consisting of 50,000 frames. **b** A close picture of the first 100 frames showing on (grey regions) and off states of different time durations. **c** The occurrence profile that displays the on and off states. **d** Numbers of on/off events, **e** on/off probability density distributions. Dashed straight line is added only as a reference

between $\log[P_{\mathrm{on}}(t)]$ and $\log(t)$ is more complicated. The almost linear relationships are generally known to be the power law.

The theoretical understanding of the QD blinking is still diverse and limited. Many qualitative models have been proposed, but the major difficulty is that there has not been a good numerical match between theoretical simulations and experimental data. A quantum mechanical surface-state associated blinking model was published recently that quantitatively bridging the model and experiments [10]. As mentioned before, the surfaces of the as-grown QDs are commonly coated with ODA molecules, which are termed as surface ligands. The as-grown QDs can be well dissolved in chloroform or toluene. For biological applications, ODA surface ligands are exchanged to, for example, 3MPA molecules. A careful quantum chemistry calculation shows that the highest occupied molecular orbital (HOMO) and lowest-unoccupied molecular orbital (LUMO) of Zn_6S_6-ODA/3MPA are closely aligned with the ground hole state E_{v0} in the valence and the ground electron state E_{c0} in the conduction band in the QD CdSe core, see Fig. 7.16a. (Note that the valence band of an intrinsic semiconductor is the highest occupied band, while the conduction band is the lowest unoccupied band, at low temperature.) This close alignment provides means for the photogenerated electron and/or hole in the QD core to tunnel to the surface LUMO and/or HOMO state.

The quantum mechanical surface-state associated blinking model adds the tunneling between the QD core and surface states to the fluorescence model of Fig. 7.9, resulting in Fig. 7.16b. To simplify narrations, we focus only on the photogenerated electron. HOMO and LUMO are formed when ODA or 3MPA attaches perfectly to the ZnS surface atoms. Imperfections occur so surface electron states are introduced, e.g., electronic traps at the QD surface [11], they replace LUMO states. LUMO states and surface electron trap states on the QD surface are modeled to distribute randomly in a 2D square grid. The photogenerated electron, after tunneling from the QD core to one LUMO state, randomly walks on the 2D grid until it reaches a nearby LUMO state where it can tunnel back into the QD core. The random walk is parameterized by the necessary migration distance Δ, see Fig. 7.16b. The rate equations of (7.5) are subsequently revised accordingly

$$\begin{aligned}
\frac{\mathrm{d}n_n}{\mathrm{d}t} &= \frac{n_0(1-n_n)}{\alpha} - \frac{n_n(1-n_0)}{\beta} - \frac{n_n(1-n_s)}{\tau'} \\
\frac{\mathrm{d}n_s}{\mathrm{d}t} &= \frac{n_n(1-n_s)}{\tau'} - \frac{n_s(1-n_s^*)}{\eta} \cdots \\
\frac{\mathrm{d}n_s^*}{\mathrm{d}t} &= \frac{n_s(1-n_s^*)}{\eta} - \frac{n_s^*(1-n_0)}{\tau'} \\
\frac{\mathrm{d}n_0}{\mathrm{d}t} &= -\frac{n_0(1-n_n)}{\alpha} + \frac{n_n(1-n_0)}{\beta} + \frac{n_s^*(1-n_0)}{\tau'}
\end{aligned} \tag{7.16}$$

where $1/\eta$ is the walking rate between two neighboring surface states $\hbar\omega_s$ and $\hbar\omega_s^*$.

Numerical simulations based on the rate equations of (7.16) with input parameters from quantum mechanical calculations quantitatively agree with experimental data [10], showing that the QD exciton is first generated by an excitation photon; It radiatively recombines to give QD's fluorescence response, i.e., the on-state, which displays the upwards-bended on-state probability density distribution profile (red stars in Fig. 7.15e); The electron and/or the hole of the photoexcited exciton in the QD core, after tunneling to the QD surface, randomly walks through the two-dimensional network of the QD surface states, resulting in the off-state probability density distribution profile of the inverse power law (black squares in Fig. 7.15e). The model explains many experimental data mentioned at the beginning of the section. Surface modifications modify the QD surface-state network, in turn modifying the on/off probability density distribution profiles. A thicker QD shell will reduce the transport probabilities of the photogenerated electron/hole to the surface state, thus reducing the off-state probability. Lattice strain and stress in gradient composited QDs are low, implying lower transport probabilities of the photogenerated electron/hole to the surface state, leading to the reduced QD blinking.

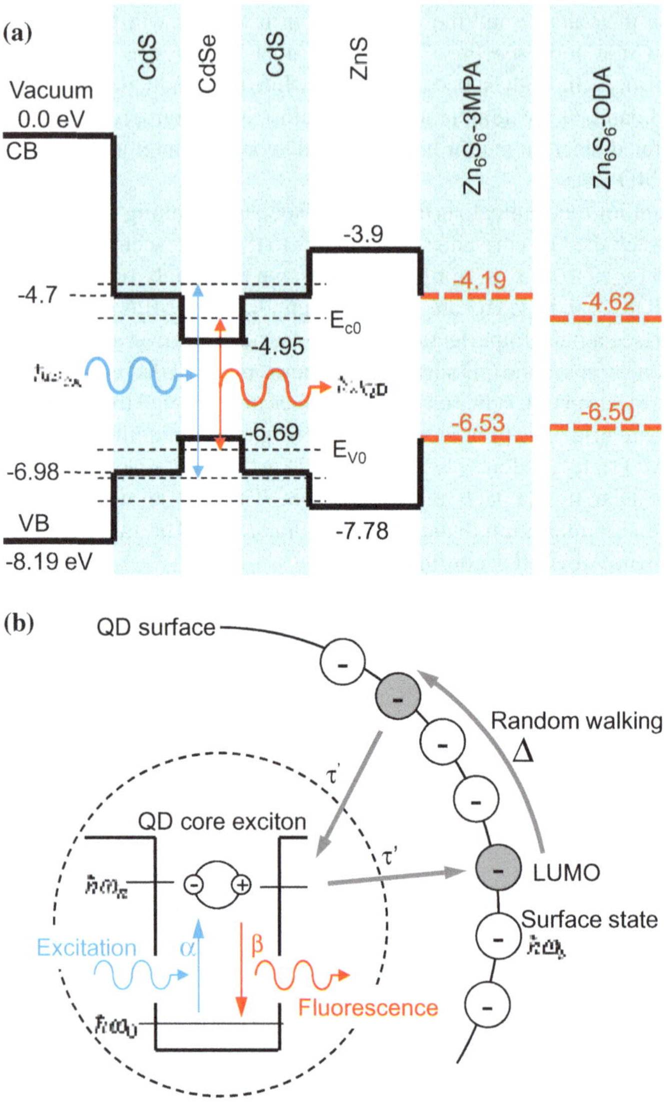

Fig. 7.16 **a** Energy diagram of a colloidal CdSe-based quantum dot. Numerical values mark various energy states in units of eV, e.g., in CdSe core, −6.69 eV is the VBO of CdSe, while −4.95 eV is the conduction bandedge of CdSe. E_{c0} is the ground electron state in the conduction band while E_{v0} is the ground hole state in the valence band. Highest occupied molecular orbital (HOMO) in Zn_6S_6-3MPA is −6.53 eV and the lowest unoccupied molecular orbital (LUMO) is −4.19 eV. **b** Electron transition pathways in the colloidal CdSe-CdS/ZnS core-multishell QD and two-dimensional QD surface-state network

7.4 Fluorescence Spectrum Unravels QD-Ion Interaction

Let us study the interaction between QDs and ions in the QD solution using various fluorescence characterization methods discussed in this chapter. We focus on water-soluble 3-MPA coated CdSe-$(CdS)_2/(ZnS)_{1.5}$ core-multishell QDs having a fluorescence peak at 607 nm at room temperature, consisting of a CdSe core, a CdS shell of 2 monolayers, and another shell of 1.5 monolayer ZnS. These QDs are dispersed to a 4-(2-hydroxyethyl)-1-piperazineethanesulfonic acid (HEPES) buffer solution containing 50 mM HEPES and 23 mM NaOH with a pH value of 7.2 at the QD concentration of 37 nM.

As schematically shown in Fig. 7.3, 3MPA surface ligand has a linear chemical formula $HSCH_2CH_2COOH$. The left end of it is attached to the surface atom Cd of the QD. Figure 7.17a shows the geometry of a model Zn_6S_6-3MPA obtained from density functional theory (DFT) calculation [12]. In water, the carboxylic group (-COOH) of the 3MPA molecule is expected to be acid dissociated. In other words, the outmost proton marked by an arrow in Fig. 7.17a will dissociate so that the acid dissociated (also known as deprotonated) QD-3MPA become negatively charged in water. This is experimentally confirmed that QD-3MPA dispersed in water migrates towards the anode in an electrophoresis experiment [13]. The negatively charged QD-3MPA is expected to be able to bind with a positive ion. Among many ions of biological relevance, Ca^{2+} is unique that it has an ion chelator, ethylene glycol tetraacetic acid (EGTA), see its molecular structure in Fig. 7.17c, which is able to capture one free Ca^{2+} ion in a one-to-one relationship. Thus, Ca^{2+}–EGTA interaction provides an excellent tool to study the interaction between QDs and ions that we can add free Ca^{2+} ions to the QD solution then extract them away in a controllable way.

Figure 7.18 shows the fluorescence spectra of (1) 3MPA coated QDs in HEPES solution, (2) after injection of 2 mM Ca^{2+} ions, and (3) after further injection of 5 mM EGTA. Clearly, interactions between QDs and Ca^{2+} ions diminish the fluorescence of the QDs, which is recovered after EGTA is added to capture Ca^{2+} ions from QDs.

Time-resolved fluorescence spectra of QDs with various Ca^{2+} concentrations (0.1 μM, 1.0 μM, 100 μM, 1 mM, 2 mM Ca^{2+}) are shown in Fig. 7.19a. EGTA (5 mM) recovers the time-resolved fluorescence spectrum (though not totally).

Finally we take a look at the Ca^{2+} effects on QDs' blinking properties by analyzing the fluorescence trajectories of many single QDs in the three different QD solutions. The average on-off probability density distributions of single-QD fluorescences (22 single QDs in HEPES, 38 in HEPES+2 mM Ca^{2+}, and 27 in HEPES+2 mM Ca^{2+}+5 mM EGTA, respectively) are presented in Fig. 7.20. It is shown here that after adding 2 mM Ca^{2+}, the on-state probability of the single-QD fluorescence is decreased, and the off-state probability is increased. After adding EGTA to chelate Ca^{2+} ions from QDs, the on-off probability densities recover.

All these results are well understood by the microscopic picture of the acid-dissociated QD binding one Ca^{2+} ion to its surface, see Fig. 7.17b. The appearance of an ion at the QD surface modifies the QD fluorescence properties through modifying the spatial distributions of the electron and the hole photoexcited in the QD

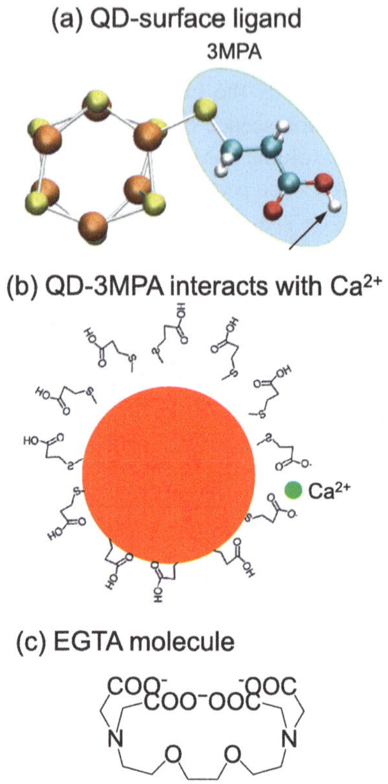

Fig. 7.17 **a** Theoretical molecular structure of a model QD-3MPA from density functional theory. The arrow indicated proton is expected to be acid dissociated when QD-3MPA is dispersed in water. **b** A QD with multiple partially deprotonated 3MPA surface ligands interacting with one Ca^{2+} ion. **c** Molecular structure of EGTA that is capable of capturing Ca^{2+} ion in a one-to-one relationship

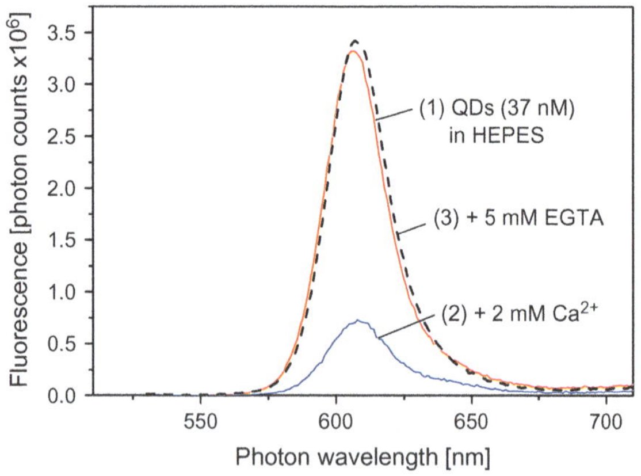

Fig. 7.18 Fluorescence spectra of (1) 3MPA coated QDs in HEPES solution, (2) after injection of 2 mM Ca^{2+} ions, and (3) after further injection of 5 mM EGTA (black dashed line)

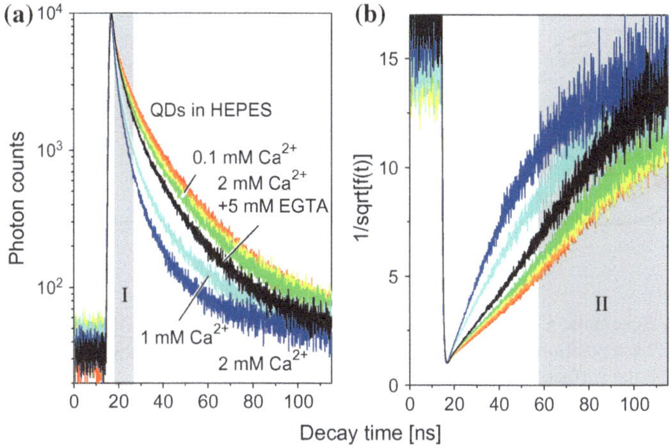

Fig. 7.19 **a** Time-resolved fluorescence spectra of QDs under the influence of Ca^{2+} and EGTA. Rainbow color from red to blue: QDs in HEPES, subsequently 0.1 μM, 1.0 μM, 100 μM, 1 mM, 2 mM Ca^{2+}, and finally 5 mM EGTA (black). **b** $1/\sqrt{f(t)}$. Ca^{2+} significantly modifies the fluorescence decay of QDs while EGTA recovers (black lines), though the recovery is not total

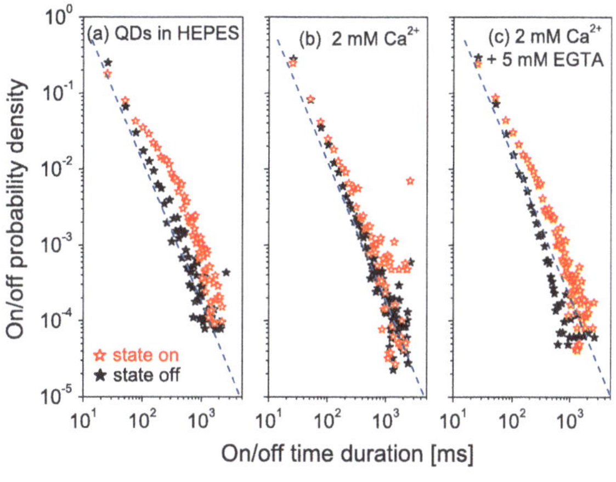

Fig. 7.20 Averaged on-off probability densities of single QD fluorescence in three different QD solutions

core by the Coulomb potential of the ion. Figure 7.21 shows the wave functions of the conduction-band ground state $\psi_{c0}(\boldsymbol{r})$ and the hole at valence-band ground state $\psi_{v0}(\boldsymbol{r})$ when a Ca^{2+} ion is present at the QD surface [12]. The electron is attracted to the Ca^{2+} ion while the hole is pushed away, resulting in a reduced wave function overlapping between the electron and the hole. Since the radiative recombination probability between the electron occupying conduction-band ground state $\psi_{c0}(\boldsymbol{r})$ and the hole at valence-band ground state $\psi_{v0}(\boldsymbol{r})$ is given by (7.14), the reduced wave function overlapping between the electron and the hole implies directly a reduced fluorescence strength, as confirmed in Fig. 7.18.

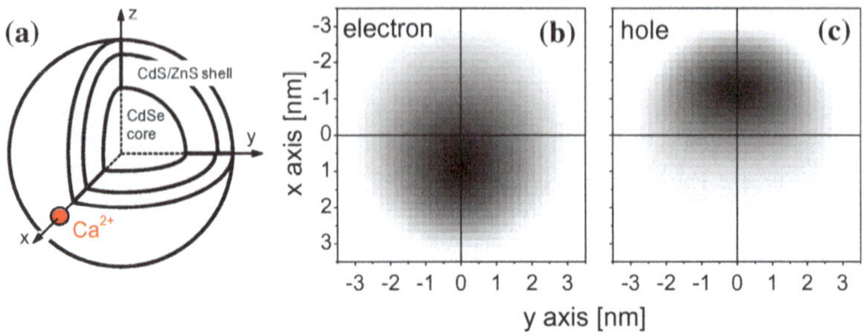

Fig. 7.21 **a** Schematic CdSe-CdS/ZnS core-multishell QD centred at (0, 0, 0) and a positively charged Ca^{2+} ion positioned at (3.7 nm, 0, 0), **b**, **c** wave functions of the electron $\psi_{c0}(\boldsymbol{r})$ and the hole $\psi_{v0}(\boldsymbol{r})$ at the surface of $z = 0$

Using the exciton transition model of Fig. 7.9, we convert the time-resolved fluorescence spectra in Fig. 7.19a to $1/\sqrt{f_n(t)}$ versus t which are presented in Fig. 7.19b. It is shown here that the slope in time region II is decreased due to interactions with Ca^{2+}. By (7.11), the decreased slope ($\propto 1/\sqrt{\beta}$) indicates a longer radiative recombination lifetime β of the exciton ground state, which agrees with the reduced $|\langle\psi_0|H'|\psi_1\rangle|$ due to the increased spatial separation between the electron and the hole, resulting in a reduced radiative recombination probability of the exciton, thus a reduced fluorescence strength, and thereafter an increased β.

Time region I in Fig. 7.19a shows that the lifetime τ of the non-radiative energy relaxation from excited exciton state to the ground exciton state is decreased due to Ca^{2+}-QD interactions. The most possible physical understanding about this is that due to the displacements of the photoexcited electron and hole towards QD surface (the electron is attracted to the surface Ca^{2+} ion while the hole is pushed towards the opposite surface region), both the electron and hole have closer interactions with surface states. This increases the numbers of energy relaxation channels, i.e., the number of V_i terms in (7.14), resulting in a shorter energy relaxation time τ.

This QD-surface-related energy relaxation channel model is supported by the blinking data. Figure 7.20b shows a reduced on-state probability due to Ca^{2+} as compared with the situation of pure QDs in HEPES in Fig. 7.20b. The on-state probability is recovered when EGTA is added to chelate Ca^{2+} in the QD solution. By the surface-state associated blinking model briefly introduced in Sect. 7.3, the introduction of Ca^{2+} at the QD surface attracts the photogenerated electron and repels the hole, facilitating the transport of the electron and the hole between exciton states in the QD core and surface states.

7.5 Quantum Dot Bioimaging

Highly fluorescent semiconductor QDs have already been introduced successfully in many fluorescence-based optical imaging applications in biomedical sciences. Polymer-coated QDs are water soluble, noncytotoxic and innocuous to normal cell physiology. To achieve targeting abilities, polymer-coated QDs are conjugated to bio-affinity ligands such as monoclonal antibodies, peptides, or oligonucleotides using several approaches including passive adsorption, multivalent chelation, or covalent-bond formation. In this last section of the book we briefly describe one example of QD's biomedical application.

High cholesterol, high blood sugar, blood pressure fluctuations, folic acid reduction and high homocysteine are all closely related to arteriosclerosis development. Early diagnosis of arteriosclerosis is therefore essential in preventing cardiovascular and cerebrovascular diseases. Vascular cell adhesion molecule 1 (VCAM1) protein expression is known to be the major performance in inflamed cells of early arteriosclerosis. VCAM1 binding peptide, see Fig. 7.22a, can target bind with VCAM1 molecule [14]. We bind fluorescent QD with VCAM1 binding peptide by forming stable amide bonds between COO^- on QD surface and NH_2^+ on VCAM1 binding peptide, denoted as VQDs, so that we can visualize VCAM1 molecule expression since we can "see" QDs, while VCAM1 molecule is not directly visible. This is known as target bioimaging. The basic protocol is to incubate VQDs with cells for a certain time then wash the cells. If the cells have VCAM1 expression on their surfaces, VQDs will target bind there and fluoresce in a fluorescence microscope.

Refer to Fig. 7.22, a more advanced application is to use QD as a target drug delivery that is loaded with both the targeting (VCAM1 binding peptide) and the drug molecules (e.g., protoporphyrin IX) for target drug delivery (aiming at VCAM1 molecule on inflamed cell surface) at molecular level [15]. Here we describe one photodynamic therapy (PDT) application. Photodynamic therapy is widely applied for treating tumors that tumor-localizing photosensitizers are first administrated by either intravenous injection or external application; Photosensitizers are then target-activated by irradiation of a specific wavelength. Protoporphyrin IX (PpIX) is a common photosensitizer, see Fig. 7.22b.

In the following case study, we use human umbilical vein endothelia cells (HUVECs) which express typical endothelial phenotypes; VCAM-1 expression in HUVEC cells can be induced by tumor necrosis factor α (TNFα) treatment; Control HUVECs (no TNFα treatment) and TNFα-treated HUVECs are thus chosen here to demonstrate the target drug delivery. The general experimental protocol is to conjugate 3MPA-QD (Fig. 7.22c) first with VCAM1 binding peptide then with PpIX, forming the PVQs, see Fig. 7.22d. Deposit PVQs to TNFα-treated HUVECs, i.e., Fig. 7.22e.

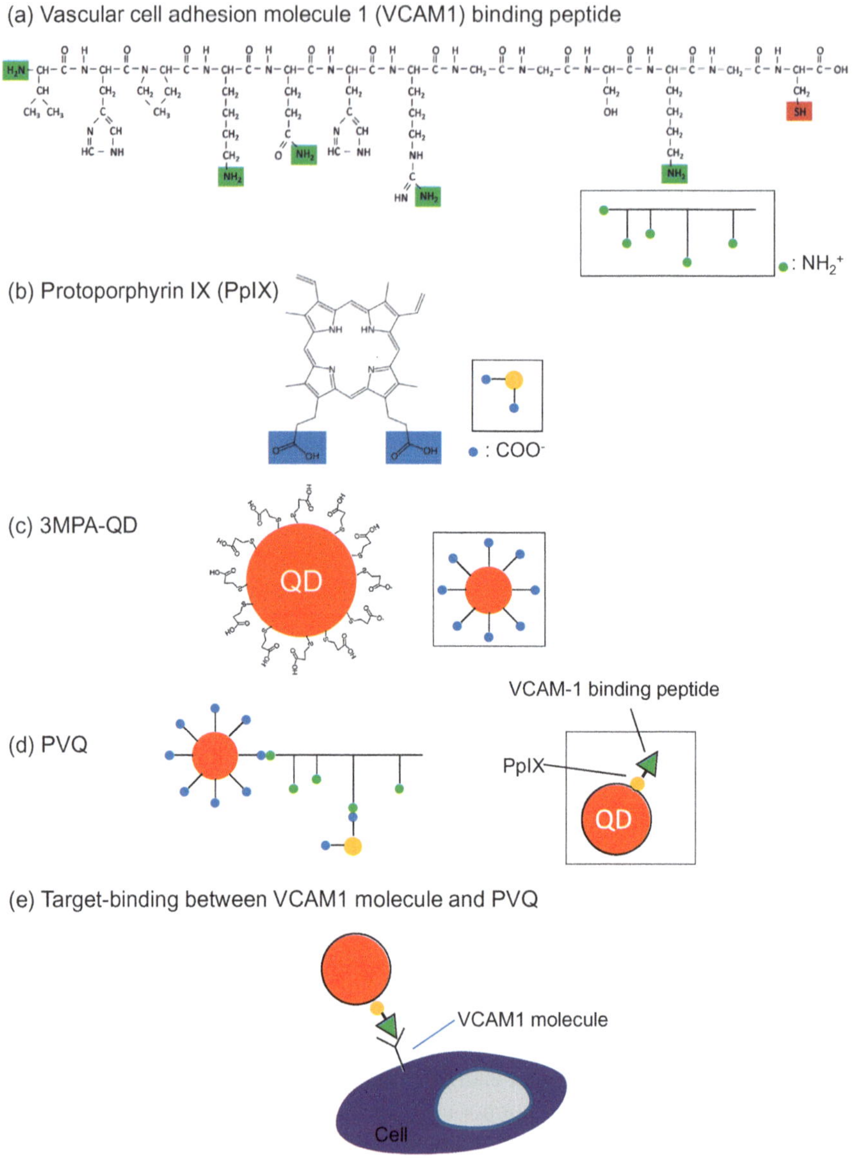

Fig. 7.22 Schematics of molecular structures of **a** vascular cell adhesion molecule 1 (VCAM1) binding peptide, **b** protoporphyrin IX (PpIX), **c** 3MPA-QD, **d** conjugate of PpIX-VCAM1 binding peptide-QD (PVQ), and **e** target-binding between VCAM1 molecule and PVQ

Figure 7.23a, b show the fluorescent images of control and TNFα treated HUVEC cells after 24 h incubation with PVQs. The images are excited by 405-nm laser and acquired by a microscopic objective 40$\times$ in lambda mode, meaning that the microscope takes a series of images of different wavelength channels, similar to the operation of the three RGB channels in common camera, though more spectrally resolved. More specifically, the two images of Fig. 7.23a, b are of 1024 $\times$ 1024 pixels at a spatial resolution of 0.208 $\times$ 0.208 μm^2. The spectral range covers from 414 to 691 nm with a spectral resolution of 8.9 nm such that each image of Fig. 7.23a, b is composed of 32 sub-images acquired at wavelength 414, 423, ... 691 nm. In other words, the microscope measures one fluorescence spectrum per one spatial pixel in optical range $\in$ (414, 691) nm at a spectral resolution of 8.9 nm. We can summarize the spectra of pixels enclosed in a special region of interest (ROI) in order to spatially obtain the fluorescence spectrum of a certain object confined in this ROI. For example, Fig. 7.23c show three fluorescence spectra of three ROIs. The spectrum of ROI1 includes both the yellow QDs emitting at 587 nm and the blue signals are auto-fluorescence from live cells, which dominates in ROI2. To make sure that the 587-nm peak does originate from QDs, we can zoom in to ROI3, where the 587-nm peak becomes predominant, while the blue-green signals are still there (so we see a sub-region in ROI3 which is green since blue plus yellow appears green).

We observe a few QD signals in Fig. 7.23a which is not surprising since there is no way to make sure that all cells not TNFα treated are totally healthy. Figure 7.23a, b thus show that TNFα treatment induces VCAM1 expression which is target imaged by PVQs.

To assess photodynamic therapeutic effects of PVQs, HUVEC cells of no-TNFα- and NTFα-treatments were irradiated by a 630-nm LED for 3 min at an optical power of 28.6 mW/cm^2. After further incubation for 2 h or 6 h, the cells were stained with Annexin V-FITC and propidium iodide (PI) for 10 min at room temperature in darkness. Annexin V-FITC is known to bind to cell membranes at early apoptosis, while PI stains nuclei at later apoptosis or death cells. The results are presented in Fig. 7.23d–g. Here the numbers of cells studied in each groups are denoted in (d), which are large enough to obtain statistical significance. Three phenomena are clearly observed: (1) Both (d) and (f) show that the percentages of PI and Annexin V-FITC positive cells are significant in TNFα treated HUVECs 6 h after PDT irradiation (though 2 h is not long enough for apoptosis expression). (2) Furthermore, though Annexin signal strengths in PI/Annexin positive cells in TNFα-treated HUVECs are ca 75, they also reach ca 50 in not-TNFα-treated HUVECs, see (e) and (g). The latter is largely due to the non-specific cellular phototoxicity effect, i.e., live cells can become damaged by irradiation. (3) The PDT effect is significant 6 h after PDT irradiation.

In a brief summary, Fig. 7.23 show clearly the target PDT effect of PVQs in TNFα treated HUVECs 6 h after PDT irradiation.

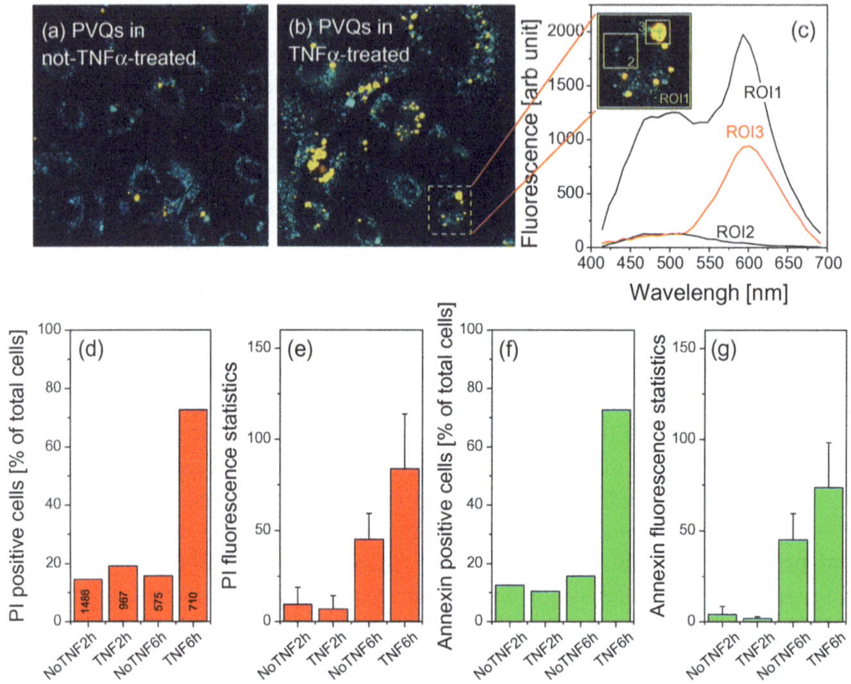

Fig. 7.23 **a**, **b** Fluorescence images of PVQs in not-TNFα-treated and TNFα-treated HUVECs. The images are excited by 405-nm laser and acquired by objective 40× in lambda mode. The image colors are true colors that the blue-green signals are auto-fluorescence from live cells while the yellow are QDs emitting at 587 nm. **c** Fluorescence spectra of three regions of interest (ROI) marked in inset. **d**, **f** Percentages of propidium iodide (PI) and Annexin positive cells in different cell groups. The numbers in the columns in **d** denote the numbers of cells studied in each cell groups. **e**, **g**: PI/Annexin signal strengths in PI/Annexin positive cells. **d–g** show clearly the target PDT effect of PVQs in TNFα treated HUVECs 6 h after PDT irradiation

References

1. X. Peng, Green chemical approaches toward high-quality semiconductor nanocrystals. Chem. Eur. J. **8**, 334–339 (2002)
2. Z. Ning, H. Tian, H. Qin, Q. Zhang, H. Ågren, L. Sun, Y. Fu, Wave function engineering of CdSe-CdS core-shell quantum dots for enhanced electron transfer to a TiO_2 substrate. J. Phys. Chem. C **114**, 15184–15189 (2010)
3. Z.-H. Chen, S. Hellström, Z.-J. Ning, Z.-Y. Yu, Y. Fu, Exciton polariton contribution to the Stokes shift in colloidal quantum dots. J. Phys. Chem. C **115**, 5286–93 (2011)
4. Y. Fu, T.-T. Han, H. Ågren, L. Lin, P. Chen, Y. Liu, G.-Q. Tang, J. Wu, Y. Yue, N. Dai, Design of semiconductor CdSe-core ZnS/CdS-multishell quantum dots for multiphoton applications. Appl. Phys. Lett. **90**, 173102(3) (2007)
5. R.M. Dickson, A.B. Cubitt, R.Y. Tsien, W.E. Moerner, On/off blinking and switching behaviour of single molecules of green fluorescent protein. Nature **388**, 355–358 (1997)

6. M. Haase, C.G. Hubner, E. Reuther, A. Herrmann, K. Müllen, T. Basche, Exponential and power-law kinetics in single-molecule fluorescence intermittency. J. Phys. Chem. B **108**, 10445–10450 (2004)
7. B. Mahler, P. Spinicelli, S. Buil, X. Quelin, J.-P. Hermier, B. Dubertret, Towards non-blinking colloidal quantum dots. Nat. Mater. **7**, 659–664 (2008)
8. Y. Chen, J. Vela, H. Htoon, J.L. Casson, D.J. Werder, D.A. Bussian, V.I. Klimov, J.A. Hollingsworth, Giant multishell CdSe nanocrystal quantum dots with suppressed blinking. J. Am. Chem. Soc. **130**, 5026–5027 (2008)
9. X. Wang, X. Ren, K. Kahen, M.A. Hahn, M. Rajeswaran, S. Maccagnano-Zacher, J. Silcox, G.E. Cragg, A.L. Efros, T.D. Krauss, Non-blinking semiconductor nanocrystals. Nature **459**, 686–689 (2009)
10. H. Xu, H. Brismar, Y. Fu, Influence of surface states on blinking characteristics of single colloidal CdSe-CdS/ZnS core-multishell quantum dot. J. Colloid Interface Sci. **505**, 528–536 (2017)
11. C. Pu, X. Peng, To battle surface traps on CdSe/CdS core/shell nanocrystals: shell isolation versus surface treatment. J. Am. Chem. Soc. **138**, 8134–8142 (2016)
12. L. Li, Y. Chen, G. Tian, V. Akpe, H. Xu, L.-M. Gan, S. Skrtic, Y. Luo, H. Brismar, Y. Fu, Reversible modification of CdSe-CdS/ZnS quantum dot fluorescence by surrounding Ca^{2+} ions. J. Phys. Chem. C **118**, 10424–33 (2014)
13. N. Shambetova, Y. Chen, H. Xu, L. Li, J. Solandt, Y. Zhou, J. Wang, H. Su, H. Brismar, Y. Fu, Acid dissociation of 3-mercaptopropionic acid coated CdSe-CdS/$Cd_{0.5}Zn_{0.5}S$/ZnS core-multishell quantum dot and strong ionic interaction with Ca^{2+} ion. J. Phy. Chem. C **120**, 3519–3529 (2016)
14. Y. Chen, M. Molnár, L. Li, P. Friberg, L.-M. Gan, H. Brismar, Y. Fu, Characterization of VCAM-1-binding peptide-functionalized quantum dots for molecular imaging of inflamed endothelium. PLoS ONE **8**(12), e83805 (2013). https://doi.org/10.1371/journal.pone.0083805
15. H. Yin, X. Shi, W. Jin, Y. Li, Y. Fu, Photodynamic therapy targeting VCAM-1 expressed human umbilical vein endothelial cell using photosensitizer-VCAM-1 binding peptide-quantum dot conjugate. RSC Adv. **7**, 50562–50570 (2017)

Index

W. Lu and Y. Fu, *Spectroscopy of Semiconductors*, Springer Series in Optical Sciences 215, https://doi.org/10.1007/978-3-319-94953-6

The manufacturer's authorised representative in the EU is Springer Nature Customer Service Centre GmbH, Europaplatz 3, 69115 Heidelberg, Germany. If you have any concerns regarding our products, please contact ProductSafety@springernature.com

Printed and bound by CPI Group (UK) Ltd, Croydon, CR0 4YY
30/09/2026
02225654-0008